SPRINGER SERIES IN PHOTONICS 3

Springer
Berlin
Heidelberg
New York
Barcelona
Hong Kong
London
Milan
Paris
Singapore
Tokyo

SPRINGER SERIES IN PHOTONICS

The Springer Series in Photonics covers the entire field of photonics, including theory, experiment, and the technology of photonic devices. The books published in this series give a careful survey of the state-of-the-art in photonic science and technology for all the relevant classes of active and passive photonic components and materials. This series will appeal to researchers, engineers, and advanced students.

1 **Advanced Optoelectronic Devices**
By D. Dragoman and M. Dragoman

2 **Femtosecond Technology**
Editors: T. Kamiya, F. Saito, O. Wada, H. Yajima

3 **Integrated Silicon Optoelectronics**
By H. Zimmermann

Horst Zimmermann

Integrated Silicon Optoelectronics

With 257 Figures and 18 Tables

Dr.-Ing. habil. Horst Zimmermann, Dipl.Phys.
Christian-Albrechts-Universität, Technische Fakultät, Halbleitertechnik
Kaiserstrasse 2, 24143 Kiel, Germany
E-Mail: horst.zimmermann@ieee.org

Series Editors:

Professor Takeshi Kamiya
Department of Electronic Engineering, Faculty of Engineering
University of Tokyo
7-3-1 Hongo, Bunkyo-ku, Tokyo 113-8656, Japan

Professor Bo Monemar
Department of Physics and Measurement Technology, Materials Science Division
Linköping University
58183 Linköping, Sweden

Dr. Herbert Venghaus
Heinrich-Hertz-Institut für Nachrichtentechnik Berlin GmbH
Einsteinufer 37
10587 Berlin, Germany

Library of Congress Cataloging-in-Publication Data

Zimmermann, Horst, 1957-
Integrated silicon optoelectronics / Horst Zimmermann.
p. cm. -- (Springer series in photonics ; 3)
Includes bibliographical references and index.
ISBN 3540666621 (alk. paper)
1. Optoelectronic devices. 2. Integrated circuits. 3. Semiconductors. 4. Silicon-on-insulator technology. I. Title. II. Springer series in photonics ; v. 3.

TK8304 .Z56 2000
621.381'045--dc21

99-088579

ISSN 1437-0379

ISBN 3-540-66662-1 Springer-Verlag Berlin Heidelberg New York

Typesetting: Camera-ready from the author using a Springer TEX macro package
Cover concept: eStudio Calamar Steinen
Cover production: *design & production* GmbH, Heidelberg

Printed on acid-free paper SPIN 10701200 57/3144/di 5 4 3 2 1 0

Preface

This book is intended as a bridge between microelectronics and optoelectronics. Usually, optoelectronics plays a minor role in electrical engineering courses at universities. Physicists are taught optics but not very much semiconductor technology and chip design. This book covers the missing information for engineers and physicists who want to know more about integrated optoelectronic circuits (OEICs) in silicon technologies and about their emerging possibilities.

Optoelectronics usually implies that III/V semiconductor materials are involved. This is the case when ultra-high-speed photodetectors or efficient light emitters are needed. For other applications, the price of III/V photodetectors and OEICs is simply too high. Silicon photodectectors and receiver OEICs, therefore, are the only choice when high volumes are needed and the price has to be low as, for instance, in consumer electronics. Such high volumes of silicon OEICs are, for example, needed in optical storage systems like audio CD, magneto-optical disk, CD-ROM, and Digital-Video-Disk or Digital-Versatile-Disk (DVD) systems. The market for DVD systems and therefore for DVD OEICs is estimated to be 120 million pieces in the year 2001.

OEICs are key devices for advanced optical storage systems and for the enhancement of their speed and data rate. This importance of OEICs is due to the following advantages of monolithic optoelectronic integrated circuits: (i) good immunity against electromagnetic interference (EMI) because of very short interconnects between photodetectors and amplifiers, (ii) reduced chip area due to the elimination of bondpads, (iii) improved reliability due to the elimination of bondpads and bond wires, (iv) cheaper mass production compared to discrete circuits, wire-bonded circuits, and hybrid integrated circuits, and (v) larger −3dB bandwidth compared to discrete circuits, wire-bonded circuits, and some hybrid integrated circuits due to the avoidance of parasitic bondpad capacitances.

Even in the domain of light emission, silicon is being investigated intensively to make silicon a competitor of III/V semiconductor materials. Much effort is made to let silicon emit light, and these attempts will be described in this book.

This book was written in parallel with the development of OEICs in CMOS and in BiCMOS technologies for optical storage systems and for optical interconnect technologies. It describes the state of the art in OEIC design and the approaches to this topic reported recently in the literature.

Parts of the book have their origin in an 'Optoelectronics' lecture I have given since 1994. It, however, dives much deeper into the topic. The possibilities of integrated silicon optoelectronics are investigated thoroughly and I have tried to initiate a link between microelectronics and photonics. The term *photonics* has come into use more and more in the last decade. This term, which was coined in analogy with electronics, reflects the growing link between optics and electronics forged by the increasing role of semiconductor materials and devices in optical systems.

As the term *electronics* already expresses, it is based on the control of electrons and of electric charge flow. Photonics is based on the control of photons, and the term photonics reflects the importance of the photon nature of light in describing the operation of many optical devices. The overlap between the two disciplines is obvious, since electrons often control the flow of photons and, conversely, photons control the flow of electrons. The term photonics is used broadly to encompass: (i) the generation of light by LEDs and lasers; (ii) the transmission of light in free space, through conventional optical lenses, apertures, and imaging systems, and through optical fibers and waveguides; (iii) the modulation, switching, and scanning of light by the use of electrically, acoustically, or optically controlled devices; (iv) the amplification and frequency conversion of light by the use of wave interaction in nonlinear materials; (v) the detection of light.

These areas have found steadily increasing applications in optical communication, signal processing, computing, sensing, display technology, printing, and energy transport. Items (i) to (iii) and (v) can be covered by silicon and will be discussed in this book.

Integrated optoelectronic receiver circuits have already made their way into microelectronics, which is dominated by silicon technology. I think it is only a question of time before true microphotonic silicon-based circuits with electronic circuits, light detectors, waveguides, grating couplers, holographic lenses, and efficient light emitters are developed and enter the market.

I would like to thank Prof. Dr.-Ing. P. Seegebrecht for the generous possibility to develop OEICs independently and for several helpful comments concerning a part of the text used to acquire the title 'habilitatus'. I am also indebted to Prof. Dr. H. Föll, who offered a waferprober for the characterization of the OEICs. The work of the OEIC group members, A. Ghazi, T. Heide, K. Kieschnick, and G. Volkholz, is highly appreciated. Three students, N. Madeja, F. Sievers, and U. Willecke, carefully performed simulations and measurements. M. Wieseke and F. Wölk helped with the preparation of numerous drawings. Special thanks go to R. Buchner from the Fraunhofer-Institute for Solid-State Technology in Munich and H. Pless from Thesys

Microelectronics in Erfurt for their engagement in the fabrication of CMOS OEICs and BiCMOS OEICs, respectively. Last but not least I would like to gratefully acknowledge the funding of the projects by the German ministry for education, science, research, and technology (BMBF) within the leading project 'optical memories'.

I extend my sincere thanks to Dr. Ascheron and his team at Springer for the good cooperation and their technical support with the text processor. My deepest gratitude, however, is directed to my wife and my daugthers, Luise and Lina, who supported this book project with their encouragement and patience during many evenings and weekends.

Kiel, January 2000 *Horst Zimmermann*

Contents

List of Symbols

Symbol	Description	Unit
a	Lattice constant	nm
A	Area	cm^2
A_0	Low-frequency open loop gain	
$A(\omega)$	Frequency-dependent gain	
c	Speed of light in a medium	cm/s
c_0	Speed of light in vacuum	cm/s
c_{bd}	Small-signal bulk-drain capacitance	F
c_{bs}	Small-signal bulk-source capacitance	F
c_{cs}	Small-signal collector-substrate capacitance	F
c_{gb}	Small-signal gate-bulk capacitance	F
c_{gd}	Small-signal gate-drain capacitance	F
c_{gs}	Small-signal gate-source capacitance	F
c_{ws}	Small-signal well-substrate capacitance	F
c_{je}	Small-signal emitter-base capacitance	F
c_{μ}	Small-signal base-collector capacitance	F
C_{e}	Doping concentration of epitaxial layer	cm^{-3}
C_{B}	Capacitance of bondpad	F
C_{BE}	Base-emitter capacitance	F
C_{C}	Base-collector space-charge capacitance	F
C_{D}	Depletion capacitance of photodiode	F
C_{E}	Total base-emitter capacitance of bipolar transistor	F
C_{F}	Feedback capacitance of transimpedance amplifier	F
C_{I}	Input capacitance of amplifier	F
C_{L}	Load capacitor	F
C_{P}	Capacitance of chip package	F
$C_{\mathrm{R_F}}$	Parasitic capacitance of feedback resistor	F
C_{S}	Parasitic capacitance of signal line	F
C_{SE}	Base-emitter space-charge capacitance	F

Symbol	Description	Unit
d_e	Thickness of epitaxial layer	μm
d_I	Thickness of intrinsic region	μm
d_p	Thickness of P-type region	μm
D	Diffusion coefficient	cm^2/s
D_n	Diffusion coefficient of electrons	cm^2/s
D_p	Diffusion coefficient of holes	cm^2/s
DR	Data rate	Mb/s
E_C	Bottom of conduction band	eV
E_F	Fermi energy level	eV
E_g	Energy bandgap	eV
E_p	Phonon energy	eV
E_t	Energy level of recombination center	eV
E_V	Top of valence band	eV
E	Photon energy	eV
$\boldsymbol{E}$	Electric field	V/cm
f	Frequency	Hz
f_g	Bandwidth, −3dB frequency	Hz
f_T	Transit frequency (gain-bandwidth product)	Hz
f_{GP}	Frequency of gain-peak	Hz
G	Photogeneration (e-h-p)	cm^{-3}/s
$G(\omega)$	Frequency response function	V/A
g_m	Transconductance	A/V
h	Planck constant	Js
$\hbar$	$h/2\pi$	Js
$h\nu$	Photon energy	eV
I	Current	A
I_{ph}	Photocurrent	A
I_{th}	Threshold current	A
I_B	Base current	A
I_C	Collector current	A
I_D	Drain current	A
I_E	Emitter current	A
I_S	Source current	A
j	Current density	A/cm^2
$\boldsymbol{k}$	Wave vector	cm^{-1}
k_B	Boltzmann constant	J/K
k_BT	Thermal energy	eV
L	Length	μm
L_n	Diffusion length of electrons	μm

Symbol	Description	Unit
L_{p}	Diffusion length of holes	µm
L_{B}	Inductance of bond wire	H
L_{G}	Gate length	µm
L_{W}	Inductance of lead wire	H
$\bar{n}$	Refractive index	
$\bar{n}_{\mathrm{s}}$	Refractive index of surroundings	
n_{sc}	Refractive index of semiconductor	
$\bar{n}_{\mathrm{ARC}}$	Refractive index of antireflection coating	
n	Density of free electrons	cm^{-3}
n_{i}	Intrinsic density	cm^{-3}
N	Impurity concentration	cm^{-3}
N_{A}	Acceptor concentration	cm^{-3}
N_{D}	Donor concentration	cm^{-3}
N_{t}	Concentration of recombination centers	cm^{-3}
p	Density of free holes	cm^{-3}
Q_{E}	Minority-carrier charge in the base	As
Q_{pix}	Charge on pixel storage capacitor	As
QE	Quantum efficiency	%
$\boldsymbol{p}$	Momentum	Js/m
P_{opt}	Incident optical power	W
$\bar{P}$	Optical power in a semiconductor	W
q	Magnitude of electronic charge	As
r_{b}	Base series resistance	Ω
r_{o}	Small-signal output resistance	Ω
r_{c}	Small-signal collector series resistance	Ω
r_{ex}	Small-signal emitter series resistance	Ω
r_{d}	Small-signal drain series resistance	Ω
r_{s}	Small-signal source series resistance	Ω
R	Responsivity	A/W
R_{bb}	Responsivity to black-body radiation	A/W
R_{D}	Parallel resistance	Ω
R_{F}	Feedback resistance	Ω
R_{S}	Series resistance	Ω
R_{I}	Input resistance of amplifier	Ω
R_{L}	Load resistance	Ω
$\bar{R}$	Reflectivity	
t	Time	s
t_{d}	Drift time	s
t_{diff}	Diffusion time	s
t_{f}	Fall time	s

Symbol	Description	Unit
t_{r}	Rise time	s
t_{gd}	Group delay	s
T	Absolute temperature	K
U	Voltage	V
U_{BE}	Base-emitter voltage	V
U_{D}	Built-in voltage	V
U_{DS}	Drain-source voltage	V
U_{Ea}	Early voltage	V
U_{GS}	Gate-source voltage	V
U_{T}	Thermal voltage $\mathrm{k_B T/q}$	V
U_{th}	Thermal generation/recombination rate	$\mathrm{cm^{-3}s^{-1}}$
U_{Th}	Threshold voltage	V
V_{det}	Detector bias	V
V_{o}	Output voltage	V
V_{rev}	Reverse voltage	V
v	Carrier velocity	cm/s
v_{s}	Saturation velocity	cm/s
v_{th}	Thermal velocity	cm/s
W	Width of space-charge region	μm
W_{B}	Base thickness	μm
W_{G}	Gate width	μm
Z_{F}	Feedback impedance of transimpedance amplifier	Ω
x	x direction	μm
y	y direction	μm
α	Absorption coefficient	$\mathrm{\mu m^{-1}}$
β	Current gain of bipolar transistor	
η_{e}	External (total) quantum efficiency	%
η_{i}	Internal quantum efficiency	%
η_{o}	Optical quantum efficiency	%
η_{mc}	Emission enhancement factor in micro-cavity	
ϵ_0	Permittivity in vacuum	F/cm
ϵ_{d}	Relative permittivity of passivation layer	
ϵ_{r}	Relative permittivity	
ϵ_{s}	Semiconductor permittivity	F/cm
$\bar{\epsilon}$	Dielectric function	
σ	Carrier capture cross section	$\mathrm{cm^{-2}}$
τ	Lifetime	s
τ_{n}	Electron lifetime	s
τ_{p}	Hole lifetime	s
τ_{B}	Base transit time	s

Symbol	Description	Unit
$\bar{\kappa}$	Extinction coefficient	
λ	Wavelength in a medium	nm
λ_0	Wavelength in vacuum	nm
λ_c	Wavelength corresponding to E_g	nm
λ_d	Design wavelength of a DBR	nm
λ_{ch}	Channel length modulation parameter	V^{-1}
ν	Frequency of light	Hz
μ	Mobility	cm^2/Vs
μ_n	Electron mobility	cm^2/Vs
μ_p	Hole mobility	cm^2/Vs
ω	Angular frequency	s^{-1}
ω_T	$2 \cdot \pi \cdot f_T$	s^{-1}
ω_{GP}	$2 \cdot \pi \cdot f_{GP}$	s^{-1}
ρ	Charge density	As/cm^3
Φ	Photon flux density	$cm^{-2}s^{-1}$
Φ_B	Barrier height	eV
Φ_{bi}	Built-in potential	V
Ψ	Potential	V
Θ	Angle	$^\circ$
Θ_c	Critical angle for total reflection	$^\circ$
Θ_i	Incidence angle	$^\circ$

1. Basics of Optical Emission and Absorption

Optical emission and absorption are fundamental processes which are exploited when electrical energy is converted into optical energy and vice versa. Optoelectronics is based on these energy conversion processes. Light emitters such as light-emitting diodes (LEDs) and diode lasers convert electrical energy into optical energy. Photodetectors convert optical energy into electrical energy. In this chapter, the most important factors needed for the comprehension of light emitters and photodetectors will be summarized in a compact form. For a detailed description of the basics of optical emission and absorption, the book [1] can be recommended. Here, emphasis will, of course, be placed on silicon devices. First we will introduce photons and the properties of light. Then, the consequences of the band structure for optical emission and absorption will be dealt with. Photogeneration will be defined. Furthermore, optical reflection and its consequences on the efficiency of photodetectors will be described.

1.1 Properties of Light

Due to the work of Max Planck and Albert Einstein it is possible to describe light not only by a wave formalism but also by a quantum-mechanical particle formalism. The smallest unit of light intensity is a quantum-mechanical particle called a photon. The photon, consequently, is the smallest unit of optical signals. Photons are used to characterize electromagnetic radiation in the optical range from the far infrared to the extreme ultraviolet spectrum. The velocity of photons c in a medium with an optical index of refraction $\bar{n}$ is

$$c = \frac{c_0}{\bar{n}}, \tag{1.1}$$

where c_0 is the velocity of light in vacuum. Photons do not possess a quiescent mass and, unfortunately for the construction of purely optical computers, cannot be stored. Photons can be characterized by their frequency ν and by their wavelength λ:

$$\lambda = \frac{c}{\nu}. \tag{1.2}$$

The frequency of a photon is the same in vacuum and in a medium with index of refraction $\bar{n}$. The wavelength λ in a medium, therefore, is shorter than the vacuum wavelength λ_0 ($\lambda = \lambda_0/\bar{n}$). As a consequence, the vacuum wavelength is used to characterize light sources like light-emitting diodes (LEDs) or semiconductor lasers, because it is independent of the medium in which the light propagates.

Photons can also be characterized by their energy E (h is Planck's constant):

$$E = h\nu = \frac{hc}{\lambda} = \frac{hc_0}{\lambda_0}. \tag{1.3}$$

A useful relation is given next which allows a quick calculation of the energy for a certain wavelength and vice versa:

$$E = \frac{1240}{\lambda_0} \tag{1.4}$$

where E is in eV and λ_0 in nm. Let us define the flux density Φ as the number of photons incident per time interval on an area A. The optical power P_{opt} incident on a detector with a light sensitive area A, then, is determined by the photon energy and by the flux density:

$$P_{\text{opt}} = E\Phi A = h\nu\Phi A. \tag{1.5}$$

The magnitude of the momentum of a photon p in a medium is determined by its wavelength in this medium:

$$p = \frac{h}{\lambda}. \tag{1.6}$$

The wave number k is the magnitude of the wave vector $\boldsymbol{k}$, which defines the direction of the motion of the photon:

$$k = \frac{2\pi}{\lambda}. \tag{1.7}$$

The momentum of a photon $\boldsymbol{p}$ is proportional to the wave vector

$$\boldsymbol{p} = \hbar\boldsymbol{k}, \tag{1.8}$$

where $\hbar = h/(2\pi)$.

1.2 Energy Bands of Semiconductor Materials

The energy-band structure of a semiconductor determines not only its electrical properties but also its optical properties such as the absorption of photons and the probability of radiative transitions of electrons from the conduction band to the valence band. The absorption of photons, which is important for photodetectors, will be dealt with in the next section. Here we will just point out the consequences of the kind of energy-band structure for the applicability of a semiconductor material when we want to obtain efficient light emitters.

There are two requirements for transitions of electrons between the valence and the conduction band and vice versa: (i) the energy has to be conserved; (ii) the momentum has to be conserved. The conservation of energy usually is not a problem in direct and in indirect semiconductors. For an electron transition between the maximum of the valence band and the minimum of the conduction band, or vice versa, the conservation of momentum, however, cannot be fulfilled with the absorption or emission of a photon alone in an indirect semiconductor, because the magnitude of the momentum of a photon is several orders of magnitude smaller than that of an electron in a semiconductor. The same large difference holds between the wave vectors of a photon and an electron in a crystal. It is, therefore, possible to compare the momentums or the wave vectors. The energy bands in dependence on the wave vector are calculated from the Schrödinger equation with a periodic potential that is characteristic for a certain semiconductor. The wave vector of an electron in a crystal is between approximately 0 and $2\pi/a$, i. e. between the k values at the boundaries of the first Brillouin zone, where a is the lattice constant. The minimum of the conduction band in the first Brillouin zone in silicon for instance is at $0.85 \times 2\pi/a$ [1]. The momentum of an electron in the minimum of the conduction band of Si with $a = 0.543\,\text{nm}$, therefore, can reach $0.85 \times 2h/a = 2 \times 10^{-24}\,\text{Js/m}$, whereas the momentum of a photon with $E = 1\,\text{eV}$ is only $5.3 \times 10^{-28}\,\text{Js/m}$. In addition to a photon, a phonon has to be absorbed or emitted in order to conserve the momentum. Phonons are quantized lattice vibrations. They possess small energies (up to approximately 100 meV) and a momentum of the order of that of an electron in a semiconductor. Many phonons are present in crystals like semiconductors at room temperature.

Let us first consider GaAs, which has a direct bandgap (Fig. 1.1), where the minimum of the conduction band (CB) is above the maximum of the valence band (VB). No phonon is needed for the conservation of the momentum when an electron makes a transition from the minimum of the conduction band to the maximum of the valence band (recombining with a hole) and when a photon with the energy of the bandgap is emitted.

Silicon is known to have an indirect bandgap. A phonon is needed for the conservation of the momentum in order to enable the radiative transition of an electron from the minimum of the conduction band to the maximum of the valence band. A phonon has to be consumed or created when a photon with an energy of approximately E_g is emitted (see Fig. 1.1). The photon energy is exactly (E_p is the phonon energy)

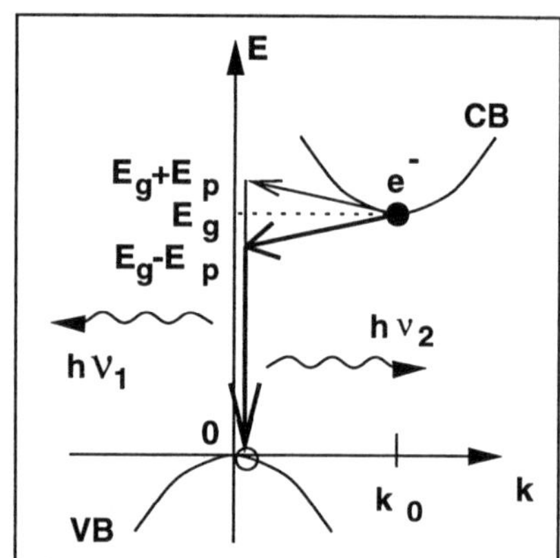

Fig. 1.1. Electron transitions in direct (*left*) and indirect (*right*) semiconductors as for instance GaAs and Si, respectively

$$E = h\nu = \frac{hc}{\lambda} = E_{\mathrm{g}} \pm E_{\mathrm{p}}. \tag{1.9}$$

The generation of a phonon possesses a larger probability than the consumption of a phonon in connection with an electron transition in an indirect semiconductor. Therefore, more photons will have the energy $E_{\mathrm{g}} - E_{\mathrm{p}}$ than $E_{\mathrm{g}} + E_{\mathrm{p}}$. From quantum mechanics it is, however, known that the probability of all electron transitions combined with phonon transitions is very small. The probability of radiative transitions in indirect semiconductors in fact is four to six orders of magnitude lower than that in direct semiconductors. Silicon devices, therefore, are usually very poor light emitters. This work will summarize the attempts to obtain more efficient silicon light emitters and it will, of course, focus on silicon photodectors.

1.3 Optical Absorption of Semiconductor Materials

The energy of a photon can be transferred to an electron in the valence band of a semiconductor, which is brought to the conduction band, when the photon energy is larger than the bandgap energy E_{g}. The photon is absorbed during this process and an electron–hole pair is generated. Photons with an energy smaller than E_{g}, however, cannot be absorbed and the semiconductor is transparent for light with wavelengths longer than $\lambda_{\mathrm{c}} = hc_0/E_{\mathrm{g}}$.

The optical absorption coefficient α is the most important optical constant for photodetectors. The absorption of photons in a photodetector to produce carrier pairs and thus a photocurrent, depends on the absorption coefficient α for the light in the semiconductor used to fabricate the detector. The

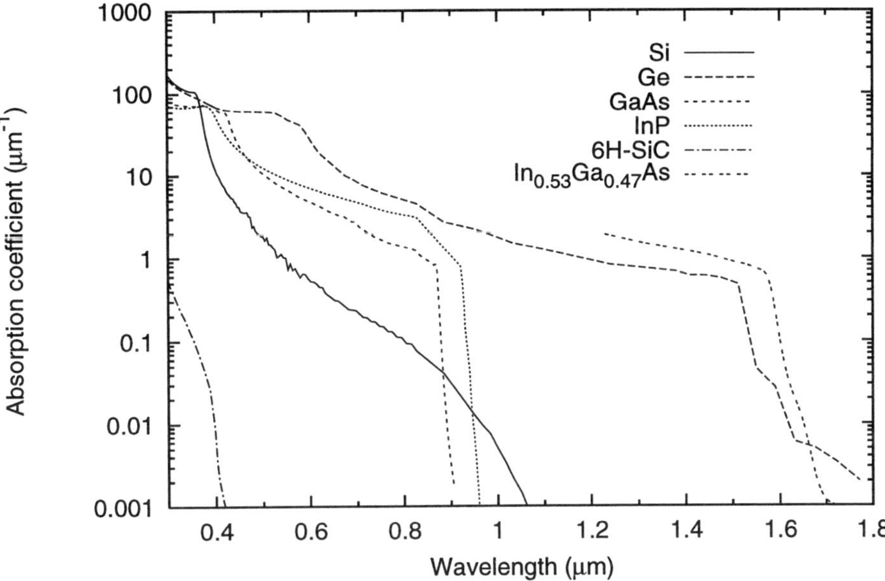

Fig. 1.2. Absorption coefficients of important semiconductor materials versus wavelength

absorption coefficient determines the penetration depth $1/\alpha$ of the light in the semiconductor material according to Lambert–Beer's law:

$$I(\bar{y}) = I_0 \exp(-\alpha \bar{y}). \tag{1.10}$$

The optical absorption coefficients for the most important semiconductor materials are compared in Fig. 1.2. The absorption coefficients strongly depend on the wavelength of the light. For wavelengths shorter than λ_c, which corresponds to the bandgap energy ($\lambda_c = hc_0/E_g$), the absorption coefficients increase rapidly according to the so-called *fundamental absorption*. The steepness of the onset of absorption depends on the kind of band–band transition. This steepness is large for direct band–band transitions as in GaAs ($E_g^{dir} = 1.42$ eV at 300 K), in InP ($E_g^{dir} = 1.35$ eV at 300 K), in Ge ($E_g^{dir} = 0.81$ eV at 300 K) and in $In_{0.53}Ga_{0.47}As$ ($E_g^{dir} = 0.75$ eV at 300 K). For Si ($E_g^{ind} = 1.12$ eV at 300 K), for Ge ($E_g^{ind} = 0.67$ eV at 300 K) and for the wide bandgap material 6H-SiC ($E_g^{ind} = 3.03$ eV at 300 K) the steepness of the onset of absorption is small.

$In_{0.53}Ga_{0.47}As$ and Ge cover the widest wavelength range including the wavelengths 1.3 and 1.54 µm which are used for long distance optical data transmission via optical fibers. The absorption coefficients of GaAs and InP are high in the visible spectrum (≈ 400–700 nm). Silicon detectors are also appropriate for the visible and near infrared spectral range. The absorption coefficient of Si, however, is one to two orders of magnitude lower than that of the direct semiconductors in this spectral range. For Si detectors, therefore, a much thicker absorption zone is needed than for the direct semiconductors. We will, however, see in this work that with silicon photodiodes GHz operation is nevertheless possible. Silicon is the economically most important semiconductor and it is worthwhile to investigate silicon optoelectronic devices and integrated circuits in spite of the nonoptimum optical absorption of silicon.

Table 1.1. Absorption coefficients α of silicon and intensity factors I_0 (ehp/cm^3 means electron–hole pairs/cm^3) for several important wavelengths for a constant photon flux density of $\Phi = I_0/\alpha = 1.58 \times 10^{18}$ photons/cm^2

Wavelength (nm)	α (μm^{-1})	I_0 (ehp/cm^3)
980	0.0065	1.03×10^{20}
850	0.06	9.50×10^{20}
780	0.12	1.89×10^{21}
680	0.24	3.79×10^{21}
635	0.38	6.00×10^{21}
565	0.73	1.16×10^{22}
465	3.6	5.72×10^{22}
430	5.7	9.00×10^{22}

The absorption coefficients of silicon for wavelengths which are the most important ones in practice are listed in Table 1.1 [2, 3]. In order to compare the quantum efficiencies of photodetectors for different wavelengths, it is advantageous to use the same photon flux for the different wavelengths. The photocurrents of photodetectors are equal for the same fluxes of photons with different energy, i. e. for different light wavelengths, when the quantum efficiency of the photodetector is the same for the different photon energies or wavelengths, respectively. According to the Lambert–Beer law, different intensity factors I_0 result for a constant photon flux. As an example, intensity factors are listed for the most important wavelengths in Table 1.1 for a certain arbitrary photon flux density.

1.4 Photogeneration

The Lambert–Beer law can be formulated for the optical power $\bar{P}$ analogously to (1.10):

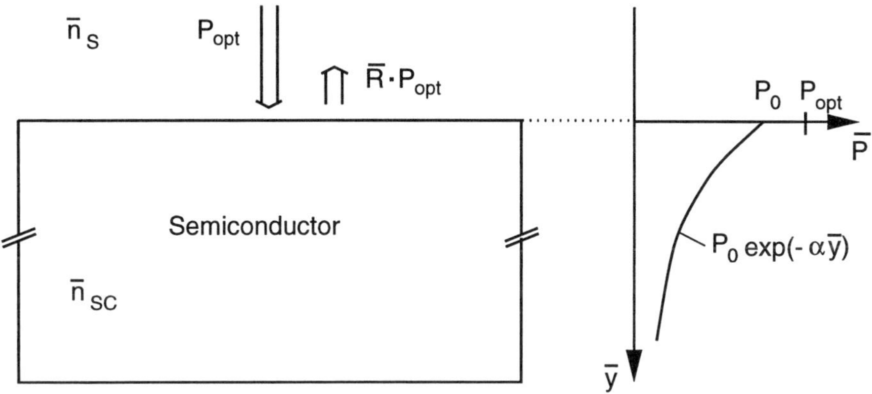

Fig. 1.3. Reflection at a semiconductor surface and decay of the optical power in the semiconductor ($P_0 = (1-\bar{R})P_{\mathrm{opt}}$)

$$\bar{P}(\bar{y}) = P_0 \exp(-\alpha\bar{y}). \tag{1.11}$$

The optical power at the surface of the semiconductor $\bar{P}(\bar{y}=0)$ is $P_0 = (1-\bar{R})$ (see Fig. 1.3). The optical power of the light penetrating into a medium decreases exponentially with the penetration coordinate $\bar{y}$ in the medium (compare Fig. 2.1).

The absorbed light generates electron–hole pairs in a semiconductor due to the internal photoeffect provided that $h\nu > E_{\mathrm{g}}$. Therefore, we can express the generation rate per volume $G(\bar{y})$ as:

$$G(\bar{y}) = \frac{\bar{P}(\bar{y}) - \bar{P}(\bar{y}+\triangle\bar{y})}{\triangle\bar{y}} \frac{1}{Ah\nu}. \tag{1.12}$$

In this equation, A is the area of the cross section for the light incidence and $h\nu$ is the photon energy. For $\triangle\bar{y} \to 0$, we can write $(\bar{P}(\bar{y}) - \bar{P}(\bar{y}+\triangle\bar{y}))/\triangle\bar{y} = -d\bar{P}(\bar{y})/d\bar{y}$. From (1.11), $d\bar{P}(\bar{y})/d\bar{y} = -\alpha\bar{P}(\bar{y})$ then follows and the generation rate is

$$G(\bar{y}) = \frac{\alpha P_0}{Ah\nu} \exp(-\alpha\bar{y}). \tag{1.13}$$

1.5 External Quantum Efficiency and Responsivity

The external or overall quantum efficiency η is defined as the number of photogenerated electron–hole pairs, which contribute to the photocurrent, divided by the number of the incident photons. The external quantum efficiency can be determined, when the photocurrent of a photodetector is measured for a known incident optical power.

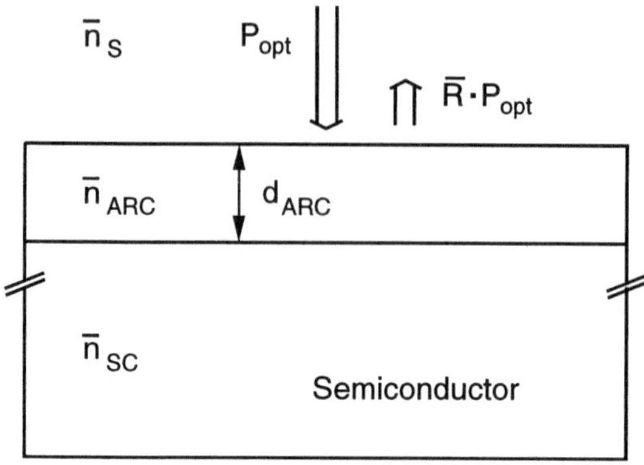

Fig. 1.4. Semiconductor with an antireflection coating

A fraction of the incident optical power is reflected (see Fig. 1.3) due to the difference in the index of refraction between the surroundings $\bar{n}_{\mathrm{s}}$ (air: $\bar{n}_{\mathrm{s}} = 1.00$) and the semiconductor $\bar{n}_{\mathrm{sc}}$ (e. g. Si, $\bar{n}_{\mathrm{sc}} \approx 3.5$). The reflectivity $\bar{R}$ depends on the index of refraction $\bar{n}_{\mathrm{sc}}$ and on the extinction coefficient $\bar{\kappa}$ of an absorbing medium, for which the dielectric function $\bar{\epsilon} = \bar{\epsilon}_1 + \mathrm{i}\bar{\epsilon}_2 = (\bar{n}_{\mathrm{sc}} + \mathrm{i}\bar{\kappa})^2$ is valid ($\bar{n}_{\mathrm{s}} = 1$) [2].

$$\bar{R} = \frac{(1 - \bar{n}_{\mathrm{sc}})^2 + \bar{\kappa}^2}{(1 + \bar{n}_{\mathrm{sc}})^2 + \bar{\kappa}^2}. \tag{1.14}$$

The extinction coefficient is sufficient for the description of the absorption. The absorption coefficient α can be expressed as:

$$\alpha = \frac{4\pi\bar{\kappa}}{\lambda_0}. \tag{1.15}$$

The optical quantum efficiency η_{o} can be defined in order to consider the partial reflection:

$$\eta_{\mathrm{o}} = 1 - \bar{R}. \tag{1.16}$$

The reflected fraction of the optical power can be minimized by introducing an antireflection coating (ARC) with thickness d_{ARC} (see Fig. 1.4):

$$d_{\mathrm{ARC}} = \frac{\lambda_0}{4\bar{n}_{\mathrm{ARC}}}. \tag{1.17}$$

The index of refraction of the ARC-layer can be calculated:

$$\bar{n}_{\mathrm{ARC}} = \sqrt{\bar{n}_{\mathrm{s}}\bar{n}_{\mathrm{sc}}}. \tag{1.18}$$

The optimum index of refraction of the ARC-layer is determined by the refractive index $\bar{n}_{\mathrm{s}}$ of the surroundings and by the refractive index $\bar{n}_{\mathrm{sc}}$ of the semiconductor. A complete suppression of the partial reflection, however, is not possible in practice. For silicon photodetectors, SiO_2 ($\bar{n}_{sc} = 1.45$) and Si_3N_4 ($\bar{n}_{sc} = 2.0$) ARC-layers are most appropriate.

Because of the partial reflection, it is useful to define the internal quantum efficiency η_i as the number of photogenerated electron–hole pairs, which contribute to the photocurrent, divided by the number of photons which penetrate into the semiconductor.

The external quantum efficiency is the product of the optical quantum efficiency η_o and of the internal quantum efficiency η_i:

$$\eta = \eta_o \eta_i . \tag{1.19}$$

The internal quantum efficiency η_i will be discussed in Sect. 2.2.3 after carrier diffusion and drift have been introduced.

For the development of photoreceiver circuits, and especially of transimpedance amplifiers, it is interesting to know how large the photocurrent is for a specified power of the incident light with a certain wavelength. The responsivity R is a useful quantity for such a purpose:

$$R = \frac{I_{ph}}{P_{opt}} = \frac{q\lambda_0}{hc}\eta = \frac{\lambda_0 \eta}{1.243} \frac{A}{W}, \tag{1.20}$$

where λ_0 is in μm. The responsivity is defined as the photocurrent I_{ph} divided by the incident optical power. R depends on the wavelength, therefore wavelength has to be mentioned if a responsivity value is given. The dependence of the responsivity R on wavelength λ_0 is shown in Fig. 1.5 for a quantum efficiency $\eta = 1$.

The line shown in Fig. 1.5 represents the maximum responsivity of an ideal photodetector with $\eta = 1$. The responsivity of real detectors is always

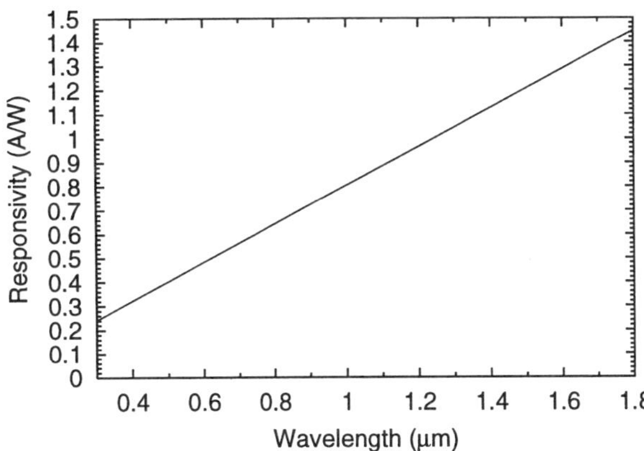

Fig. 1.5. Responsivity of an ideal photodetector with a quantum efficiency $\eta = 1$ versus wavelength ($\lambda_c = 1.8\,\mu m$). The responsivity of real detectors possesses smaller values

lower due to partial reflection of the light at the semiconductor surface and due to partial recombination of photogenerated carriers in the semiconductor or at its surface.

2. Theory

After the collection of the most important optical and optoelectronic definitions in the last chapter, we will now summarize the fundamentals of device physics and modeling of solid-state electron devices including photodetectors in a compact form. For a detailed description of the principles of photodetectors, the book [1] is recommended. A detailed review on modeling of solid-state electron devices can be found in [4]. Here, the semiconductor equations with implemented photogeneration and the models for carrier mobility used in device simulators will be listed first. Carrier drift and diffusion as well as their consequences for the speed and the quantum efficiency of photodetectors will be explained. Furthermore, the equivalent circuit of a photodiode will be discussed in order to show further aspects concerning the speed of photoreceivers.

2.1 Semiconductor Equations

The physics of semiconductor devices like diodes, MOSFETs (Metal-Oxide-Silicon Field-Effect Transistors), bipolar transistors, and photodetectors is well known. The equations describing the behavior of these devices in most important cases are the semiconductor equations. These equations have already been implemented in many device simulation programs. The physical models for the parameters used in the semiconductor equations were discussed thoroughly, for instance, in [4].

Device simulation programs have been valuable tools for the development of semiconductor devices for many years. Much time and money can be saved with their help for such a purpose. They are also valuable for the development of photodetectors. We will take the two-dimensional device simulator MEDICI as an example [5]. The drift-diffusion model is implemented in this simulator. MEDICI solves the Poisson equation (2.1), the transport equations (2.3) and (2.4), and the continuity equations (2.6) and (2.7) for electrons and holes. Furthermore, photogeneration is implemented. Due to the photogeneration for the internal photoeffect, electron–hole pairs are created, i. e. the corresponding generation terms for electrons and holes are equal.

The potential Ψ in the device, for which the simulation is performed, is calculated according to the Poisson equation:

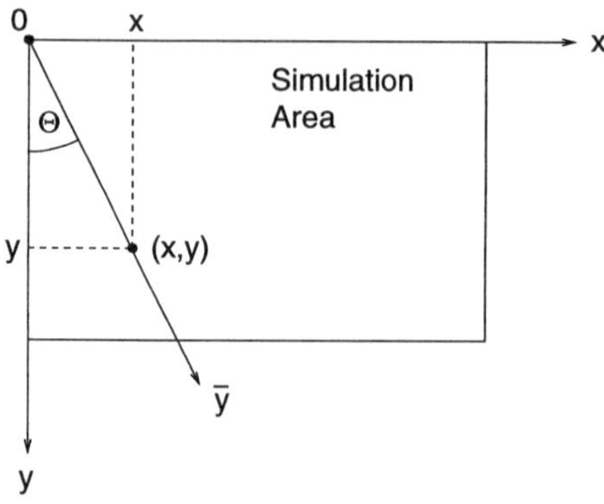

Fig. 2.1. Coordinate transformation from the penetration coordinate $\bar{y}$ to the coordinates (x,y) used in the two-dimensional device simulator for non-perpendicular light incidence

$$\triangle\Psi = -\frac{\rho}{\epsilon} \,. \tag{2.1}$$

The quantity ϵ is the product of the relative and absolute dielectric constants: $\epsilon = \epsilon_r\epsilon_0$. The symbol ρ represents the charge density, which can be further broken apart into the product of the elementary charge q times the sum of the hole concentration p (positively charged), of the electron concentration n (negatively charged), of the donor concentration N_D (positively charged), and of the acceptor concentration N_A (negatively charged):

$$\rho = q(p - n + N_D - N_A). \tag{2.2}$$

The current densities for electrons and holes are the sum of the drift and diffusion current densities:

$$\boldsymbol{j}_n = qn\mu_n\boldsymbol{E} + qD_n \,\mathbf{grad}\, n, \tag{2.3}$$

$$\boldsymbol{j}_p = qp\mu_p\boldsymbol{E} - qD_p \,\mathbf{grad}\, p. \tag{2.4}$$

The total current density results from the electron and hole current densities:

$$\boldsymbol{j} = \boldsymbol{j}_n + \boldsymbol{j}_p. \tag{2.5}$$

The continuity equations with the inclusion of photogeneration $G(x,y)$ due to the penetration of light into the semiconductor can be written as:

$$\frac{\partial n}{\partial t} = \frac{\mathrm{div}\,\boldsymbol{j}_n}{q} + U_{th} + G(x,y), \tag{2.6}$$

$$\frac{\partial p}{\partial t} = -\frac{\mathrm{div}\,\boldsymbol{j}_p}{q} + U_{th} + G(x,y). \tag{2.7}$$

U_{th} is the thermal generation/recombination term. The photogeneration $G(\bar{y})$ was derived above. The simulator allows us to define nonperpendicular incidence for the light. Then, $G(x, y) = G(\bar{y} = (x^2 + y^2)^{1/2})$ has to be used (compare with Fig. 2.1).

The electric field is determined by (2.1) and obeys:

$$\boldsymbol{E} = -\mathbf{grad}\, \Psi. \tag{2.8}$$

The electric field, for instance, is important for the calculation of the drift velocity $\boldsymbol{v}$ of photogenerated carriers in photodetectors:

$$\boldsymbol{v} = \mu \boldsymbol{E}. \tag{2.9}$$

The mobilities of electrons and holes differ strongly and μ_n or μ_p has to be used for μ in order to calculate the electron drift velocity or the hole drift velocity, respectively. The mobilities μ_n and μ_p are only constant for a low electric field. It will be shown below that the drift velocities of electrons and holes saturate for large values of the electric field.

For the calculation of the speed of photodetectors we not only need the mobilities and the electric field but also the width of the space-charge region, where an electric field is present. In the so-called depletion approximation, i. e. with n and p approximately equal to zero in the space-charge region and $\rho = 0$ outside the space-charge region, the distribution of the electric field and the width of the space-charge region W can be calculated analytically for an abrupt PN junction:

$$W = \sqrt{\frac{2\epsilon_\mathrm{r}\epsilon_0}{q} \frac{N_\mathrm{A} + N_\mathrm{D}}{N_\mathrm{A} N_\mathrm{D}} \left(U_\mathrm{D} - U - \frac{2k_\mathrm{B}T}{q} \right)}, \tag{2.10}$$

with

$$U_\mathrm{D} = \frac{k_\mathrm{B}T}{q} \ln \frac{N_\mathrm{A} N_\mathrm{D}}{n_\mathrm{i}^2}. \tag{2.11}$$

Let us assume that one side of the PN junction is doped to a much larger extent than the other, for instance $N_\mathrm{D} = 10^{20}\,\mathrm{cm}^{-3} \gg N_\mathrm{A} = 10^{16}\,\mathrm{cm}^{-3}$ or vice versa. We want to define the lower doping concentration as N_I. Equation (2.10) then can be simplified to

$$W = \sqrt{\frac{2\epsilon_\mathrm{r}\epsilon_0}{qN_\mathrm{I}} \left(U_\mathrm{D} - U - \frac{2k_\mathrm{B}T}{q} \right)}. \tag{2.12}$$

For the reverse voltage U, here, a negative value always has to be used. The space-charge region only spreads into the low doped side of the PN junction due to the simplification. This behavior, however, is a good approximation for most semiconductor diodes and for photodiodes, accordingly. The width of the space-charge region as a function of the doping concentration of the low doped side of the PN junction is shown in Fig. 2.2.

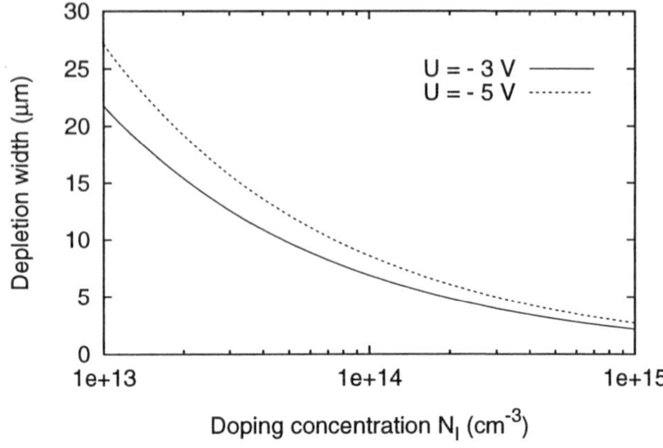

Fig. 2.2. Depletion layer width versus doping concentration of the lower doped side of a PN junction

It should be mentioned for the sake of completeness that for the modeling of very small devices with large electric fields, for which the drift velocities of electrons and holes usually saturate in devices with larger dimensions, and with locally varying electric fields, the drift-diffusion model will not be sufficient [6]. For these very small devices with drift times on the picosecond time scale, the so-called energy balance equations have to be solved self-consistently together with the drift-diffusion equations [6, 7]. The resulting model is called the hydrodynamic model. It allows the carrier temperature distribution and, therefore, the carrier velocity distribution inside a device to be determined for both electrons and holes. It allows us to consider transient effects on a picosecond time scale and carrier cooling due to impact ionization. Energy relaxation times for carriers in silicon are on the order of 0.1 ps [8]. Drift times usually are at least 1000 times larger in photodetectors of optoelectronic integrated circuits. The drift-diffusion model, therefore, is sufficient for the modeling of silicon photodetectors described in this work.

The hydrodynamic model is important for devices wherein the electric field varies significantly within the length of one carrier mean-free path. The mean-free path is of the order of 10 nm for the maximum carrier mobilities and decreases with the mobilities [9]. The hydrodynamic model has been used successfully in small bipolar devices for reducing spurious velocity overshoot effects [5]. The hydrodynamic model is necessary for the simulation of compound semiconductor devices, for which negative derivatives of the mobility in dependence on the electric field may occur. For the photodetectors described in this work with thicknesses of the drift regions in the range from 1–10 µm, none of these aspects is relevant and the drift-diffusion model is sufficient for the modeling of these photodetectors. We, therefore, continue with the description of the models used together with the drift-diffusion equations.

2.2 Important Models for Photodetectors

The most important processes for the characterization of photodetectors with respect to their speed are carrier drift and minority carrier diffusion. The carrier drift in conventional semiconductor materials is a much faster process than the minority carrier diffusion. For the calculation of the frequency response of photodiodes, series resistances and PN junction capacitances are also important. An even more complete equivalent circuit for photodiodes considering wiring capacitances will be discussed. Minority carrier diffusion, series resistances and capacitances reduce the speed of photodiodes.

2.2.1 Carrier Drift

The carrier mobilities in the drift terms (see (2.3) and (2.4)) are the parameters which are responsible for the obtainable speed of photodetectors. The carrier mobilities depend on the doping concentration and on the electric field. An empirical expression is available for the dependence of the mobility on the total impurity concentration N_{total}, which also considers the influence of temperature T in K [10, 11]:

$$\mu_{0\mathrm{n}} = \mu_{\mathrm{n,min}} + \frac{\mu_{\mathrm{n,max}}(T/300)^{\nu_{\mathrm{n}}} - \mu_{\mathrm{n,min}}}{1 + (T/300)^{\chi_{\mathrm{n}}} (N_{\mathrm{total}}/N_{\mathrm{ref,n}})^{\alpha_{\mathrm{n}}}}, \tag{2.13}$$

$$\mu_{0\mathrm{p}} = \mu_{\mathrm{p,min}} + \frac{\mu_{\mathrm{p,max}}(T/300)^{\nu_{\mathrm{p}}} - \mu_{\mathrm{p,min}}}{1 + (T/300)^{\chi_{\mathrm{p}}} (N_{\mathrm{total}}/N_{\mathrm{ref,p}})^{\alpha_{\mathrm{p}}}}. \tag{2.14}$$

N_{total} is the sum of the concentrations of acceptors and donors. The other parameters for silicon are listed in Table 2.1.

The dependence of the carrier mobilities in silicon on the doping concentration according to (2.13) and (2.14) is shown in Fig. 2.3 for $T = 300$ K.

Table 2.1. Parameter values for the approximation of the electron and hole mobilities in silicon with (2.13) and (2.14)

Parameter	Electrons	Holes
μ_{min} (cm^2/Vs)	17.8	48.0
μ_{max} (cm^2/Vs)	1350	495
N_{ref} (cm^{-3})	1.072×10^{17}	1.606×10^{17}
$\nu_{\mathrm{n,p}}$	−2.3	−2.2
$\chi_{\mathrm{n,p}}$	−3.8	−3.7
$\alpha_{\mathrm{n,p}}$	0.73	0.70

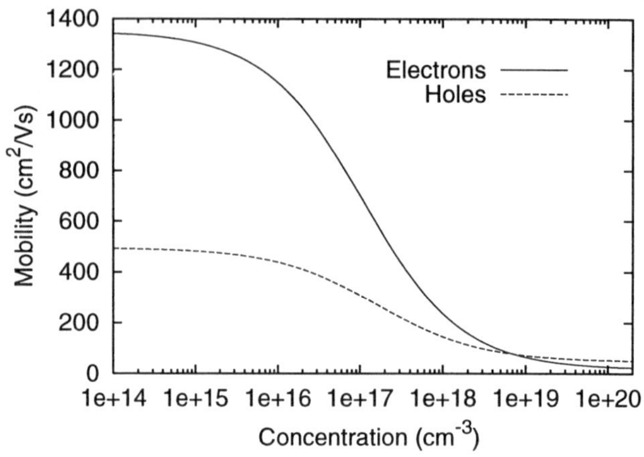

Fig. 2.3. Carrier mobilities in silicon versus total doping concentration [10, 11]

According to [10], the dependence of the carrier mobilities in silicon on the electric field can be approximated by

$$\mu_{\rm n} = \frac{\mu_{0\rm n}}{1 + (\mu_{0\rm n} E / v_{\rm n}^{\rm sat})^2}, \tag{2.15}$$

$$\mu_{\rm p} = \frac{\mu_{0\rm p}}{1 + (\mu_{0\rm p} E / v_{\rm p}^{\rm sat})^1}. \tag{2.16}$$

The values for the saturation velocities (in cm/s) can be computed from the expression [12] with T in K:

$$v_{\rm n}^{\rm sat}(T) = v_{\rm p}^{\rm sat}(T) = \frac{2.4 \times 10^7}{1 + 0.8 \exp(T/600)}. \tag{2.17}$$

The curves calculated with (2.15) and (2.16) are shown in Fig. 2.4 for a very low doping concentration, i. e. for the 'intrinsic' zone of a PIN photodiode. The carrier mobilities begin to degrade for electric fields in excess of approximately 2000 V/cm.

The drift velocities, therefore, are proportional to the electric field only for smaller values of the electric field (Fig. 2.5). For values of the electric field larger than approximately 10 000 V/cm the electron drift velocity saturates. For holes, the electric field has to exceed a value of approximately 100 000 V/cm in order to obtain the hole saturation velocity.

The dependence of the carrier mobilities on the doping concentration is implemented in MEDICI with the CONMOB model [5]. The model in MEDICI for the dependence of the carrier mobilities, i. e. of the drift velocities on the electric field is called FLDMOB.

One quantity, which should be mentioned here, is the drift time, i. e. the time needed by carriers to drift through the whole width W of the space-charge region or drift zone. The drift time is determined by the drift velocity

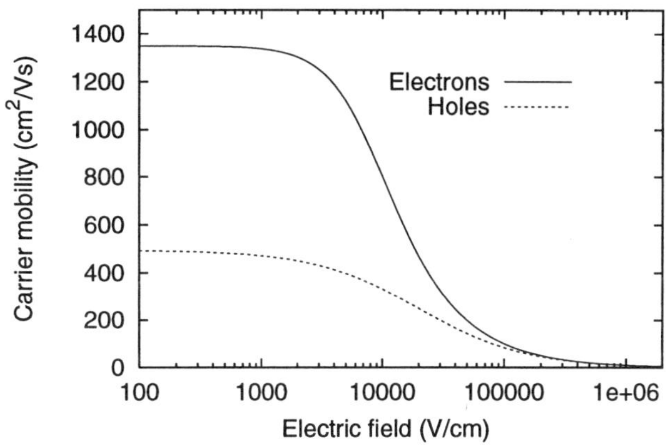

Fig. 2.4. Carrier mobilities versus electric field for weakly doped silicon

v and by the width of the drift zone W. The drift velocity v depends on μ and E. Furthermore, μ depends on the doping concentration. The doping concentration and the bias voltage applied to the photodiode determine the distribution of the electric field. It is, therefore, very difficult to calculate the drift time analytically, because v, μ and E in a real photodetector depend on the space coordinates. Numerical process and device simulations, in general, are necessary in order to compute the transient response of photodetectors.

When we choose the ideal PIN photodiode with a constant electric field for instance (see Fig. 2.6), it is possible to give some analytical expressions for the drift time $t_{\rm d}$ and for the rise and fall times $t_{\rm r}$ and $t_{\rm f}$ of the photocurrent (time difference between the 10 % and 90 % values of the stationary photocurrent) [13]:

$$t_{\rm d} = t_{\rm r} = t_{\rm f} = \frac{d_{\rm I}}{v(E_0)}. \qquad (2.18)$$

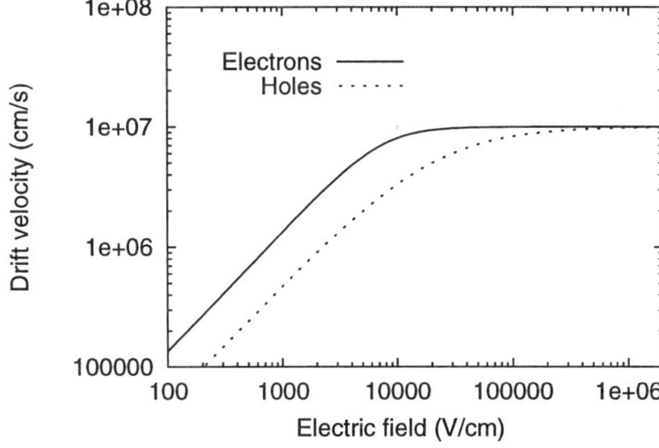

Fig. 2.5. Carrier drift velocities in silicon versus electric field calculated according to (2.9), (2.15) and (2.16) for a low doping concentration

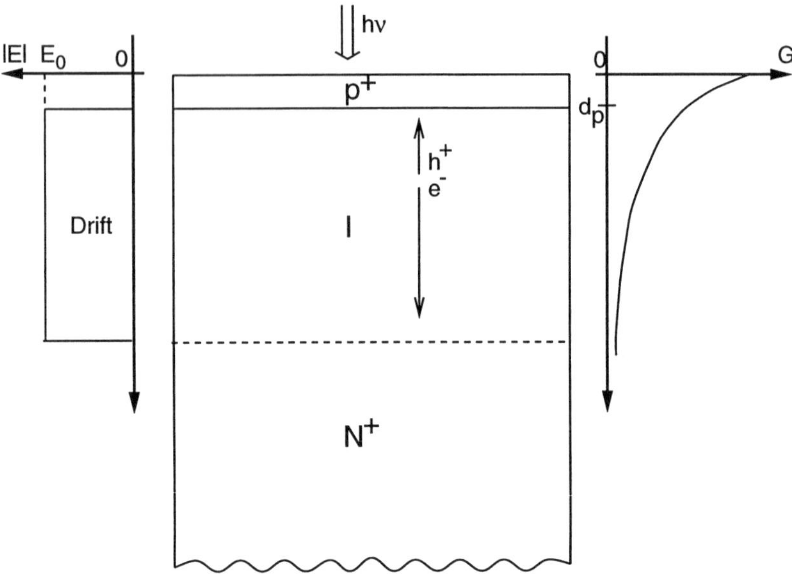

Fig. 2.6. Drift region in an ideal PIN photodiode with an undoped intrinsic I-layer

Instead of W, the thickness of the intrinsic I-layer of the ideal PIN photodiode d_{I} is used in (2.18). For a large reverse voltage, which results in a large value of the electric field, the drift velocities may be approximated by the saturation velocity v_{s} and

$$t_{\mathrm{d}} = t_{\mathrm{r}} = t_{\mathrm{f}} = \frac{d_{\mathrm{I}}}{v_{\mathrm{s}}} \tag{2.19}$$

results. The rise and fall times limit the maximum data rate DR of optical receivers. The maximum data rate of a photodiode (in the non-return-to-zero transmission mode) can be estimated in a conservative way to be the lower value of DR$= 1/(3t_{\mathrm{r}})$ and DR $= 1/(3t_{\mathrm{f}})$. The mean value of t_{r} and t_{f} may lead to DR$= 2/(3(t_{\mathrm{r}} + t_{\mathrm{f}}))$. In a more aggressive way, instead of the factor 3, the factor 2 can be used, however, in practice the bit-error rate of optical receivers determines the usable data rate. The relation between the rise time and the -3dB bandwidth should also be mentioned [14,15]:

$$f_{3\mathrm{dB}} = \frac{2.4}{2\pi t_{\mathrm{r}}} \approx \frac{0.4 v_{\mathrm{s}}}{W}. \tag{2.20}$$

This equation was obtained for a time-dependent sinusoidal photogeneration at the surface of the PIN photodiode after integrating the conduction current density and the displacement current density over the drift zone thickness [15]. The amplitude of the short circuit photocurrent density dropped to $1/\sqrt{2}$ of its low-frequency value for $\omega t_{\mathrm{d}} = 2.4$ leading to (2.20) when assuming that the transit time t_d is equal to the rise time t_r. When the thickness of the intrinsic I-layer is chosen as $W = d_I = 1/\alpha$, a simple expression is obtained for the −3dB frequency of a PIN photodiode:

$$f_{3dB} \approx 0.4\alpha v_s. \tag{2.21}$$

Another case, for which an analytical solution may be justified, is a PN photodiode with a very large electric field. Then $v \approx v_s$ and $t_d = W/v_s$ may be used.

The electric field in most real photodiodes, however, is much lower than would be necessary in order to justify the assumption of the saturation velocities. The electric field in real PIN photodiodes also is not constant even at the low 'intrinsic' doping level of $10^{13}\,\mathrm{cm}^{-3}$ [16] and, besides, a rather limited validity range of $\alpha d_I > 3$ for an analytical model, process and device simulations for the computation of the rise and fall times are necessary for real PIN photodiodes. These process and device simulations are also capable of considering the influence of carrier diffusion, which will be discussed next, on the transient behavior of photodetectors.

2.2.2 Carrier Diffusion

The diffusion of minority carriers is important for the calculation of the speed of photodetectors, when electron–hole pairs are generated in the quasineutral regions of the semiconductor where no electric field is present. There can be two such regions in a photodiode. The first region is the heavily doped surface region (P^+ in Fig. 2.7), which, however, in most cases is not very thick. The second and usually more critical region is in the depth of the semiconductor below the edge of the space-charge region SCR (see Fig. 2.7).

The carrier diffusion coefficients D_n and D_p depend on the carrier mobilities according to the Einstein relation

$$D_{n/p} = \mu_{n/p}\frac{k_B T}{q} = \mu_{n/p} U_T. \tag{2.22}$$

The value of the diffusion coefficient of electrons in Si is only approximately $35\,\mathrm{cm^2/s}$ for low doping concentrations. The value reduces considerably for high doping concentrations. The diffusion coefficient of holes in Si is less than $12.5\,\mathrm{cm^2/s}$ and also depends on the doping concentration.

The distance over which minority carriers diffuse is called the carrier diffusion length L_n or L_p for electrons or holes, respectively. The carrier diffusion lengths depend on the carrier diffusion coefficients and on the carrier lifetimes τ_n and τ_p:

$$L_n = \sqrt{D_n \tau_n}, \tag{2.23}$$

$$L_p = \sqrt{D_p \tau_p}. \tag{2.24}$$

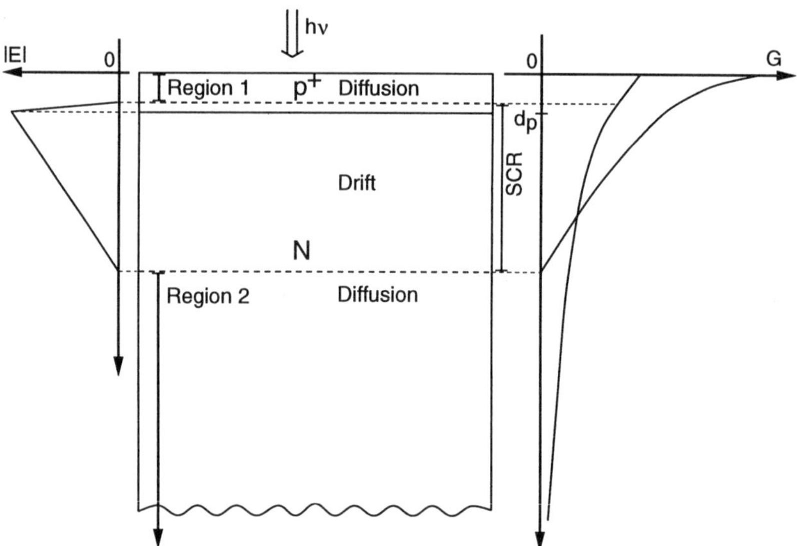

Fig. 2.7. Drift and diffusion regions in a photodiode

The carrier lifetimes are due to dopants and unwanted impurities in the semiconductor, which are unavoidably introduced in quite small but noticeable amounts in electronic devices during the fabrication process. These impurities may act as recombination centers. Recombination centers introduce deep energy levels in the bandgap, which enhance the recombination rate of electrons and holes [17–19]. Short carrier lifetimes result in a large recombination rate. The Shockley–Read–Hall (SRH) generation/recombination rate U_{th} is implemented in MEDICI in this form [5]:

$$U_{\mathrm{th}} = U_{\mathrm{n}} = U_{\mathrm{p}} = \frac{pn - n_{\mathrm{i}}^2}{\tau_{\mathrm{p}}[n + n_{\mathrm{i}} \exp((E_{\mathrm{t}} - E_{\mathrm{i}})/k_{\mathrm{B}}T)] + \tau_{\mathrm{n}}[p + n_{\mathrm{i}} \exp((E_{\mathrm{i}} - E_{\mathrm{t}})/k_{\mathrm{B}}T)]} . \tag{2.25}$$

The recombination rate U_{th} depends on the densities of electrons and holes, on the intrinsic carrier density n_{i}, on the carrier lifetimes τ_{n}, τ_{p}, and on the difference between the energy level of the recombination center E_{t} and the intrinsic Fermi level E_{i}. The carrier lifetimes themselves depend on the total impurity density N_{total} [5]:

$$\tau_{\mathrm{n}} = \frac{\tau_{0,\mathrm{n}}}{1 + N_{\mathrm{total}}/N_{\mathrm{SRH,n}}} , \tag{2.26}$$

$$\tau_{\mathrm{p}} = \frac{\tau_{0,\mathrm{p}}}{1 + N_{\mathrm{total}}/N_{\mathrm{SRH,p}}} . \tag{2.27}$$

In N_{total}, dopants and recombination centers have to be considered. The carrier lifetimes for low doping concentrations and low densities of recombination centers are $\tau_{0,\mathrm{n}}$ and $\tau_{0,\mathrm{p}}$, respectively. The reference parameters $N_{\mathrm{SRH,n}}$ and $N_{\mathrm{SRH,p}}$ for silicon are both equal to $5 \times 10^{16}\,\mathrm{cm}^{-3}$.

Equation (2.25) can be simplified for $\tau_\mathrm{n} = \tau_\mathrm{p}$ and it can be seen that the recombination rate approaches a maximum for $E_\mathrm{t} = E_\mathrm{i}$. The carrier lifetimes are inversely proportional to the recombination rate for low injection conditions, i. e., when the densities of injected carriers $\triangle n = \triangle p$ are much smaller than the density of majority carriers.

The minority carrier lifetime (hole lifetime) in an N-type semiconductor then is

$$\tau_\mathrm{p} = (\sigma_\mathrm{p} v_\mathrm{th} N_\mathrm{t})^{-1}. \tag{2.28}$$

The electron lifetime in a P-type semiconductor similarly is

$$\tau_\mathrm{n} = (\sigma_\mathrm{n} v_\mathrm{th} N_\mathrm{t})^{-1}. \tag{2.29}$$

The carrier capture cross sections σ_n, σ_p, the thermal carrier velocity v_th and the density of unwanted impurities N_t determine the minority carrier lifetimes in low doped silicon.

The minority carrier lifetimes in as grown silicon can reach several milliseconds, whereas the fabrication process of photodetectors or integrated circuits can reduce the carrier lifetimes to the order of microseconds. Usually the minority carrier lifetimes are important only when carrier diffusion is the dominating transport mechanism. Drift times are usually of the order of nanoseconds and carrier lifetimes of the order of microseconds do not deteriorate the quantum efficiency of photodetectors when carrier drift is dominating.

Carrier diffusion is the dominant aspect in solar cells, which do not need a very quick transient response to changing light intensities [20]. The spectrum of the sun contains rather long wavelengths, which penetrate several tens of micrometers into silicon. Therefore, many electron–hole pairs are generated in region 2 (see Fig. 2.7), where carrier diffusion occurs. In order to obtain a high quantum efficiency of a solar cell, a long carrier diffusion length and, therefore, a long carrier lifetime is necessary in the semiconductor. The long carrier lifetimes in silicon with a highly developed crystal growth and process technology are the reason why silicon is the dominant material in solar cell production. The interested reader will find the state of the art in solar cell technology in [21–24], for instance.

Other optoelectronic devices, where carrier diffusion is important, are bipolar phototransistors. The diffusion of minority carriers in the base of bipolar transistors limits their switching speed and their transit frequency [25]. We will return to photodiodes here.

We want to estimate the time t_diff for the diffusion of electrons through a P^+ region with thickness d_p [26, 27]

$$d_\mathrm{p} = \sqrt{2 D_\mathrm{n} t_\mathrm{diff}}. \tag{2.30}$$

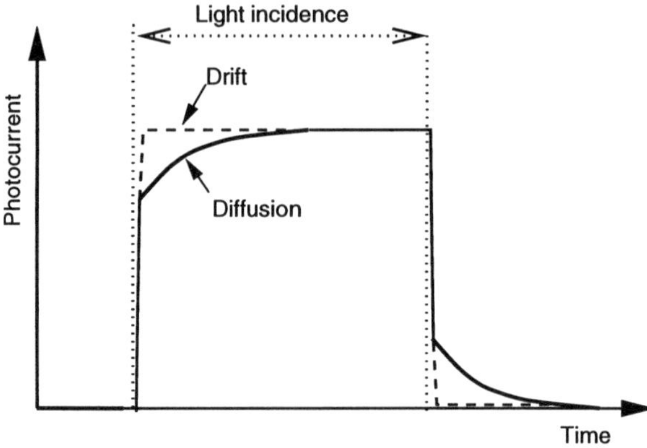

Fig. 2.8. Transient behavior of the photocurrent for carrier drift and diffusion

This equation has been derived for a time-dependent sinusoidal electron density due to photogeneration in the P^+ layer from the electron diffusion equation [27]. The time response of a low-pass filter with time constant $\tau_r = d_p^2/(2D_n)$ was obtained and this time constant was interpreted as the time t_{diff} that electrons need to diffuse to the space-charge region at the P^+N junction. With (2.22), t_{diff} can be obtained:

$$t_{diff} = \frac{d_p^2 q}{2\mu_n k_B T}. \tag{2.31}$$

A much slower contribution to the photocurrent usually results from carriers being generated in the second diffusion region (see Fig. 2.7), because the holes have to diffuse a much longer distance than d_p from this region 2 to the space-charge region. The hole diffusion time through 10 μm of silicon is 40 ns whereas the electron diffusion time over the same distance is approximately 8 ns. The resulting shape of the photocurrent shown in Fig. 2.8 is characteristic of the diffusion of carriers from the second region into the space-charge region. The influence of carrier diffusion on the frequency response of photodetectors is shown in Fig. 2.9.

Summarizing, it can be concluded that carrier diffusion should be avoided in order to obtain fast photodetectors. One possibility is to use light with a short wavelength, i. e. with a large absorption coefficient in Si in order to avoid photogeneration in region two (see Fig. 2.7). Another possibility can be the choice of a semiconductor with larger absorption coefficients. Where neither of these measures is possible, a larger reverse voltage should be applied across the photodiode or the doping concentration in the photodiode should be reduced, in order to obtain a thicker space-charge region.

2.2.3 Internal Quantum Efficiency

Strictly speaking we have to distinguish between the stationary and the dynamical internal quantum efficiency. In the stationary case, the light inten-

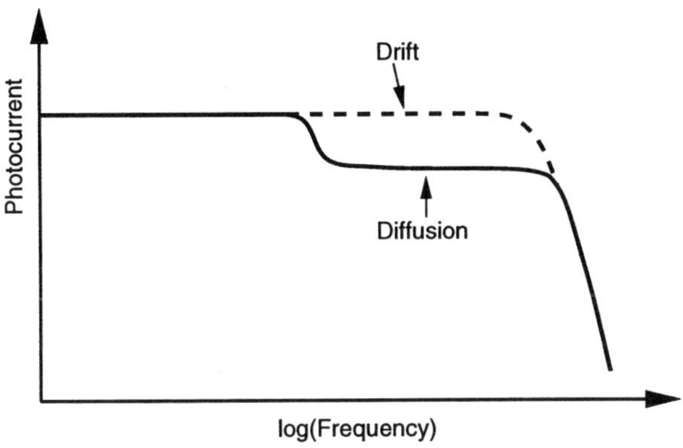

Fig. 2.9. Frequency response of the photocurrent for carrier drift and diffusion

sity and the photocurrent are constant with time. In the dynamical case both change with time. The dynamical quantum efficiency generally is lower than the stationary one.

Let us discuss the stationary case first. As explained above, practically all carriers which are photogenerated in drift regions contribute to the photocurrent. In other words, there is no negative influence of recombination on the internal quantum efficiency in the space-charge regions of photodiodes. The recombination of photogenerated carriers in region 1 and 2 (see Fig. 2.7), however, reduces the internal quantum efficiency. In the highly doped region 1, the carrier lifetime is reduced considerably according to (2.26) and (2.27). This reduces the internal quantum efficiency for short wavelengths considerably, because a large portion of the light is absorbed in region 1.

Light with long wavelengths penetrates deep into silicon and the recombination of photogenerated carriers in region 2 can reduce the internal quantum efficiency. The recombination of photogenerated carriers in region 1 is not very important for long wavelengths due to the large penetration depth and the small portion of photogenerated carriers in region 1.

In the dynamical case, carriers being photogenerated in region 1 and especially in region 2 do not have enough time to diffuse to the space-charge or drift region before the light intensity is reduced again. The diffusion tails of the photocurrent of consecutive light pulses overlap (Fig. 2.10). The photocurrent for a sine-wave light modulation reduces similarly at high frequencies (see Fig. 2.9). It should be mentioned explicitly that the dynamical quantum efficiency depends on the frequency or data rate. The higher both these are, the smaller the dynamical quantum efficiency becomes until the minimum is reached. This minimum is set by the portion of carriers being generated in the space-charge region, when we assume that the frequency is not extremely high and all drifting carriers still reach the boundary of the space-charge region and contribute to the photocurrent. For this case, we can use the expression

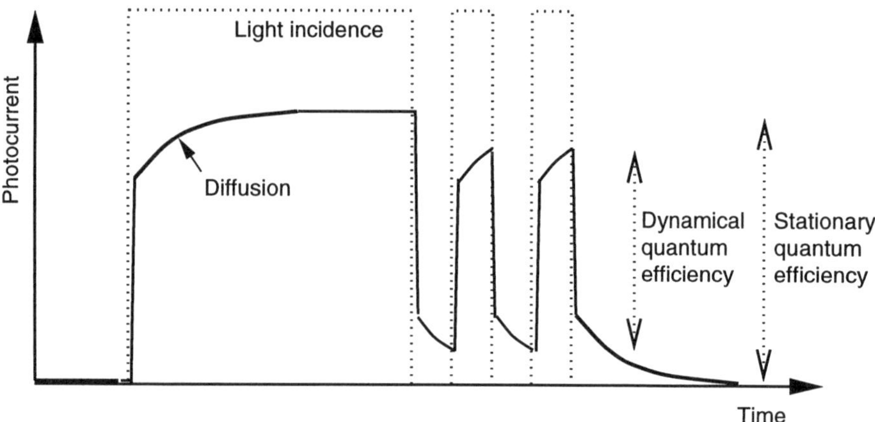

Fig. 2.10. The influence of carrier diffusion on the dynamical quantum efficiency of photodetectors at high data rates

$$\eta_{\mathrm{i}} = (1 - \exp[-\alpha(d_{\mathrm{p}} + d_{\mathrm{I}})]) \exp(-\alpha d_{\mathrm{p}}) \tag{2.32}$$

to describe the dynamical internal quantum efficiency. This expression was derived for ideal PIN photodiodes (compare with Fig. 2.6) with the thickness d_{I} of the intrinsic region, i. e. the thickness d_{I} of the drift region. We can also use (2.32) for a PN photodiode shown in Fig. 2.7 to a good approximation because the space-charge region with the thickness d_{I} does not penetrate far into the highly doped P^+ layer.

2.2.4 Equivalent Circuit of a Photodiode

A photoreceiver consists at least of a photodetector, of a load resistor R_{L}, and of an amplifier with an input capacitance C_{I} and with an input resistance R_{I} (Fig. 2.11). The input resistance R_{I} can be neglected for amplifiers with a JFET or MOS input transistor. R_{I}, however, is important for amplifiers with a bipolar input transistor.

The complete equivalent input circuit of the photoreceiver considering parasitics due to mounting is shown in Fig. 2.12. The equivalent circuit of a photodiode is obtained without mount parasitics and without R_{L}, C_{I}, and R_{I}.

The current source I_{ph} represents the photocurrent in Fig. 2.12. C_{D} is the capacitance of the space-charge region of the photodiode. C_{B} is the capacitance of the bondpad of the photodiode. L_{B} represents the inductance of the bond wire. The stray capacitance of the detector package is considered in C_{P}. A typical value for photodiode chips and integrated circuits is $C_{\mathrm{B}} = 0.2$–$0.3\,\mathrm{pF}$. The stray capacitance of package pins of discrete photodiodes and of the output pins of integrated circuits is of the order of $1\,\mathrm{pF}$. The lead wires and the wiring on electronic boards introduce the additional

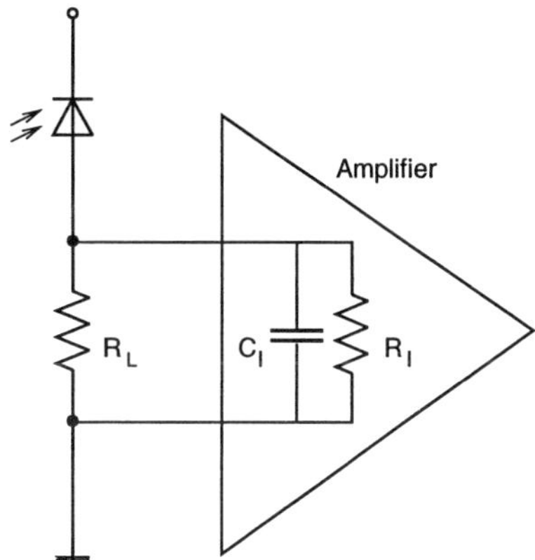

Fig. 2.11. Essential input circuit of a photoreceiver

inductance L_W. The inductances L_B and especially L_W become important at high frequencies.

The parallel resistor R_D, which models the reverse, leakage, or dark current of a photodiode usually is very large and can be neglected in most cases. The series resistance R_S may not be neglected when the photocurrent has to flow through low doped regions in the photodiode. R_S should be negligible for PIN photodiodes with highly doped P and N regions, when carrier drift dominates, i. e. when the 'intrinsic' zone is fully depleted.

The time constant t_D for a PIN photodiode with a connected amplifier is, in general, when the inductances can be neglected:

$$t_D = (R_L \| R_I)(C_D + C_B + C_P + C_I). \tag{2.33}$$

The -3dB bandwidth f_{3dB} of the input circuit of a photoreceiver is then

$$f_{3dB} = \frac{1}{2\pi (R_L \| R_I)(C_D + C_B + C_P + C_I)}. \tag{2.34}$$

Optoelectronic integrated circuits (OEICs) avoid the bondpad and package capacitances and the inductances of bond and lead wires as well as

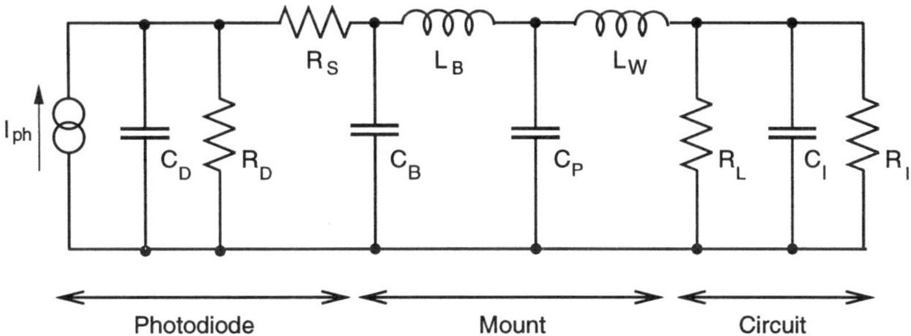

Fig. 2.12. Small-signal equivalent input circuit of a photoreceiver

the inductance of wiring on electronic boards. Optoelectronic integrated circuits, therefore, reach a larger bandwidth $f_{3\mathrm{dB}} = 1/(2\pi(R_\mathrm{L}\|R_\mathrm{I})(C_\mathrm{D} + C_\mathrm{I}))$. OEICs with a MOSFET or JFET input stage of the amplifier possess a very large input resistance and the bandwidth can be written as $f_{3\mathrm{dB}} = 1/(2\pi R_\mathrm{L}(C_\mathrm{D} + C_\mathrm{I}))$.

For a reverse-biased abrupt PN junction, for which one doping type is present in a concentration several orders of magnitude larger than the other doping type, the capacitance is given by

$$C_\mathrm{D} = A\sqrt{\frac{q\epsilon_\mathrm{r}\epsilon_0 N_\mathrm{A/D}}{2}}\frac{1}{\sqrt{U_\mathrm{D} - U - (2k_\mathrm{B}T/q)}}, \tag{2.35}$$

with

$$U_\mathrm{D} = \frac{k_\mathrm{B}T}{q}\ln\frac{N_\mathrm{A}N_\mathrm{D}}{n_\mathrm{i}^2}. \tag{2.36}$$

These prerequisites for the doping concentrations are usually fulfilled in photodiodes. For $\mathrm{P^+N}$ or $\mathrm{N^+P}$ photodiodes, an abrupt PN junction is a good approximation. Equation (2.35), therefore, approximates the capacitance of most photodiodes quite well. For photodiodes with deeply diffused doping regions like N-well to P-substrate photodiodes in CMOS technology, the capacitance can be calculated with the expression for linearly graded junctions [12]. However, deeply diffused doping regions should be avoided in silicon photodiodes at least for short wavelengths in order to keep the quantum efficiency high.

In (2.35), $N_\mathrm{A/D}$ represents the doping concentration of the low doped side of the PN junction. A negative value has to be used for the reverse bias of the photodiode U.

For a so-called PIN structure, in which high doped P and N regions are at the two sides of a low doped 'intrinsic' zone with thickness d_I, the capacitance is approximately that of a plate capacitor (neglecting the boundary capacitance):

$$C_\mathrm{D}^\mathrm{PIN} = A\frac{\epsilon_\mathrm{r}\epsilon_0}{d_\mathrm{I}}, \tag{2.37}$$

where A is the area of the $\mathrm{P^+}$-doped (PIN photodiode) or $\mathrm{N^+}$-doped (NIP photodiode) region at the surface of the device.

Figure 2.13 depicts the capacitances according to (2.37) and (2.35) as a function of the doping concentration $N_\mathrm{A/D}$. A diode area A of $2500\,\mu\mathrm{m}^2$, thereby, was assumed. When the doping level is reduced starting from $10^{15}\,\mathrm{cm}^{-3}$, the capacitance first decreases according to (2.35) until the space-charge region spreads across the whole thickness d_I of the intrinsic zone. Then C_D remains constant according to (2.37), when the doping level is reduced further. For a reverse bias of 3 V and $d_\mathrm{I} = 5\,\mu\mathrm{m}$, the capacitance finally remains

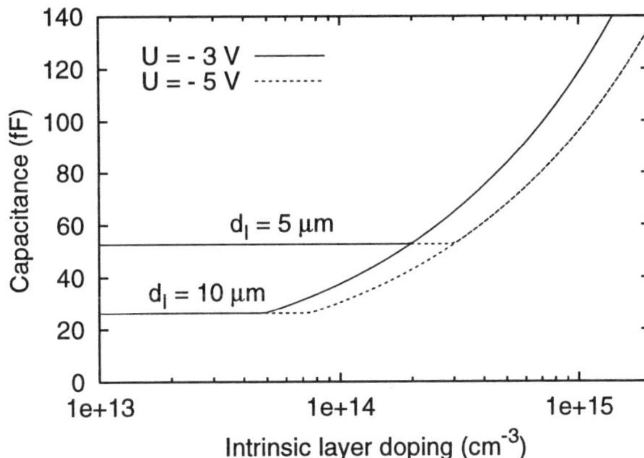

Fig. 2.13. Capacitance of PN and PIN diodes with an area of $2500\,\mu\text{m}^2$ versus doping concentration. In the constant capacitance regime, the diode can be considered as a PIN diode. For larger doping concentrations, it is not appropriate to call the diode a PIN diode, because the 'intrinsic' zone is not fully depleted

constant at a value of 53 fF, when the doping concentration drops below approximately $2 \times 10^{14}\,\text{cm}^{-3}$. For $d_{\text{I}} = 10\,\mu\text{m}$, the capacitance remains constant at a value of 26.5 fF, when the doping concentration falls below approximately $5 \times 10^{13}\,\text{cm}^{-3}$. For the larger reverse bias of 5 V, constant capacitances are already reached for slightly larger doping levels (see Fig. 2.13).

3. Silicon Technologies and Integrated Photodetectors

In this chapter the bipolar, CMOS, and BiCMOS process technologies are described. Photodetectors which are produced in these technologies without process modifications and their properties are introduced. Furthermore, the possible improvements of photodetectors resulting from small substrate and process modifications are discussed.

CMOS is the economically most important technology. The section on integrated photodetectors in CMOS technology, therefore, is more comprehensive than the sections on photodetectors in bipolar and BiCMOS technologies. Within the CMOS section, the innovative integration of vertical PIN photodiodes will be highlighted, since they allow a considerable improvement of the speed of CMOS OEICs. Furthermore, image sensors using charge-coupled-devices and active pixel image sensors will be described in some detail because of their economical importance. Within the BiCMOS section, the exploitation of double photodiodes will be mentioned as another innovation for high-speed OEICs and OPTO-ASICs in standard technology.

3.1 Bipolar Processes

In the standard buried collector (SBC) technology [28], the N^+ collector is implanted in the P-type substrate and an N-type epitaxial layer is grown, which serves as the N^- collector (Fig. 3.1).

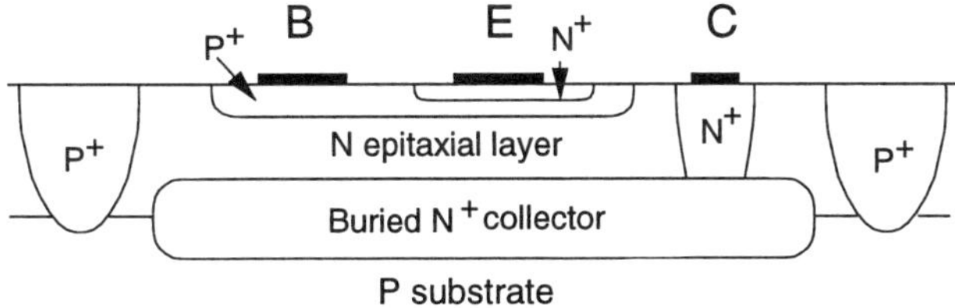

Fig. 3.1. Schematic cross section of an NPN transistor in SBC technology [28]

The P^+ isolation, the N^+-collector contact, the P-type base, and the N^+ emitter are implanted and annealed subsequently. In modern technologies, polysilicon emitters [29] and polysilicon extrinsic bases [30], oxide isolation

[31, 32], or recessed oxide isolation [33, 34] are used. Alternatively, trench isolation [35–37] is also used. Here, however, a simple process flow (Fig. 3.2) will be described, in order to demonstrate more agreement with [38], where a vertical PIN photodiode has been integrated (see Fig. 3.8).

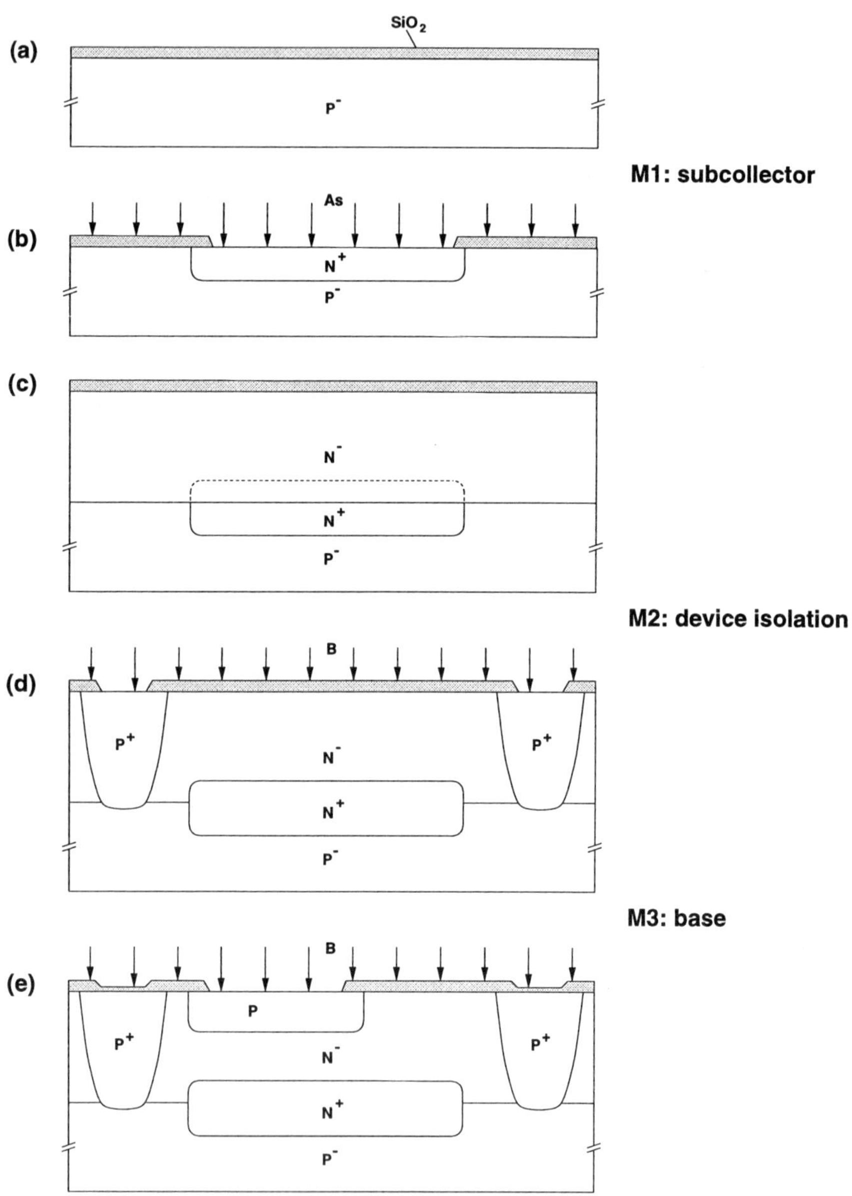

Fig. 3.2a–e. For caption see next page

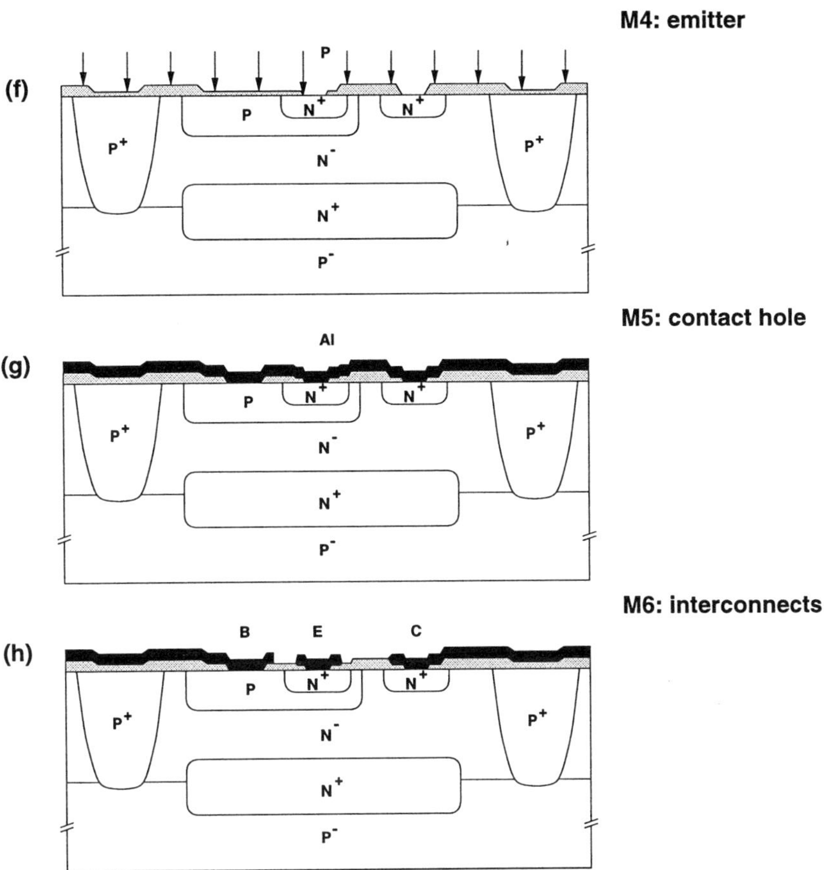

Fig. 3.2a–h. Simplified process flow of SBC bipolar technology. (**a–e**) subcollector, device isolation, and base formation, (**f–h**) emitter, contact hole, and interconnect formation [39]

The P-type substrate with a doping concentration of $\approx 10^{15}\,\mathrm{cm}^{-3}$ is oxidized (Fig. 3.2a). Then the area for the buried collectors is defined by lithography (M1). Arsenic is diffused or, more recently, antimony is implanted (Fig. 3.2b) to form the N^+ subcollector, which is needed to minimize the collector series resistance. Subsequently, the N^- collector layer is grown epitaxially (Fig. 3.2c). An oxidation is performed (Fig. 3.2c) and the areas for device isolation are defined by lithography (M2). Boron is diffused and driven in under oxidizing conditions (Fig. 3.2d). The resulting P^+ regions reach the P substrate and isolate the NPN transistor from neighboring transistors. Then lithography is applied to define the base area (M3). Boron is implanted in this area (Fig. 3.2e) and annealed partially under oxidizing conditions. The emitter area and the collector contact area are defined next (M4) and implanted with phosphorus (Fig. 3.2f) or arsenic. The annealing is again performed partially under oxidizing conditions. The contact holes are defined by lithography (M5) and aluminum is deposited (Fig. 3.2g) and structured by lithography (M6) in order to obtain interconnects (Fig. 3.2h). After oxide and nitride deposition for passivation, a seventh lithography step is needed for the opening of the bondpads. Seven steps of lithography are applied in total for this simple bipolar process with one metal layer. This simple process has often been varied and supplemented [39]. A second metal layer for instance would result in two additional lithography steps. An additional mask before M2 is required for a deep collector contact diffusion (collector plug) in order to eliminate the N^- region and the resulting collector series resistance between N^+ collector contact and N^+ subcollector resulting in a single N^+ path from the collector contact to the subcollector (see Fig. 3.1).

For our purposes, however, this simple process will be sufficient to explain the photodetectors available in bipolar processes without modifications (see Sect. 3.2). It will also be sufficient to understand the complexity of PIN photodiode integration, which will be discussed in Sect. 3.3.2.

3.2 Integrated Detectors in Standard Bipolar Technology

3.2.1 Photodiodes

An N^+/P substrate photodiode can be obtained in a bipolar OEIC, whereby the N^+ region is formed by the N^+ buried collector and the deep N^+ collector plug (Fig. 3.3).

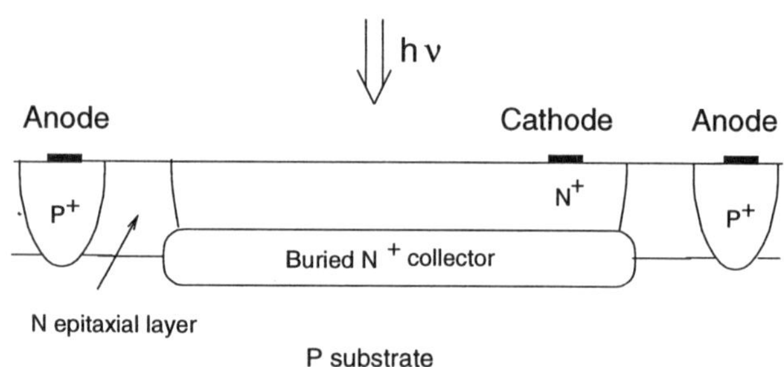

Fig. 3.3. Schematic cross section of a subcollector to substrate photodiode

Such a photodiode was realized in [40]. For an N^+ area of $100 \times 100\,\mu m^2$, a bit rate of 150 Mb/s was reported for $\lambda = 850$ nm. The speed of the photodiode was limited by carrier diffusion in the P substrate with a doping concentration of approximately $10^{15}\,cm^{-3}$. The transient behavior of a smaller photodiode with an N^+ area of $10 \times 10\,\mu m^2$ was much better. A rise time of 0.04 ns and a fall time of 0.4 ns was reported for this smaller photodiode with

$\lambda = 820$ nm. These values would allow a data rate DR of about 850 Mb/s, when we use the conservative estimate $DR = 1/(3t_f)$. The reverse bias of the photodiode was approximately 4.2 V. The quantum efficiency of the photodiode was 30 %, which was probably due to optical interferences in the isolation and passivation layers above the photodiode.

The subcollector may be omitted in Fig. 3.3. Then the photodiode is formed by the N^+ contact diffusion and the P substrate. The N^+ penetration depth, however, will be only slightly smaller. The quantum efficiency for blue and green light will be somewhat increased.

Also, without any technological modifications, the buried N^+ collector can serve as the cathode (see Fig. 3.4), the N collector epitaxial layer can serve as the 'intrinsic' layer of a PIN photodiode, and the base implant can serve as the anode in order to integrate PIN photodiodes with a thin 'intrinsic' region in bipolar technologies [41, 42]. The process of [41] was described in [43].

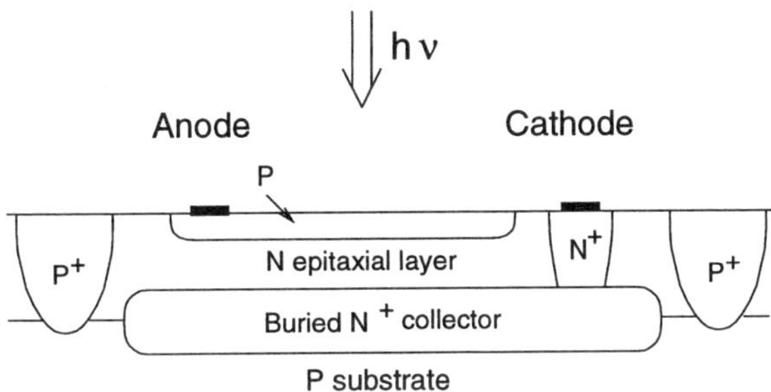

Fig. 3.4. Schematic cross section of a base-collector photodiode

The small thickness of the epitaxial layer of high-speed bipolar processes in the range of about 1 µm causes a low quantum efficiency in the yellow to the infrared spectral region (580–1100 nm). The rise and fall times of the photocurrent for light pulses are very short due to the small epitaxial layer thickness. In [41] a bit rate of 10 Gb/s for the base-collector diode and a responsivity R of merely 48 mA/W for 840 nm were reported. This low responsivity at wavelengths from 780 to 850 nm, which are widely used for optical data transmission on short fiber lengths of up to several kilometers, is a major disadvantage of standard bipolar OEICs.

The base-collector diode with a sensitive area of 100 µm^2 fabricated in a 0.8 µm silicon bipolar technology worked up to 3 Gb/s for a wavelength of 850 nm [42]. A sensitivity of 0.045 A/W was reported. A phototransistor with a very small sensitive area of 10 µm^2 reached a data rate of 5 Gb/s. With these monolithic silicon OEICs, a higher data rate was obtained than with a III/V photodetector and a silicon amplifier [44] showing the advantage of monolithic photodetector integration.

Emitter-base photodiodes may be useful for blue and green light when the emitters are implanted. The emitter-base diode, however, does not seem to be very advantageous, when polysilicon emitters are used in the bipolar process. The polysilicon then covers the N^+ emitter and, therefore, the light sensitive area of the photodiode completely, reducing the quantum efficiency by partial light absorption in the polysilicon. In particular, the quantum efficiency for blue and UV light would be very low.

3.2.2 Phototransistors

The NPN transistor with an increased base-collector junction area can, of course, be used as a phototransistor (see Fig. 3.5). Bipolar phototransistors, simply speaking, use the base-collector diode as a photodiode and amplify the photocurrent of this diode. The P type region of this photodiode and the P-type base of the NPN transistor are one P-type region; the cathode of this photodiode and the collector of the NPN transistor consist of one N-type region. The base contact can be omitted. The electron–hole pairs generated in the base-collector space-charge region are separated by the electric field. In this space-charge region the holes are swept into the base and the electrons are swept into the collector. The holes make the base potential positive and the emitter can inject electrons into the base and towards the collector. The photocurrent of the photodiode I_{pd} is amplified by the current gain β of the NPN transistor. This amplified photocurrent is available at the collector electrode ($\beta \cdot I_{\mathrm{pd}}$) and at the emitter ($(\beta+1) \cdot I_{\mathrm{pd}}$). It should be noted, however, that for long wavelengths, i. e. for large penetration depths of the light, only the fraction of the carriers photogenerated in the base-collector space-charge region will contribute to I_{pd}, which is amplified. The other fraction of the photogenerated carriers does not contribute to I_{pd} and is not amplified.

Bipolar phototransistors are known to be much slower than photodiodes. This is due to their current gain β, to the base transit time τ_{B}, for which carrier diffusion may be limiting, to the emitter-base space-charge capacitance C_{SE} and mainly to the rather large base-collector space-charge capacitance C_{C}, which results from the necessary light sensitive area. It should be mentioned that the total emitter-base capacitance C_{E} is the sum of the emitter-base diffusion capacitance C_{DE} and the emitter-base space-charge capacitance C_{SE}, i. e. $C_{\mathrm{E}} = C_{\mathrm{DE}} + C_{\mathrm{SE}} = \tau_{\mathrm{B}} g_{\mathrm{m}} + C_{\mathrm{SE}}$. The -3dB frequency of

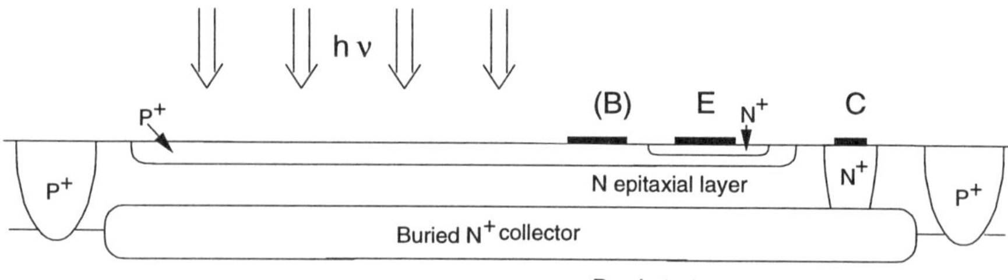

Fig. 3.5. Schematic cross section of a phototransistor in SBC technology

a bipolar phototransistor is [45]:

$$f_{3\mathrm{dB}} = \frac{1}{\beta 2\pi(\tau_{\mathrm{B}} + (k_{\mathrm{B}}T/qI_{\mathrm{E}})(C_{\mathrm{SE}} + C_{\mathrm{C}}))} \tag{3.1}$$

This equation is the same as for a common bipolar transistor [46]. The base-collector capacitance is proportional to the size of the light-sensitive area, which is much larger in a phototransistor than the base-collector junction area of a common bipolar transistor. Therefore, the bipolar phototransistor is much slower than a combination of a PIN photodiode with a small-area common bipolar transistor. Nevertheless, a phototransistor is advantageous at low frequencies and when PIN photodiodes cannot be integrated.

For the application in optical clock distribution, a NPN transistor with a self-adjusting emitter was investigated as a phototransistor receiver for the wavelength $\lambda = 840$ nm [41]. A double-polysilicon process [43] with a 0.5 μm thick epitaxial layer was used. The self-aligned emitter technique was used in order to realize a minimum emitter width of 0.6 μm with a 1.0 μm lithography. The current gain factor β of the transistors of 70 resulted in a relatively high sensitivity of 3.2 A/W, although the thickness of the epitaxial layer was much smaller than the penetration depth $1/\alpha = 16$ μm of the light with this wavelength. A data rate of 1.25 Gb/s was reported for the NPN transistor as a phototransistor with a small size.

3.3 Integrated Detectors in Modified Bipolar Technology

3.3.1 UV Sensor

An example of a photodiode integrated in a bipolar OEIC for ultraviolet (UV) detection will also be given. An industrial application for such a sensor system is, for instance, flame detection for combustion monitoring. A spectral range has to be used in which the emitted light of the flame is more intensive than the background radiation, i. e. the black-body radiation, at the temperature of 2000 K. In the UV spectral region from 250 to 400 nm, the optical emission of the flame is much larger than the thermal background radiation from the furnace walls.

Silicon photodiodes usually have a low quantum efficiency in this UV region. The need for cheap sensor systems, however, implied the monolithic integration of the detector together with the electronic circuit in a silicon technology [47]. Additional reasons for the monolithic integration were the insensitivity of OEICs to electromagnetic interference (EMI) and a larger signal-to-noise ratio. The UV sensor was implemented in a bipolar technology [48, 49]. The cross section of the silicon UV sensor is shown in Fig. 3.6. Shallow P^+ and N^+ implants form the depletion region close to the crystal surface, where UV light is absorbed and carriers are generated. The depth of the implanted P^+ and N^+ profiles was not standard and a few additional process steps with low thermal budget and rapid thermal annealing allowed the fabrication of the UV detector and of the interface circuit on a single chip without degrading the electrical performance of the analog components.

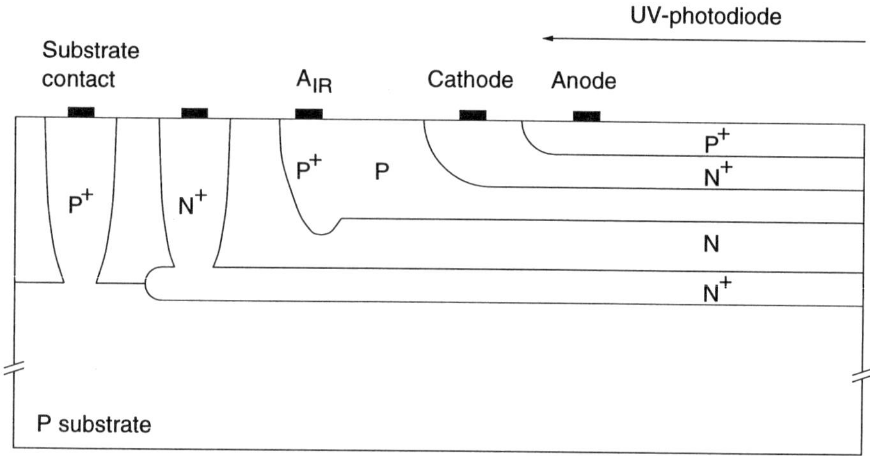

Fig. 3.6. Cross section of a bipolar-integrated UV sensitive photodiode [47]

The P^+/N^+ UV sensitive photodiode was fabricated in the base island of the NPN transistor, which is contacted by A_{IR} (see Fig. 3.6). The diode formed by the base and the added N^+ region was used to reduce contributions of deeper penetrating green, red, and infrared (IR) light to the signal current. The anode current is used as the UV signal current and the photocurrent from longer wavelengths flowing to the cathode and to the A_{IR} anode is not used by the UV sensor system.

The photocurrent of this UV detector ranged from 20 pA to 1 nA in the flame detection application. The UV-OEIC was mounted in a TO5 package with a lens focusing the UV light onto the UV sensitive photodiode with an octagonal area of 1 mm^2. A glass filter was used in addition to perform optical bandpass filtering. The electronic circuit of this UV sensor system will be described in Sect. 12.4.1.

3.3.2 PIN Photodiode Integration

For the wavelengths 780 and 850 nm, a thickness of at least 10 μm is necessary for the so-called intrinsic layer of a PIN photodiode, due to the low optical absorption coefficient of silicon for wavelengths larger than about 700 nm. Such a large I-layer thickness requires the reduction of the epitaxial layer concentration below about 10^{14} cm^{-3} for a voltage of 3 V across the PIN photodiode in order to obtain a spreading of the electric field (drift region) over the whole I-zone [16]. There is, however, the so-called Kirk effect [50] or base push-out effect, which requires a large concentration for the N^- collector of the order of $10^{16}\,cm^{-3}$ for high-speed bipolar transistors with transit frequencies in the 10 GHz range or above. According to the Kirk effect, the start and the end of the base-collector space-charge region move closer to the N^+ collector and thereby increase the width of the base drastically, when the concentration of the electrons crossing the base becomes comparable to the doping concentration N_D of the N^- collector. The Kirk effect is a high-injection effect, which was investigated intensively in [51–57]. We will not discuss the details. In short, the Kirk effect becomes active when the collector current density exceeds the critical value $j_{critical}$ [58]:

$$j_{critical} = qN_D v_s. \tag{3.2}$$

To avoid the Kirk effect, the doping level in the N^- collector should be larger than $10^{16}\,cm^{-3}$. A reduction of the doping concentration N_D in the epitaxial layer to below $10^{14}\,cm^{-3}$, as would be necessary for the integration of PIN photodiodes, therefore, causes a dramatic decrease of the critical collector current density $j_{critical}$. In turn, the current gain and the transit frequency of the NPN transistor would drop strongly for $N_D \leq 10^{14}\,cm^{-3}$.

In order to combine both of the above mentioned requirements — thick, low-doped I-layer and thin, higher doped N collector — process modifications were suggested in [38, 59].

Kyomasu [59] described a PIN photointegrated circuit sensor (PIN-PICS) on a P^+ substrate, creating a high-speed and high-optical-responsivity PIN photodiode. Figure 3.7 shows the structure of this PIN-PICS. The photodiode is isolated from the bipolar components by the combination of a P well and a trench isolation. Kyomasu states that only one additional mask (disregarding antireflection coating) is necessary for the integration of the PIN photodiode for adding the P well, because the trench isolation is part of the original bipolar process. This P well avoids the wide collector/P^{--} space-charge region, which would result from the low P^{--}-I-layer doping of about $1 \times 10^{13}\,cm^{-3}$ [59]. The P^+ substrate serves as the anode of the PIN photodiode. A surface anode contact is used, which may result in a certain series resistance across the P^{--} layer between P well and P^+ substrate. The collector N^+ contact implant and diffusion is used to form the cathode of the PIN photodiode [59], resulting in a rather thick cathode. For the integration of the PIN photodiode, two epitaxial layers had to be grown. First, the P^-/P^{--} layer was grown resulting in the intrinsic region of the PIN photodiode. Then, the P well was formed and the buried collector was implanted. Subsequently, the second epitaxial layer was grown to form the N-type collector.

The autodoping problem, which makes it usually difficult to grow very low-doped epitaxial layers was solved by the incorporation of a P^--buffer layer before the P^{--}-I-layer was grown. The process complexity of this bipolar PIN-PICS added to that of the original process is rather low. The voltage

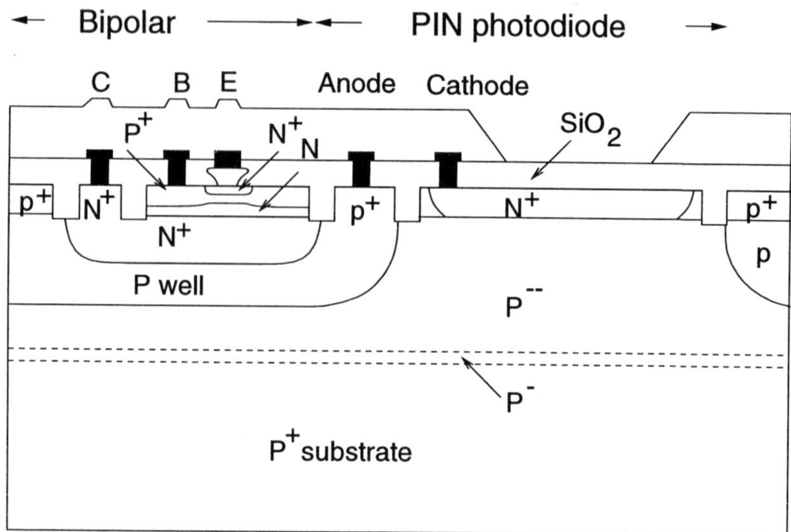

Fig. 3.7. Cross section of a PIN bipolar OEIC [59]

across the PIN photodiode is limited to the range of the circuit operating voltage ($U_{\mathrm{PIN}} < U_{\mathrm{CC}}$, when the anode is grounded (similar to Fig. 3.20). For $U_{\mathrm{PIN}} = 3\,\mathrm{V}$ and for $\lambda = 830\,\mathrm{nm}$, a $-3\mathrm{dB}$ bandwidth of 270 MHz was reported for an active photodiode area of $0.145\,\mathrm{mm}^2$. When the substrate is biased at a more negative voltage than the ground of the circuit, the voltage across the PIN photodiode can be made larger than the circuit operating voltage. U_{PIN}, then, is limited by the breakdown voltage of the P well to N^+ collector PN junction. A $-3\mathrm{dB}$ bandwidth of 680 MHz for $U_{\mathrm{PIN}} = 10\,\mathrm{V}$ and $\lambda = 830\,\mathrm{nm}$ was reported. The responsivity of the PIN photodiode was 0.5 A/W at $\lambda = 830\,\mathrm{nm}$ ($\eta = 74\,\%$). This relatively large responsivity is due to the reduced oxide thickness above the photodiode (Fig. 3.7), which can be denoted as an antireflection coating. The dark current density for the PIN photodiode was $50\,\mathrm{pA/mm}^2$.

Figure 3.8 shows the cross section of the OEIC proposed in [38]. The PIN cathode and a P isolation were implanted into the substrate. A three-step epitaxial growth (first 2 μm as a buffer layer in order to avoid autodoping during the epitaxial process, then a 9 μm I-layer, and finally a 4 μm I-layer after the buried collector implant) was performed in order to supply the thick N-type 'intrinsic' layer with a reduced doping level ($< 3 \times 10^{13}\,\mathrm{cm}^{-3}$) and a thin collector [38]. Assuming that the N^+-buried collector and the N^+-contact implant can be used for the N^+-cathode contact diffusions, that the P^+-isolation implant (compare Figs. 3.1 and 3.2) of the standard process can be used for the P isolation contact, and that the base contact implant is used for the PIN anode, the process complexity is increased by three additional lithography steps compared to the original bipolar process (without steps

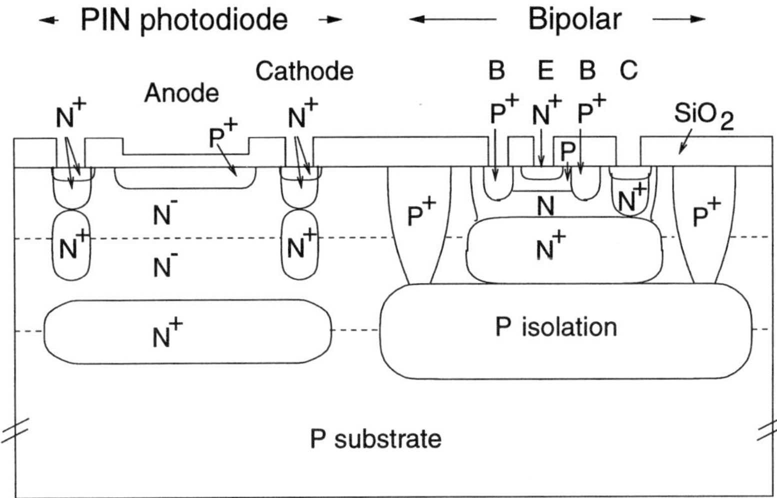

Fig. 3.8. Cross section of a PIN bipolar OEIC [38]

for antireflection coating): (1) for the N^+-buried PIN cathode; (2) for the P isolation in order to isolate the buried collectors from the N^- 'intrinsic' region and to restrict the cathode potential to the photodiode area; and (3) for the N collector implant in order to provide the doping level of about $10^{16}\,\text{cm}^{-3}$. Here, the bias voltage of the photodiode is not limited by the circuit operating voltage U_{CC} or by the breakdown voltage BV_{CE0} of the NPN transistors. Larger voltages, therefore, enable high-speed operation of the PIN photodiode. The advantage of this approach compared to the structure investigated in [59] is that here the substrate can be at the ground potential of the circuit for $U_{\text{PIN}} > U_{\text{CC}}$, because both contacts of the PIN photodiode are at the wafer surface and because both contacts do not have to be at the substrate potential.

A -3dB bandwidth of 300 MHz for $\lambda = 780$ nm was reported for the PIN photodiode with an area of 0.16 mm^2 at a bias of 3 V. The rise and fall times of the photocurrent were 1.6 ns for these wavelength and bias values. Bandwidth values for higher bias values, unfortunately were not reported in [38]. The responsivity of the PIN photodiode was 0.35 A/W ($\eta = 57\,\%$) for $\lambda = 780$ nm. This responsivity seems to be rather low, although an antireflection coating is indicated in Fig. 3.8.

3.3.3 Color Sensor

An all-silicon color sensor based on the strong wavelength dependence of the penetration depth of light in silicon [60] will be discussed in order to emphasize the large variety of silicon photodetectors. The sensor was inte-

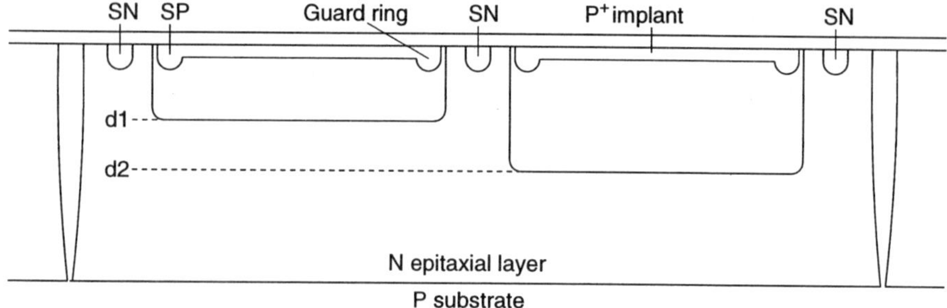

Fig. 3.9. Cross section of an integrated color indicator [60]

grated together with the required electronic functions in bipolar technology. Figure 3.9 shows the cross section of this color sensor.

The P^+N photodiodes were formed by implantation of boron in a 6 Ωcm N-epitaxial layer on a 2–5 Ωcm P substrate. The P^+ anodes of the photodiodes are surrounded by a lower doped P diffusion guard ring to increase the breakdown voltage. To form the color sensor, two identical photodiodes are differently reverse-biased. The space-charge regions, therefore, have different widths (actually depths) d_1 and d_2 (see Fig. 3.9). The operation of the color sensor is quite complex. Let us assume that blue light is completely absorbed within d_1. Then the photocurrents of the two photodiodes are equal, when the intensity of the light is equal on both photodiodes. When red light is incident into both photodiodes, the diode with the wider space-charge region d_2 has a larger photocurrent than the other diode. By varying the reverse voltage of diode 1, a lower wavelength limit can be adjusted. In such a way the wavelength of monochromatic light can be determined. In fact a more complex evaluation of the dependence of the photocurrents and of their difference on the scanned reverse voltages was implemented in a bipolar circuit. This OEIC provided a voltage proportional to the average of the optical power distribution of the incident light within an electronically tunable part of the spectrum [60]. It should be mentioned that the carrier lifetime and diffusion length must not be large for such a color sensor. Charge carriers, generated in the epitaxial layer below the space-charge region depth d_2, were prevented from diffusing to the depleted regions by a variably reverse-biased P–substrate/N–epitaxial layer diode collecting these carriers.

3.4 CMOS Processes

The CMOS technology is the economically most important technology for the fabrication of microelectronic circuits. Early processes were one-well processes. Modern processes are twin-well processes. We will briefly discuss one-well processes first and then twin-well processes extensively.

3.4.1 One-Well Processes

In [61] an N-well process with a minimum feature size of 1.2 μm is described, which uses epitaxial silicon wafers. The P-type substrate has a doping concentration of $2 \times 10^{18}\,\mathrm{cm}^{-3}$. On this substrate, a 12 μm thick epitaxial P^- layer with a doping concentration of $2 \times 10^{15}\,\mathrm{cm}^{-3}$ is deposited.

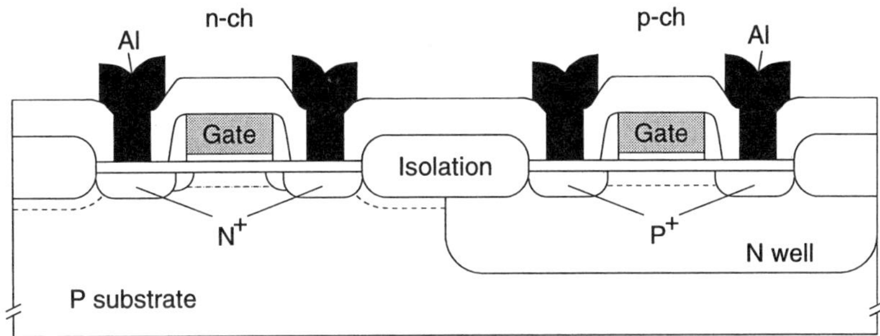

Fig. 3.10. Schematic cross section of a CMOS chip in an N-well technology

The N-channel MOSFET obtains only an anti-punchthrough implant instead of a deep P well (Fig. 3.10). The immunity against latch-up, therefore, is not very good. In order to improve the latch-up immunity of CMOS circuits and to optimize the N- and P-channel transistors independently, twin-well processes were developed.

3.4.2 Twin-Well Processes

A typical submicrometer CMOS process will be described in the following. It uses twin wells [62], LOCOS (LOCal Oxidation of Silicon) isolation [32], an N^+ polysilicon gate, and lightly doped drain (LDD) N-channel MOSFETs. As illustrated in the final device cross section (Fig. 3.11), a CMOS process consists of six main sections [63] plus passivation:

(1) Formation of the wells by deep N- and P-type diffusions, which allow the independent tailoring of the doping profiles in each well.

(2) LOCOS field isolation and the formation of channel-stopper regions, which isolate neighboring devices from each other electrically.

(3) Ion implantation for the formation of surface- and buried-channel regions for N- and P-MOSFETs, respectively, in order to adjust the threshold voltages.

(4) Gate oxidation and deposition of the N^+ polysilicon layer with subsequent gate definition.

(5) Source/drain junction formation by ion implantation including sidewall spacers for LDD formation.

(6) Deposition of oxide, contact hole opening, and metallization.

(7) Passivation.

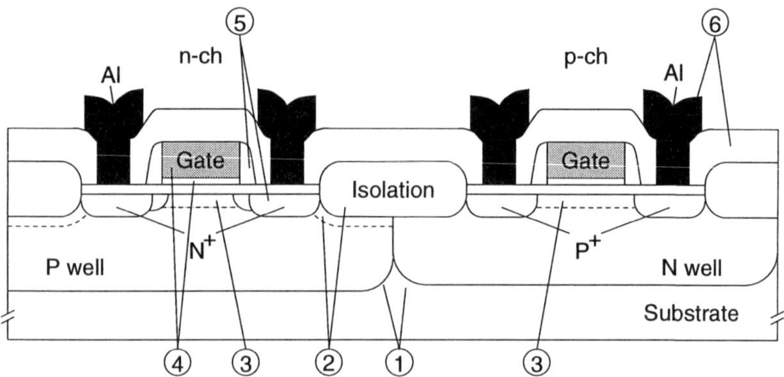

Fig. 3.11. Schematic cross section of a CMOS chip. The common CMOS process can be divided roughly into six main steps [63]

The process flow will be treated in more detail in Sect. 3.5.2. Here several other examples of modern twin-well processes will be mentioned. A 0.6 μm twin-well CMOS process of AT&T uses a 7 μm thick epitaxial layer [64]. The supply voltage is 3.3 V. The drive-in of the wells to a depth of about 1.8 μm each is performed at 1150 °C. Boron is implanted for the P^+ source/drain regions of the P-channel transistors with an energy of 20 keV using silicon-pre-amorphization by Si implantation in order to avoid channeling. The implantation of BF_2 with 30 keV was reported as an alternative to obtain shallow P^+/N-well junctions. In the first case, the crystal damage is annealed at 800 °C and the electrical activation of the boron is achieved with rapid thermal annealing (RTA) at a higher temperature. Further examples of twin-well processes were reported in [65, 66]. The first of the two processes is a 0.5 μm process with a specified supply voltage of 3.3 V. The second process is a 0.25 μm process with a specified supply voltage of 2.5 V.

3.5 CMOS-Integrated Detectors

3.5.1 Integrated Detectors in Standard CMOS Processes

PN Photodiodes. The simplest way to build CMOS OEICs is to use the PN junctions available in CMOS processes: source/drain-substrate, source/drain-well, and well-substrate diodes. These PN photodiodes, however, possess regions which are free from electric fields. In these regions, the slow diffusion of photogenerated carriers determines the transient behavior of such PN photodiodes. Published PN CMOS OEICs are characterized by bandwidths of less than 15 MHz [67–69]. Another example is described in [70], where the OEIC was optimized for a dynamic range of six decades in illumination.

The source/drain-substrate and source/drain-well photodiodes are more appropriate for the detection of wavelengths shorter than about 600 nm, whereas the well-substrate photodiode is more appropriate for long wavelengths like 780 or 850 nm. Figure 3.12, for instance, shows an N^+/P-substrate photodiode.

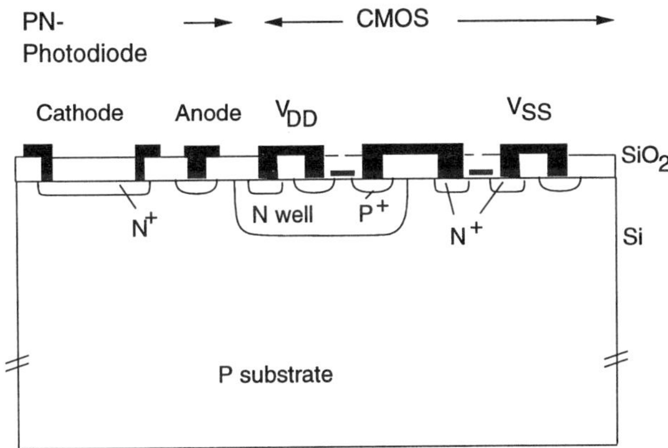

Fig. 3.12. Schematic cross section of a PN photodiode integrated in a one-well CMOS chip

In addition to carrier diffusion, the series resistance of the photodiodes due to lateral anode contacts at the silicon surface together with the relatively large junction capacitance of the photodiode may limit the dynamical PN photodiode behavior. In an N-well process, the anode of the N^+/P-substrate photodiode has to be at V_{SS} potential, which may be a restriction for circuit design.

A lateral N^+/P-substrate/P^+ photodiode was used as an optical detector, into which light was coupled by an integrated waveguide [67]. This detector was realized in a 0.8 μm N-well CMOS process. A maximum bandwidth of 10 MHz was achieved with a transimpedance amplifier for a photocurrent of 1 μA with $\lambda = 675$ nm. The speed of the detector was limited by carrier diffusion due to photogeneration outside the diode [68].

The N-well/P-substrate diode in a 2 μm N-well CMOS process was used as a photodiode in [69]. A bandwidth of 1.6 MHz with a wavelength $\lambda = 780$ nm was reported for an unoptimized system. The leakage current density of the photodiode was 15 pA/mm^2 at 5 V. The responsivity for $\lambda = 780$ nm was 0.5 A/W ($\eta = 70\,\%$). The light was coupled into the photodiode via an integrated waveguide. Therefore, the light transmission of the isolation oxide was not strongly influenced by destructive interference.

In addition to the continuous observation of the photocurrent of a PN photodiode, PN photodiodes can be operated in a storage mode. The capacitance of the photodiode can be used to integrate and store a photogenerated charge and to read it after certain periods (Fig. 3.13).

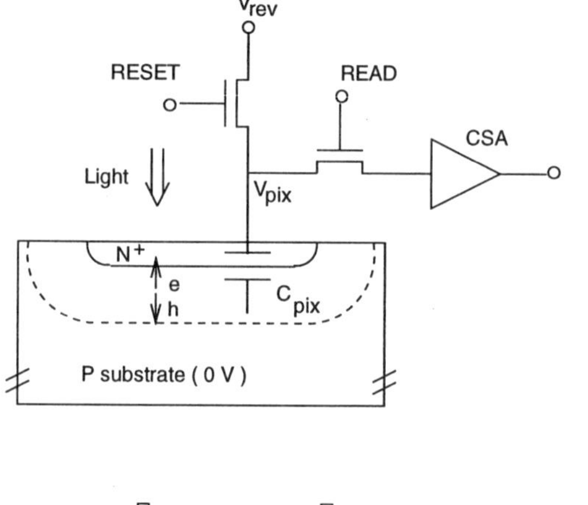

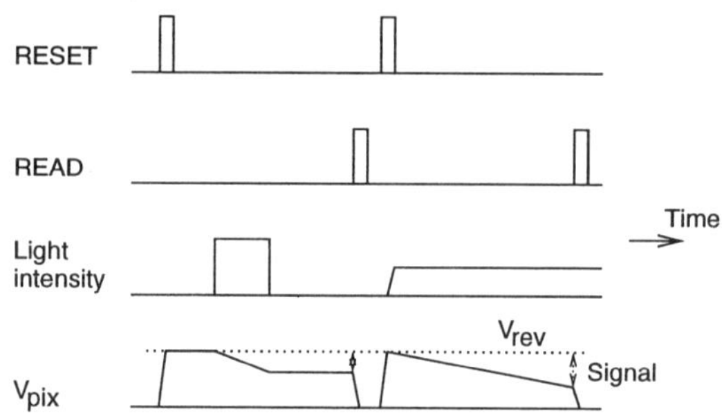

Fig. 3.13. Operation of a PN photodiode in a discharging storage mode

For such a purpose, the capacitance of the photodiode is initially set to the reverse voltage V_{rev}, when the reset-MOS switch is ON. Then the reset-MOS switch is opened to the OFF state and the photodiode floats for the light integration period. During this time, the capacitance is being discharged with a rate which depends on the light intensity. The remaining charge on the capacitance Q_{pix} is switched by the read-MOSFET to the input of a charge sensitive amplifier (CSA) at the end of the integration period. The smaller this

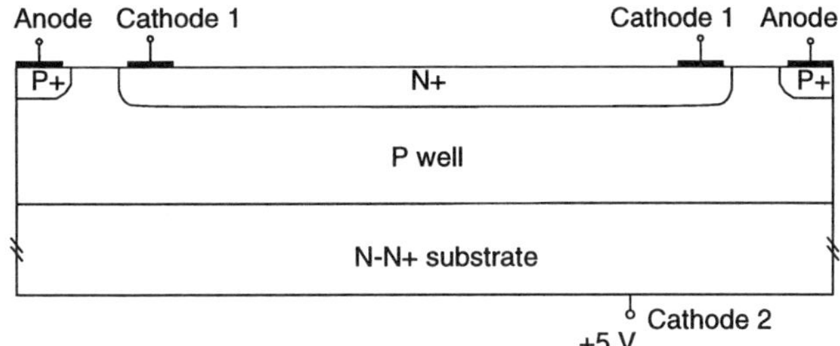

Fig. 3.14. Schematic cross section of a double photodiode

charge being read, the more intensive was the incident light. This discharging method is advantageous compared to a charging method, because the electric field due to $V_{\rm rev}$ in the photodiode helps to collect the photogenerated charge effectively. With a charging method, the space-charge region would be thinner and charge collection would not be as effective. N^+/P-substrate photodiodes in a 512 one-dimensional array were, for instance, integrated in a 1.2 μm analog CMOS technology and used in the charge storage mode [71].

Double Photodiodes. The double photodiode shown in Fig. 3.14 does not need any process modifications. It consists of the N^+/P-well diode and of the P-well/N^-N^+-substrate diode. The load resistor has to be connected to the common P-well anode in order to collect the photocurrents of both photodiodes.

Figure 3.15 shows the frequency response of a double photodiode (DPD) on a N^-N^+ substrate, which was fabricated with a 1 μm CMOS process. The doping concentration of the N epitaxial layer was approximately $1 \times 10^{15}\,\text{cm}^{-3}$. The integrated 500 Ω resistor was connected to the P-well anode, which was biased at +2 V. The N^+ cathode and the N-substrate cathode were at +5 V. The reverse bias of the double photodiode accordingly was 3 V. The size of the photodiode was approximately $50 \times 50\,\mu\text{m}^2$. A −3dB frequency

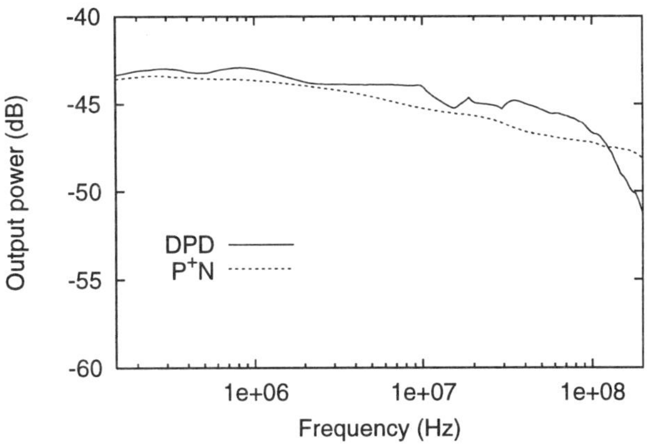

Fig. 3.15. Comparison of the frequency responses of a double photodiode (DPD) and of a P^+N photodiode fabricated in a 1 μm CMOS process and measured with integrated 500 Ω gate polysilicon resistors and a picoprobe for a wavelength of 638 nm

of approximately 90 MHz was measured for this double photodiode with a picoprobe.

In Fig. 3.15 the frequency response of a P^+N photodiode with the same size and fabricated on the same wafer as the double photodiode is shown for comparison. The −3dB frequency of the P^+N photodiode is only 32 MHz and the photocurrent already begins to decrease at a frequency of 2 MHz due to slow diffusion of photogenerated carriers from a depth of more than approximately 2 µm. In the double photodiode two space-charge regions are present. The first space-charge region is formed at the N^+/P-well junction and the second space-charge region is present at the P-well/N-substrate junction at a depth of approximately 4 µm. Furthermore, there is an electric field between these two space-charge regions due to the doping gradient of the N well [72]. The field-free region in the substrate below the P-well/N-substrate junction where a small portion of carriers is photogenerated with red light, therefore, is thinner in the double photodiode and carrier diffusion is much less important resulting in a much larger −3dB bandwidth.

The rise and fall times of 2.34 ns and 2.48 ns, respectively, for the double photodiode were determined with a picoprobe and a digital oscilloscope (Fig. 3.16). The frequency and transient responses of the double photodiode were limited by the *RC* time constant of its large capacitance of more than 1 pF and the 500 Ω resistor. The capacitance of the picoprobe of 0.1 pF is rather small and can be neglected.

In addition to the application of the double photodiode as a fast photodetector, it is possible to use the DPD as a color detector [73]. The diode with the shallow junction preferentially collects carriers created by shorter wave-

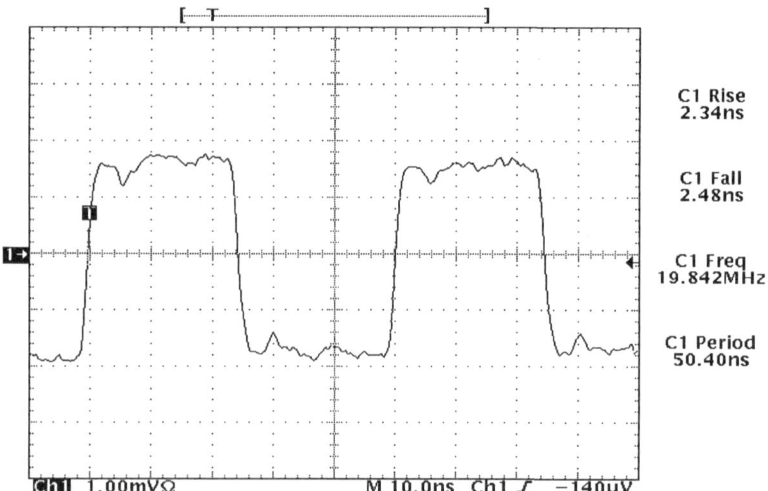

Fig. 3.16. Transient response of a CMOS-integrated double photodiode measured with an integrated 500 Ω polysilicon resistor and a picoprobe for $\lambda = 638$ nm

length light, while the diode with the deeper junction collects carriers created by longer wavelength light. The shorter wavelengths predominantly generate a photocurrent in the source/drain-well diode, while the longer wavelengths predominantly generate a photocurrent in the well-substrate diode. The quantum efficiency of the source/drain-well diode peaked at about 530 nm and the quantum efficiency of the well-substrate diode showed a maximum at about 710 nm [73]. The ratio of the photocurrents of the source/drain-well diode and the well-substrate diode versus wavelength was a monotonously decreasing function in the range from 450 to 900 nm. Measuring the photocurrents of the two diodes, computing their ratio, and looking up the wavelength for this ratio value from a list stored in an EEPROM for instance allows us, therefore, to determine the wavelength in monochromatic or narrow bandwidth applications.

Interrupted-P-Finger Photodiode. Another example of a high-speed photodiode in a 0.35 μm standard CMOS technology [74] will be given in the following. The photodiode uses the P^+ source/drain implant of the CMOS process for the anode and the N well for the cathode (Fig. 3.17).

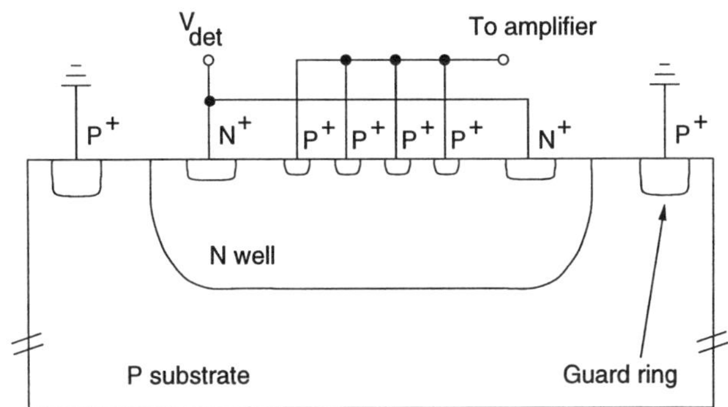

Fig. 3.17. Schematic cross section of an N-well / interrupted - P - finger photodiode [74]

The photoreceiver (see Fig. 12.43) uses the anode current for amplification of the optical signal (see Fig. 3.17). The slow diffusion of carriers generated by light with a wavelength of 850 nm in the P substrate, therefore, does not contribute to the photocurrent. This slow current is collected by the N well and shorted to the power supply V_{det}. To enhance the speed of the photodiode further, an interdigitated network of P^+ fingers is employed instead of a continuous P^+ region for maximizing the depletion regions available for carrier collection, particularly near the surface of the device. These P^+ fingers are connected outside the light sensitive area and form the anode of the photodiode. The N well had a size of $16.5 \times 16.5\,\mu m^2$. With a detector bias of 10 V, a bit rate of 1 GBit/s was reported with a bit error rate of less than 10^{-9}. The responsivity of the photodiode with the reported values of 0.01–0.04 A/W was very low due to the shallow P^+/N-well junction. It should be mentioned that this wide range of 0.01–0.04 A/W probably is due to optical interference in the CMOS isolation and passivation stack.

3.5.2 PIN Photodiode Integration

Lateral PIN Photodiode. A lateral PIN photodiode was fabricated together with NMOS transistors in a 1.0 μm technology using a nominally undoped substrate, which was actually P-type with $N_A = 6 \times 10^{12}\,\text{cm}^{-3}$ [75,76]. The schematic cross section of the NMOS OEIC is shown in Fig. 3.18. The PIN photodiode was fabricated directly on the high-resistivity substrate. It was stated that the thickness of the depletion layer of the PIN photodiode was approximately equal to the penetration depth of 850 nm light in silicon. The lateral PIN photodiode had a circular form with the cathode in the center surrounded by the anode. The gap between anode and cathode had a width of 10 μm. A dark current of 20 nA at 5 V, a breakdown voltage in excess of 60 V, and a capacitance of 40 fF at 5 V were reported. The external quantum efficiency of the photodiode at 870 nm was 67 % without an antireflection coating.

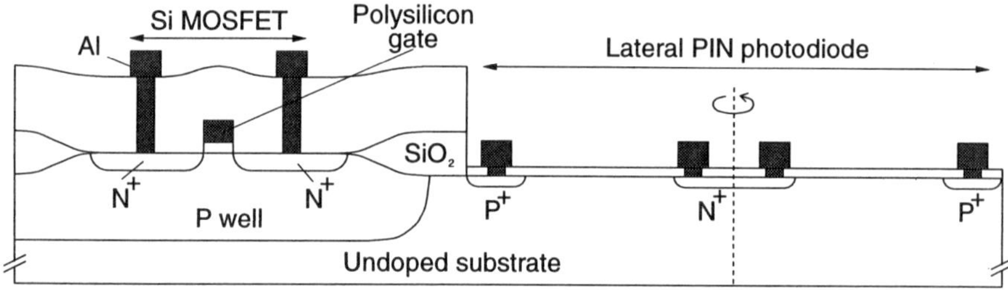

Fig. 3.18. Schematic cross section of an NMOS fiber receiver OEIC with a circular lateral PIN photodiode [75]

The −3dB bandwidth of the photodiode was determined by measuring the spectral content of the photodiode response to an optical input approximating a comb of "delta" functions in the time domain. Pulses with a duration of 200 fs and a separation of 13.2 ns from a mode-locked Ti-sapphire 850 nm laser were used as the optical input for the photodiode for this measurement. The output spectrum of the photocurrent was a comb function in the frequency domain with a peak separation of 75 MHz (1/(13.2 ns)) and a pulse height determined by the frequency response of the photodiode. A −3dB bandwidth of the photodiode equal to approximately 1.3 GHz was indicated by this measurement technique at a photodiode bias of 5 V. This large value, however, seems questionable for central light incidence due to the low electric field beneath the wide N^+ cathode.

Vertical PIN Photodiode. Our goal is the integration of a vertical PIN photodiode in a twin-well CMOS process. We will use Fig. 3.19 to illustrate the CMOS process sequence and the simplicity of PIN photodiode integration by means of the 'third' column on the right of this figure.

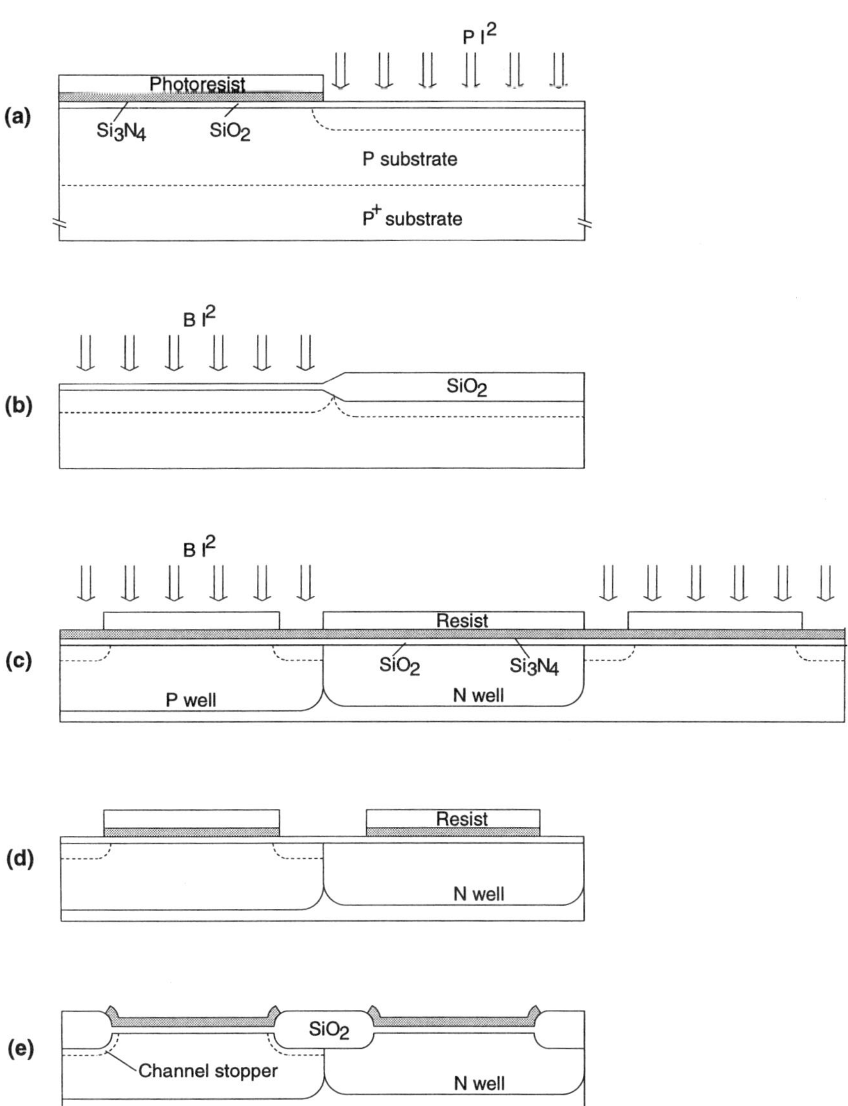

Fig. 3.19a–e. For caption, see the page after next page

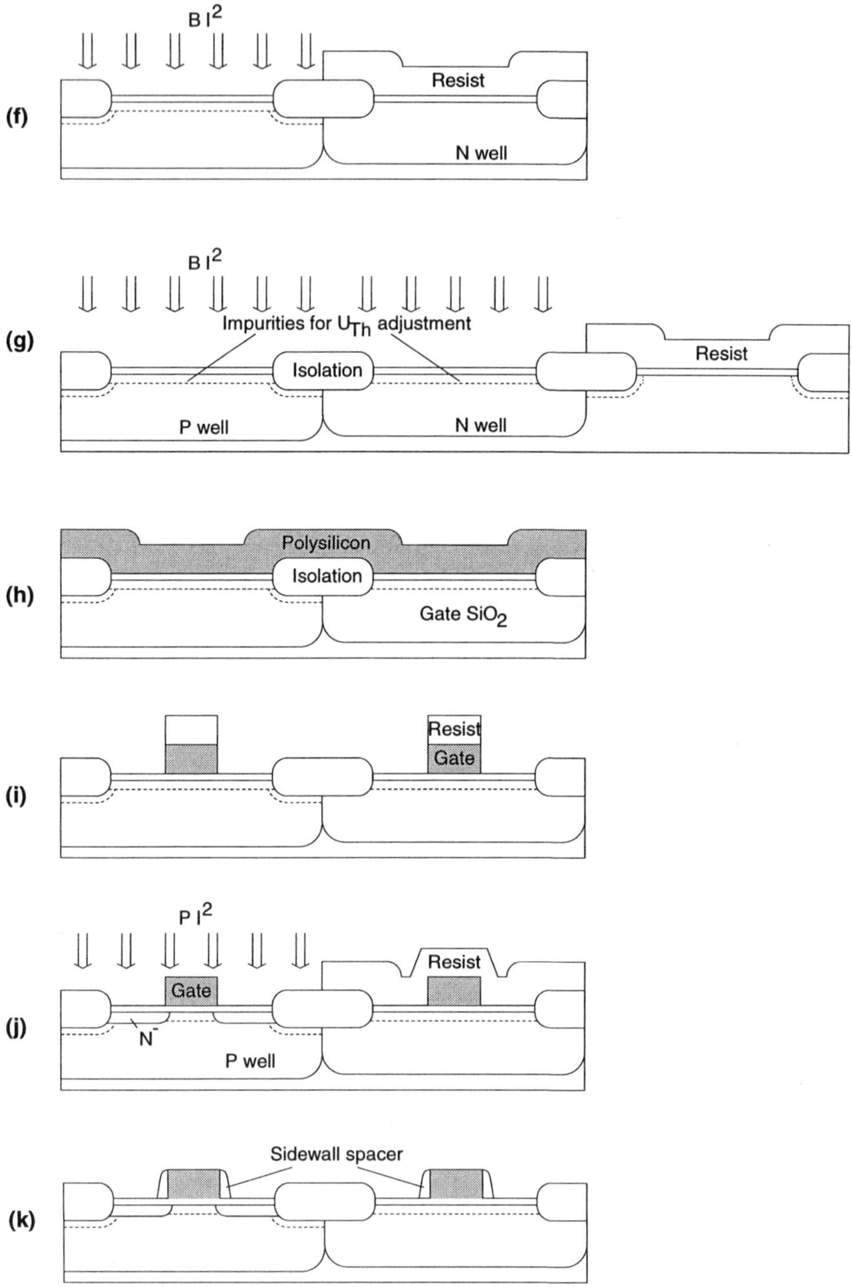

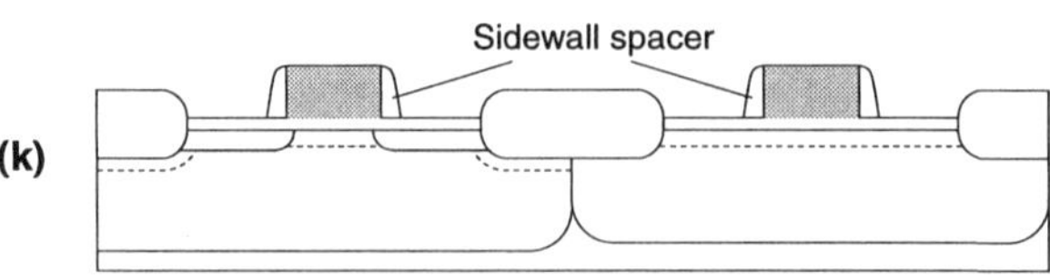

Fig. 3.19f–k. For caption see next page

(l)

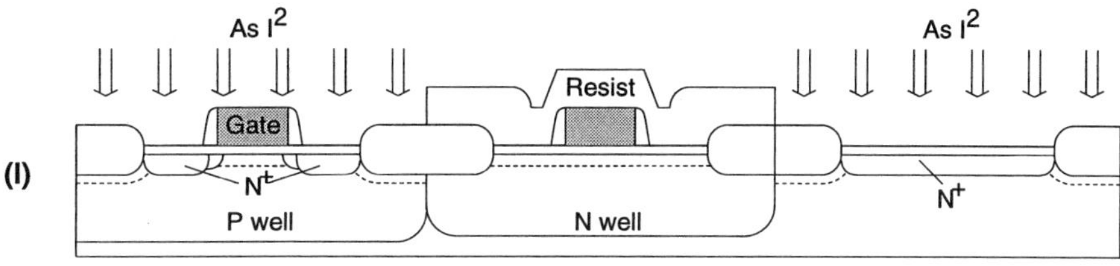

(m)

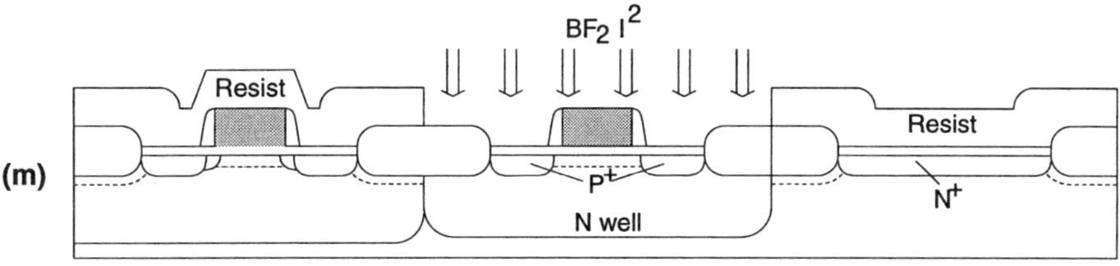

(n)

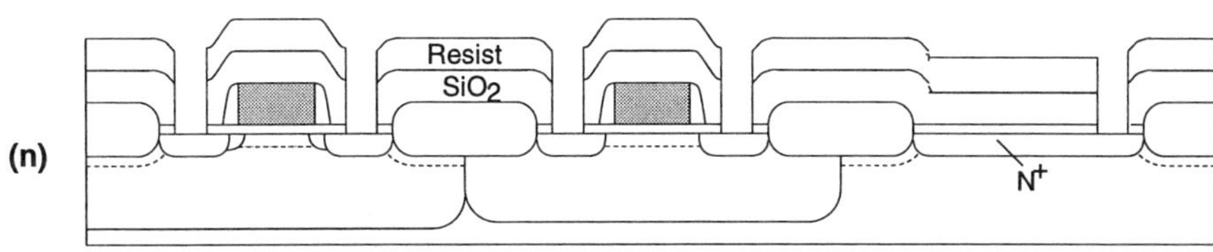

Fig. 3.19a–n. Process flow of a modern CMOS process [63] extended by the right column for PIN photodiode integration: (**a, b**) well formation, (**c–e**) isolation formation with channel stoppers (**f, g**) threshold voltage adjustment, (**h, i**) gate-oxide and gate-electrode formation, (**j**) LDD implantation, (**k–m**) LDD sidewall and source/drain formation, (**n**) contact-hole formation

The starting material is usually a (100)-oriented silicon substrate — P-type in this case — with an epitaxial layer having a doping concentration of $\approx 10^{15}\,\mathrm{cm}^{-3}$. Steps (a) and (b) show the self-aligned twin-well formation with only one lithographic mask step.

(a) The structured photoresist and nitride layer on a thin oxide layer are used for doping with phosphorus by ion implantation (I^2) into the N-well area selectively.

(b) After removing the photoresist, the wafers are selectively oxidized over the N-well region (LOCOS). The nitride layer is then etched off and boron is implanted to form the P well, whereby the LOCOS oxide prevents the boron from being implanted into the N-well area. The two wells are then driven in by high-temperature annealing to a depth of several micrometers and all oxides are etched off. The final impurity density of the wells at the silicon surface should be higher by at least one order of magnitude than that in the epitaxial layer to ensure independence of the device parameters of tolerances in the doping level of the epitaxial layer. The final impurity density of the wells at the silicon surface is typically of the order of $10^{16}\,\mathrm{cm}^{-3}$.

There are, however, also processes which do not use self-aligned twin-wells. These processes, needing two masks for the twin-well formation, are more appropriate for PIN photodiode integration, because no additional mask, i. e. no process modification is required in order to block out one of the two well implantations in the photodiode area. Although both types of processes end up with two masks for twin-well formation and PIN photodiode integration, for the process using self-aligned twin-wells the additional mask for PIN photodiode integration is a process modification. We will assume a process not using self-aligned twin-wells and start with step (c) for PIN photodiode integration.

The steps (c–e) show how LOCOS field isolation with an underlying channel-stopper region is formed.

(c) A nitride layer is deposited on a thin thermal oxide, called *pad oxide*. It is necessary for less mechanical stress concentrated at the LOCOS edge [77]. A photoresist mask is used to implant boron selectively through the oxide/nitride layer into the silicon, where the channel-stopper region is desired. The P-type channel-stopper is needed to increase the impurity density beneath the field isolation in P-well regions. Without this channel stopper, a parasitic N-channel MOSFET between two neighboring N-channel MOSFETs might be normally ON due to a large amount of positive fixed charge induced by subsequent LOCOS oxidation.

(d) The nitride layer is defined by dry etching and the pad oxide is exposed over the isolation region.

(e) The resist is removed and a thick field oxide is thermally grown to a thickness of several hundred nanometers over the isolation region. The nitride protects the remaining *active region* from being oxidized. The channel-stopper implant is driven in during this LOCOS oxidation. The nitride and thin oxide are then removed.

(f) The threshold implantation for the N-MOSFET can be done in two steps making two different threshold voltages available. The first threshold implantation is shown in Fig. 3.19f. A thin oxide is grown prior to the implantation to prevent ion channeling. The doping of the surface channel for the N-MOSFET is performed by boron or BF_2 implantation.

(g) The second threshold implantation for the N-MOSFET is shown in Fig. 3.19g. This boron or BF_2 implantation simultaneously results in the formation of a buried-channel P-MOSFET. The buried-channel P-MOSFET has the advantage of a lower 1/f noise compared to a surface-channel P-MOSFET. The second threshold implantation is carried out without a mask into all active regions. The photodiode region, therefore, has to be protected from being implanted by a resist mask. Let us call this mask a photodiode protection mask. The resist and all thin oxides are then removed to obtain the cross-sectional view before gate oxidation.

(h) The gate oxide is grown thermally in order to obtain a high oxide quality. Subsequently polysilicon for the gate electrode is deposited. The intrinsic polysilicon is then heavily doped with phosphorus in a gas ambient of $POCl_3$ at a temperature between 850 and 950 °C. In a polycide gate process, a silicide layer is then deposited on top of the polysilicon.

(i) The N^+-polysilicon layer is defined by lithography and dry etching. The resist is removed and the gate electrode is exposed to the so-called gate reoxidation or sidewall oxidation to prevent gate-oxide integrity near the gate edges from degradation [78].

The steps (j–m) show how the N^- regions for the LDD N-MOSFET, sidewall spacers, and N^+ as well as P^+ source/drain regions are formed:

(j) Phosphorus — or more recently arsenic — is implanted as the N^- LDD dopant. The implantation energy is low so that the phosphorus or arsenic implantation is masked by the thick field oxide, the gate electrode, and the photoresist mask above the opposite type of well.

(k) After removing the resist, oxide is deposited everywhere on the wafer and anisotropically etched to form sidewall spacers.

(l) Arsenic is implanted for N^+ source/drain formation of the N-MOSFET. A resist mask protects the P-MOSFETs from being implanted. This implant is used for the cathode formation of the PIN photodiode.

(m) BF_2 is implanted selectively into the P-MOSFET source/drain regions using another resist mask.

(n) Two oxide layers are then deposited, from which the first is undoped oxide (TEOS) and the second is boron and/or phosphorus doped for improving the glass flow property, which is needed for planarization. Contact holes are then defined by lithography and opened by dry etching.

Finally, the first aluminum layer for contact and interconnect formation is deposited and defined by lithography and dry etching. Then intermetal dielectric layers and further metal depositions may follow. In industrial practice, passivating oxide, nitride, or oxynitride layers are added by plasma-assisted deposition to protect the chip from impurities and humidity coming from the surroundings. Finally, bondpad areas have to be defined and opened.

Summarizing, we can state that only one additional mask is necessary for PIN photodiode integration on P^-P^+-substrate in order to block out an originally unmasked P-type threshold implantation, when the CMOS process does not use self-adjusting wells. Without this additional mask an N^+/P junction at the surface would result instead of the desired N^+/P^- junction. There would be a high electric field at the N^+/P junction and this would lead to a strongly reduced electric field in the 'intrinsic' P^--zone of the PIN photodiode and to longer rise and fall times.

On an N^-/N^+-substrate, the P-type PIN anode would be at the surface and the unmasked P-threshold implant increases the P^+/N^- junction depth by a small amount. In this case the electric field in the 'intrinsic' zone of the PIN photodiode is not reduced. The quantum efficiency for blue light may be reduced due to the larger junction depth. If this can be tolerated, no additional mask is required for PIN photodiode integration on an N^-/N^+-substrate.

As mentioned above, the doping concentration of the epitaxial layer is usually approximately $10^{15}\,\mathrm{cm}^{-3}$. Considering circuits with integrated photodiodes, a reasonable reverse bias of the photodiodes is approximately 3 V, when a single-supply voltage of 5 V is used. We know from Fig. 2.2 that the width of the space-charge region for a reverse bias of 3 V is only approximately 2 µm for a doping concentration of $10^{15}\,\mathrm{cm}^{-3}$. Due to the larger penetration depth of red and infrared light, slow diffusion of photogenerated carriers results for the epitaxial layer doping level of $10^{15}\,\mathrm{cm}^{-3}$. The doping level of the epitaxial layer has to be reduced in order to avoid this slow carrier diffusion [79]. It should be mentioned that this reduction of the doping concentration in the epitaxial layer is not a process modification. The CMOS manufacturer only has to buy wafers with a lower doping concentration in the epitaxial layer.

For the optical wavelengths of 780 and 850 nm being used in optical data transmission, the thickness of the epitaxial layer of a standard twin-well CMOS process of 7 µm [64], for instance, is too small due to the small absorption coefficients of Si for these wavelengths. In addition to the reduction of the doping concentration in the epitaxial layer, therefore, its layer thickness should be increased.

Figure 3.20 shows the structure of a PIN-CMOS-OEIC on a P^+ substrate. An antireflection coating (ARC) is drawn in the light sensitive area over the PIN cathode. It will be shown below that the integration of an ARC is highly recommended.

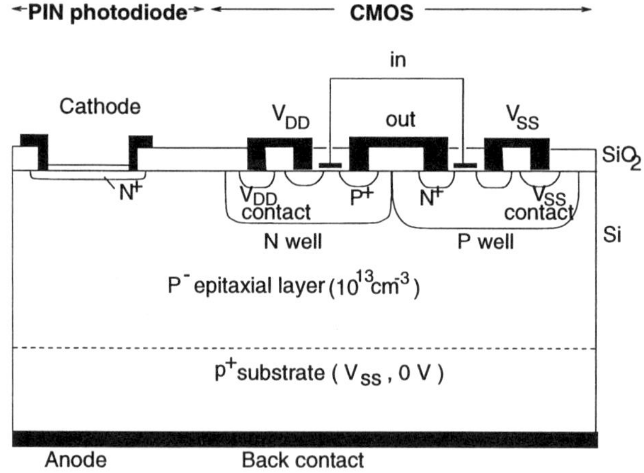

Fig. 3.20. Cross section of a P^-P^+-PIN-CMOS-OEIC [79]

There are several aspects which must be investigated when the epitaxial layer of a twin-well CMOS process is modified: (i) Latch-up immunity may be reduced. (ii) Reach-through (punch-through) between wells of the some type may occur. (iii) Electrostatic discharge hardness may suffer. Before these aspects are discussed the gain in photodiode performance by this modification of the epitaxial layer will be shown. The gain in photodiode performance by the modification of the epitaxial layer, i. e. by the reduction of its doping concentration will be shown in the following.

Results for P-type Substrate. We will investigate the behavior of PIN photodiodes by simulations in order to demonstrate their behavior and properties more thoroughly than would be possible by measurements. No details for the process illustrated in Fig. 3.19 were reported in [63]. Therefore, an AT&T 0.6 μm CMOS process [64] is chosen here. Process simulations with TSUPREM4 using the process description of the AT&T 0.6 μm CMOS process [64] completed with [80] were performed. The arsenic source/drain implantation of the N-channel MOSFETs was used for the N^+ cathode of the PIN photodiode. The doping profiles (Fig. 3.21) obtained in such a way were used by the device simulator MEDICI in order to compute the properties of the integrated PIN photodiodes [16]. The transient behavior of PIN photodiodes integrated in a 0.6 μm twin-well process with a grown epitaxial layer thickness of 15 μm is listed in Table 3.1.

For the standard doping concentration of $1 \times 10^{15}\,\mathrm{cm}^{-3}$ in the epitaxial layer there is slow carrier diffusion for $\lambda = 780\,\mathrm{nm}$ and the rise and fall times of the photocurrent are greater than 7 ns and 11 ns, respectively (Table 3.1). The more the doping concentration is reduced, the larger is the width of the space-charge region (Fig. 3.22), the less becomes the contribution of slow diffusion of photogenerated carriers and the shorter are the rise and fall times (Figs. 3.23 and 3.24).

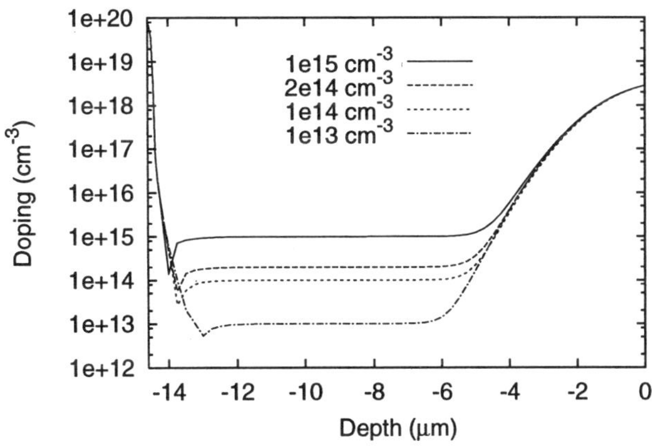

Fig. 3.21. Doping profiles of integrated $N^+P^-P^+$ PIN photodiodes. The original Si surface before epitaxy was at a depth of 0 μm. Oxidations consumed about half a micrometer of silicon from the epitaxially grown 15 μm Si

Table 3.1. Rise and fall times of the photocurrent of CMOS-integrated PIN photodiodes on P-type substrate with $U_{\mathrm{PIN}} = 3\,\mathrm{V}$ for different doping levels in the epitaxial layer with a growth thickness of 15 μm and for several wavelengths

	$\lambda = 780\,\mathrm{nm}$		$\lambda = 635\,\mathrm{nm}$		$\lambda = 565\,\mathrm{nm}$		$\lambda = 430\,\mathrm{nm}$	
I-Doping	t_{r}	t_{f}	t_{r}	t_{f}	t_{r}	t_{f}	t_{r}	t_{f}
(cm^{-3})	(ns)	(ns)	(ns)	(ns)	(ns)	(ns)	(ns)	(ns)
1×10^{15}	7.24	11.8	2.69	3.60	0.22	0.28	0.16	0.16
5×10^{14}	6.11	9.60	1.69	1.95	0.27	0.28	0.22	0.22
2×10^{14}	4.01	4.62	0.61	0.60	0.32	0.33	0.32	0.32
1×10^{14}	1.63	1.71	0.47	0.49	0.42	0.43	0.44	0.45
5×10^{13}	0.63	0.70	0.48	0.49	0.51	0.50	0.52	0.52
2×10^{13}	0.55	0.60	0.48	0.47	0.51	0.49	0.51	0.50
1×10^{13}	0.55	0.59	0.48	0.47	0.51	0.50	0.51	0.51

For the standard doping concentration of $1 \times 10^{15}\,\mathrm{cm}^{-3}$ in the epitaxial layer, the rise and fall times of the photocurrent decrease with decreasing wavelength (see Table 3.1) due to a decreasing penetration depth of the light. For the shortest wavelength, the light is absorbed completely within the space-charge region resulting in very short rise and fall times. For the short wavelengths 565 and 430 nm, the rise and fall times are shorter for the higher doping concentrations (Table 3.1), which may be surprising at first sight. This behavior, however, can easily be explained. The higher the doping concentration, the thinner the space-charge region, the thinner the drift zone, and the shorter the rise and fall times, because in the epitaxial layer below the drift zone ohmic conduction occurs, which is very fast. The lower the doping level in the epitaxial layer, the thicker the space-charge region,

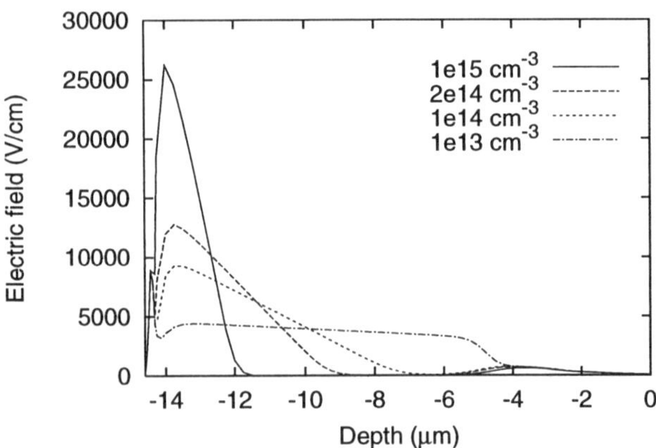

Fig. 3.22. Electric field distributions in integrated PIN photodiodes with different doping concentrations in the intrinsic layer

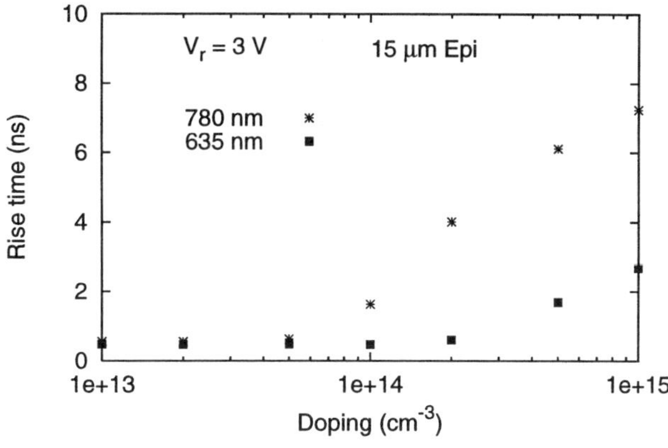

Fig. 3.23. Rise time of the PIN photocurrent for different doping concentrations of the intrinsic layer

the thicker the drift zone, the longer the drift time, and the longer the rise and fall times (Figs. 3.25 and 3.26).

For $\lambda = 780\,\text{nm}$, a reduction of the doping level in the epitaxial layer to $2 \times 10^{13}\,\text{cm}^{-3}$ is sufficient to reach the minimum rise and fall times (see Table 3.1). For $\lambda = 635\,\text{nm}$, already a reduction of the doping level in the epitaxial layer to $1 \times 10^{14}\,\text{cm}^{-3}$ is sufficient to reach the minimum rise and fall times. For this doping level, the space-charge region spreads completely through the intrinsic zone. The electric field distribution has a triangular shape. The rise and fall times for $\lambda = 635\,\text{nm}$ do not decrease when the doping concentration is reduced below $1 \times 10^{14}\,\text{cm}^{-3}$ and when the electric field distribution becomes more rectangular (Fig. 3.22). For the shorter wavelengths, no reduction of the doping concentration in the epitaxial layer is recommended. For the wavelengths 565 nm and 430 nm, the epitaxial layer should be as thin as possible in order to reduce the series resistance.

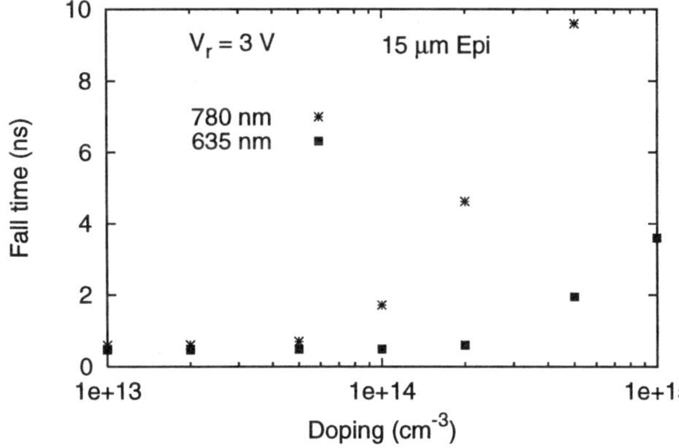

Fig. 3.24. Fall time of the PIN photocurrent for different doping concentrations of the intrinsic layer

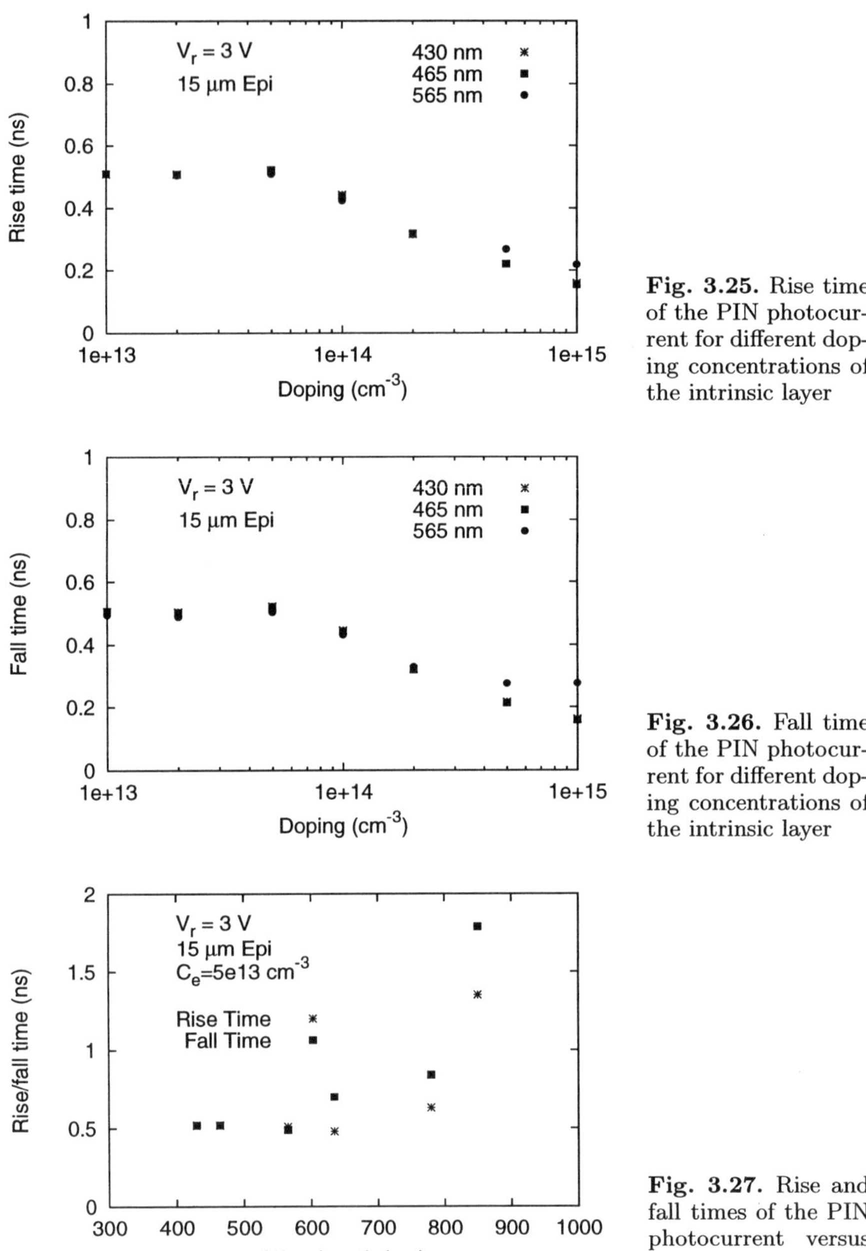

Fig. 3.25. Rise time of the PIN photocurrent for different doping concentrations of the intrinsic layer

Fig. 3.26. Fall time of the PIN photocurrent for different doping concentrations of the intrinsic layer

Fig. 3.27. Rise and fall times of the PIN photocurrent versus wavelength

Figure 3.27 shows the dependence of the rise and fall times on the wavelength for a doping concentration of $5 \times 10^{13}\,\mathrm{cm}^{-3}$ in the epitaxial layer, which was grown to a thickness of 15 µm. The rise and fall times remain constant when the wavelength is shorter than 565 nm, because the largest part of the light then is absorbed in the space-charge region (compare Figs. 3.28 and 3.22). The diffusion tail of the photocurrent is below the 10 % level of the maximum photocurrent for wavelengths shorter than 635 nm (see Fig. 3.29). With the wavelengths of 780 and 850 nm, many charge carriers are generated in the P^+ substrate. These carriers diffuse slowly to the drift zone and cause the slowly decaying tail of the photocurrent after the end of the light incidence at 13 ns shown in Fig. 3.29. These diffusing charge carriers can also diffuse laterally to the transistors in amplifier circuits or other circuit units

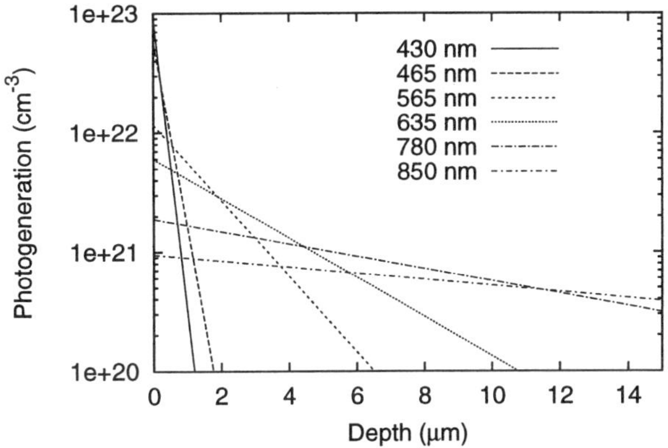

Fig. 3.28. Photogeneration of electron-hole pairs for different wavelengths

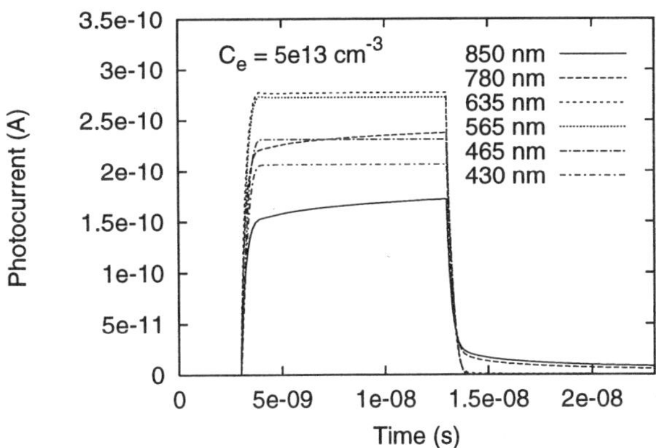

Fig. 3.29. Transient response of CMOS-integrated PIN photodiodes for different wavelengths

and cause optoelectronic cross-talk. An N^+ guard-ring (in a P-type substrate) biased in the reverse direction and surrounding the circuits or the photodiodes can attract these diffusing minority carriers (electrons) [81]. The portion of the photogenerated carriers causing the diffusion tail after the end of the light pulse is missing in the photocurrent pulse during the light pulse. This missing portion leads to a reduced dynamical quantum efficiency. Figure 3.29 illustrates this reduced dynamical quantum efficiency for long wavelengths. For short wavelengths, there is also a reduced amplitude of the photocurrent pulse. This reduction, however, is due to carrier recombination in the highly doped N^+ cathode at the surface of the silicon device. Figure 3.30 depicts the dynamical quantum efficiency as a function of wavelength.

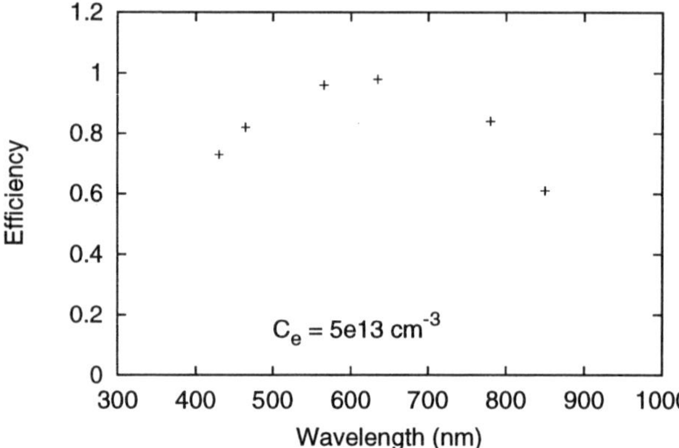

Fig. 3.30. Dynamical internal quantum efficiency for the wavelengths from Fig. 3.29 extracted just before the end of the light pulse at t = 13 ns

The lightly doped drain (LDD) structure of the N-channel transistors could lead us to the supposition that the photoresist mask in Fig. 3.19j for blocking out this LDD phosphorus implant has to be used when the quantum efficiency of the photodiodes in the green and blue spectral range shall be high. The influence of the phosphorus LDD implant on the quantum efficiency, therefore, will be investigated next. Figure 3.31 compares the doping profiles of the cathodes with and without the LDD implantation. The depth of the PN junction for a substrate doping concentration of 5×10^{13} cm^{-3} remains unchanged. The distributions of the electric field are plotted in Fig. 3.32. The peaks in the electric field at $-14.4\,\mu$m are due to the doping gradient of the N^+ source/drain region. The decrease from about $-13.7\,\mu$m to $-14.2\,\mu$m is due to the space-charge region penetrating from the position of the N/P junction at about $-13.7\,\mu$m into the N region.

The maximum of the electric field is reduced by the LDD implantation. The decrease towards the silicon surface at $-14.6\,\mu$m, however, begins at the same depth. The influence of the LDD implantation on the quantum efficiency of the PIN photodiode, therefore, should be small. Simulations calculate a photocurrent of 0.2089 nA with the LDD implant and a photocurrent

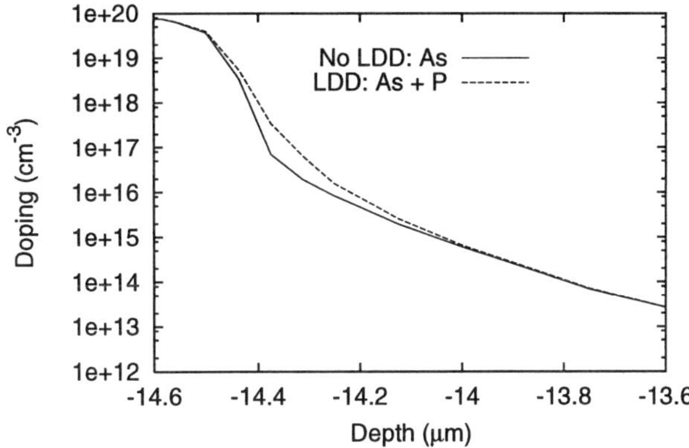

Fig. 3.31. Cathode doping profiles of CMOS-integrated PIN photodiodes

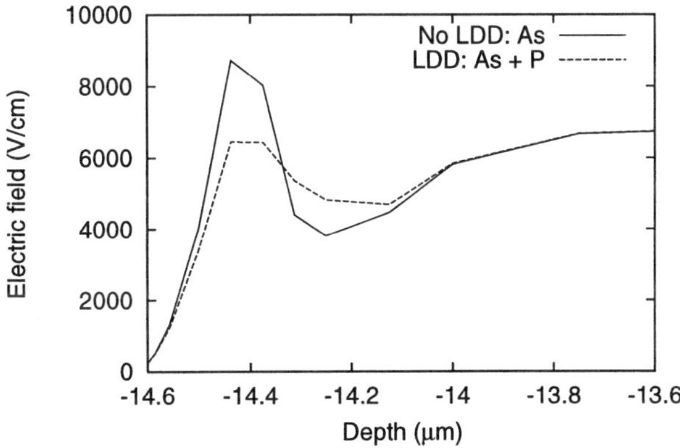

Fig. 3.32. Electric field for the different cathode doping profiles of CMOS-integrated PIN photodiodes shown in Fig. 3.31

of 0.2066 nA without LDD implant for $\lambda = 430$ nm. The difference is approximately 1 % and therefore within the numerical accuracy. Even if the doping concentration in the epitaxial layer is raised to $1 \times 10^{15}\,\mathrm{cm}^{-3}$ and when the PN junction is moved to approximately $-14\,\mu\mathrm{m}$ (see Fig. 3.31), the decrease of the electric field towards the silicon surface will not change significantly. The quantum efficiency in the blue spectral range, therefore, does not depend on the presence or absence of the LDD implant. It does not matter whether the mask for blocking out the LDD implant from the photodiode area is present or not.

Results for N-type Substrate. Figure 3.33 shows the structure of a N^-N^+ CMOS-OEIC in the twin-well approach. Here the N^+ substrate serves as the cathode and the P^+ source/drain region as the anode of the integrated PIN photodiode.

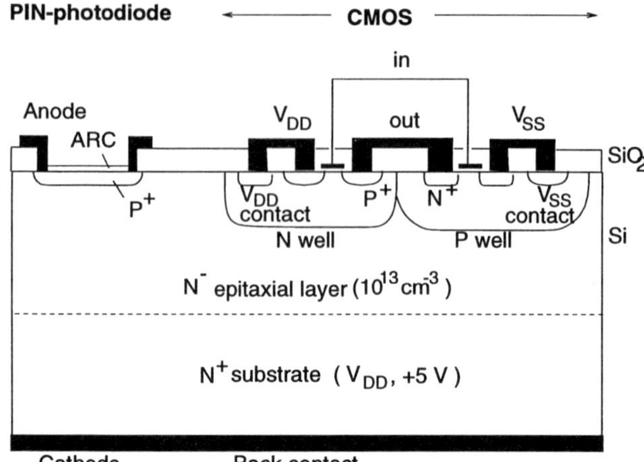

Fig. 3.33. Cross section of a N^-N^+ PIN-CMOS-OEIC [82]

PIN-CMOS photodiodes with an area of 2700 μm^2, with a standard doping concentration of 1×10^{15} cm^{-3} and with reduced doping concentrations in the epitaxial layer down to 2×10^{13} cm^{-3}, and with an integrated polysilicon resistor of 500 Ω, were fabricated in an industrial 1.0 μm CMOS process [83]. For the measurements, a laser with $\lambda = 638.3$ nm was modulated with a commercial ECL generator. The light pulses were coupled into the photodiodes on a wafer prober via a single-mode optical fiber. The rise (t_r) and fall (t_f) times of the photocurrent of the photodiodes were measured with a picoprobe (pp), which possesses a −3dB bandwidth of 3 GHz and an input capacitance of 0.1 pF, and with a 20 GHz digital sampling oscilloscope HP54750/51.

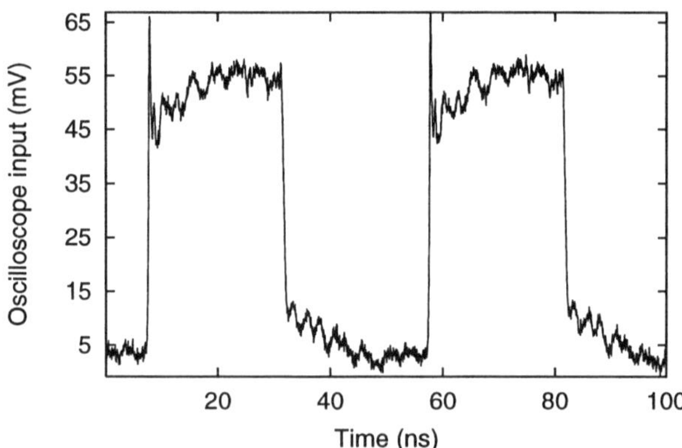

Fig. 3.34. Measured transient response of a CMOS-integrated P^+N photodiode with a standard epitaxial doping concentration of 1×10^{15} cm^{-3} for $\lambda = 638$ nm and $|V_{PD}| = 3.0$ V. The overshoot is due to direct laser modulation. The 0 % and 100 % lines for the determination of t_r and t_f are at 3.5 and 55.5 mV

Figure 3.34 contains the waveform of the photocurrent for the photodiode with the standard doping concentration of $1\times10^{15}\,\text{cm}^{-3}$ in the epitaxial layer. The values $t_r = 5.3$ ns and $t_f = 7.3$ ns were determined from the waveform for the photodiode with the standard doping concentration of $1 \times 10^{15}\,\text{cm}^{-3}$ in the epitaxial layer. These large values are due to the diffusion tail of the photocurrent, which is caused by the slow carrier diffusion in the standard epitaxial layer of the photodiode. Values up to $t_r = 15.5$ ns and $t_f = 17.6$ ns were also measured for the photodiode with the nominal doping concentration of $1 \times 10^{15}\,\text{cm}^{-3}$ in the epitaxial layer. The large ranges for t_r and t_f can be explained like this: The diffusion tail of the photocurrent is very sensitive to the actual doping concentration and thickness of the epitaxial layer. The rise and fall times, therefore, vary strongly with the position of the photodiode on a wafer and from wafer to wafer due to variations of doping concentration and epitaxial layer thickness.

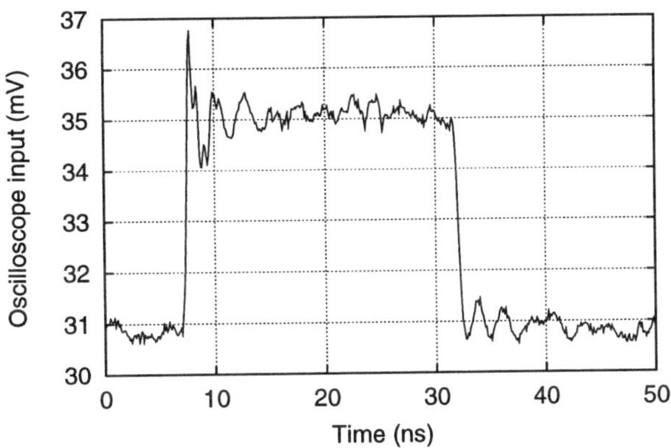

Fig. 3.35. Measured transient response of a CMOS-integrated PIN photodiode with an I-layer doping concentration of $2 \times 10^{13}\,\text{cm}^{-3}$ for $\lambda = 638$ nm and $|V_{PD}| = 3.0$ V. The overshoot in the signal is due to the direct modulation of the laser [84]

For the PIN photodiode with a doping concentration in the epitaxial layer of $2 \times 10^{13}\,\text{cm}^{-3}$, the oscilloscope extracted $t_r^{osc,disp} = 0.37$ ns and $t_f^{osc,disp} = 0.57$ ns from the waveform shown in Fig. 3.35. The evaluation according $(t_r^{PIN})^2 = (t_r^{osc,disp})^2 - (t_r^{laser})^2 - (t_r^{PP})^2 - (t_r^{osc})^2$ with $t_r^{laser} = 0.30$ ns, $t_f^{laser} = 0.51$ ns, $t_{r/f}^{PP} = 0.1$ ns and $t_{r/f}^{osc} \approx 0.02$ ns results in $t_r^{PIN} = 0.19$ ns and $t_f^{PIN} = 0.24$ ns [82]. With $f_{3dB} = 2.4/(\pi(t_r + t_f))$, the -3dB bandwidth can be estimated to be 1.7 GHz for $|V_{PD}| = 3.0$ V. With the conservative estimate BR=$1/(1.5\ (t_r + t_f))$, a bit rate, BR, of 1.5 Gb/s results for $|V_{PD}| = 3.0$ V. With an antireflection coating (ARC), the quantum efficiency η could be increased from 49 % to 94 %. To our knowledge this is the first time that such a high speed and such a high quantum efficiency have been achieved with an integrated silicon photodiode for a reverse voltage of only 3 V, whereby an only slightly modified standard CMOS process has been used.

CMOS-integrated photodiodes with doping concentrations of $5\times10^{13}\,\mathrm{cm}^{-3}$ and $1\times10^{14}\,\mathrm{cm}^{-3}$ in the epitaxial layer were also fabricated and characterized [85, 86]. The measured rise times are compared to the simulated rise times in Fig. 3.36. Measured doping profiles of the epitaxial wafers used for the fabrication of the CMOS-integrated PIN photodiodes were read into the process simulator for the doping concentrations $1\times10^{15}\,\mathrm{cm}^{-3}$, $1\times10^{14}\,\mathrm{cm}^{-3}$, and $5\times10^{13}\,\mathrm{cm}^{-3}$ in the epitaxial layers. No measured doping profiles for the lowest doping concentration of $2\times10^{13}\,\mathrm{cm}^{-3}$ in the epitaxial layer was supplied by the company from which the epitaxial wafers had been ordered. The simulated results for the lowest doping concentration in the epitaxial layer, therefore, are less reliable than for the other doping concentrations. The implantation energies and doses as well as the annealing, oxidation, and BPSG reflow temperatures and times of the 1.0 μm CMOS process used for the fabrication of the PIN photodiodes and CMOS OEICs were applied in the process simulation. The same models and parameters for the device simulations were used as for the simulation of CMOS-integrated PIN photodiodes on a P-type substrate discussed above.

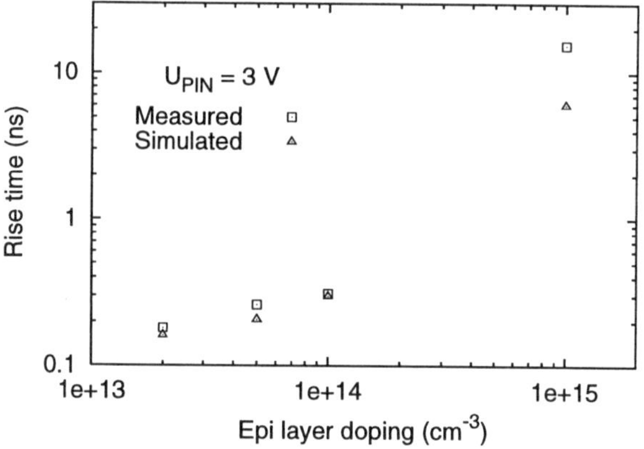

Fig. 3.36. Rise time of the photocurrent of CMOS-integrated PIN photodiodes with different doping concentrations in the epitaxial layer corrected for the laser and picoprobe rise times. The photodiodes were connected to integrated resistors of 500 Ω ($\lambda = 638\,\mathrm{nm}$, $|V_{\mathrm{PD}}| = 3.0\,\mathrm{V}$)

The measured fall times versus the doping concentration in the 'intrinsic' zone of the CMOS-integrated PIN photodiodes are compared to the simulated fall times in Fig. 3.37. There is a rather good agreement between measured and simulated rise and fall times. The deviation for $C_{\mathrm{e}} = 1\times10^{15}\,\mathrm{cm}^{-3}$ is not surprising as the 10 % point of the photocurrent is in the rather flat diffusion tail. Therefore, noise makes accurate measurements very difficult.

Furthermore, the rise and fall times strongly depend on the exact doping concentration in the epitaxial layer C_e, because C_e determines the width of the space-charge region and, consequently, the fraction of photogenerated carriers participating in slow diffusion. The amplitude of the diffusion tail and the time value of the 10 % point of the photocurrent, therefore, react very sensitively on tolerances in C_e for the highest doping concentration.

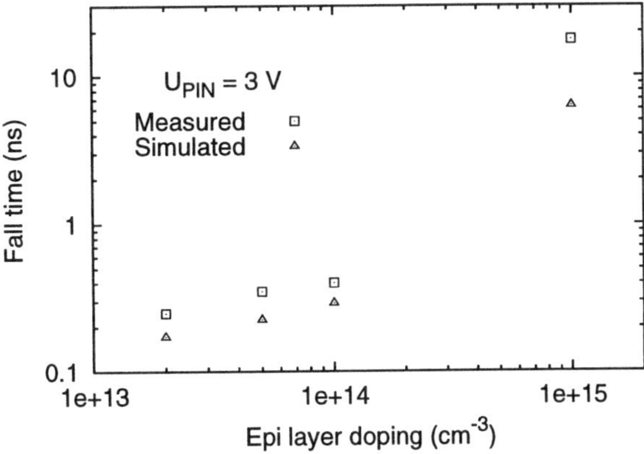

Fig. 3.37. Fall time of the photocurrent of CMOS-integrated PIN photodiodes with different doping concentrations in the epitaxial layer corrected for the laser and picoprobe fall times. The photodiodes were connected to integrated resistors of 500 Ω ($\lambda = 638$ nm, $|V_{PD}| = 3.0$ V)

The agreement between measured and simulated results seems to be rather good for $C_e \leq 1 \times 10^{14}\,\text{cm}^{-3}$, when it is considered that no measured doping profiles were available and that the dopant diffusion parameters in the process simulator, therefore, could not be adjusted to the process used for the fabrication of the CMOS-integrated PIN photodiodes and CMOS OEICs. It can be concluded, finally, that the models and the default parameters implemented in the process and device simulators used allow a rather good estimation for the transient behavior of CMOS-integrated PIN photodiodes.

Transistor Parameters. In contrast to transistors in bipolar OEICs, the electrical performance of the N- and P-channel MOSFETs is not degraded when the epitaxial layer is modified. This statement was verified by measurements (Table 3.2). The threshold voltages of the NMOS and PMOS transistors are practically independent of the doping concentration in the epitaxial layer, because these transistors are placed inside wells which possess a much higher doping level of several times $10^{16}\,\text{cm}^{-3}$ than the standard epitaxial layer with about $1 \times 10^{15}\,\text{cm}^{-3}$ and because the threshold implants produce an even higher doping level ($\approx 10^{17}\,\text{cm}^{-3}$) than the wells [87].

Table 3.2. Measured threshold voltages U_{Th} and drain leakage currents I_{D}^{r} for different doping levels in the epitaxial layer

I-Doping (cm^{-3})	$U_{\text{Th}}^{\text{NMOS}}$ (V)	$I_{\text{D}}^{\text{r,NMOS}}$ (pA)	$U_{\text{Th}}^{\text{PMOS}}$ (V)	$I_{\text{D}}^{\text{r,PMOS}}$ (pA)
Standard	0.79	2.19	−0.62	63.1
1×10^{14}	0.79	0.83	−0.60	66.1
5×10^{13}	0.79	0.81	−0.60	66.1
2×10^{13}	0.78	1.02	−0.60	67.6

The reverse, i. e. the leakage current of the drain to well diodes is also listed in Table 3.2 for the NMOS and PMOS transistors [87]. The leakage current for the NMOS transistor in an epitaxial layer with reduced doping concentrations actually seems to be smaller than for the standard concentration.

These results confirm the superiority of the PIN-CMOS integration [79] compared to the PIN-bipolar integration [38]. In contrast to the PIN-bipolar integration, the electrical transistor parameters of the standard twin-well CMOS process are completely unaffected for the PIN-CMOS integration and can be used for circuit simulations within the design of OEICs.

Latch-up Effect. At places where N and P wells are in contact, there is the danger of latch-up. Figure 3.38 shows P- and N-type regions acting as parasitic bipolar transistors in a CMOS cross section.

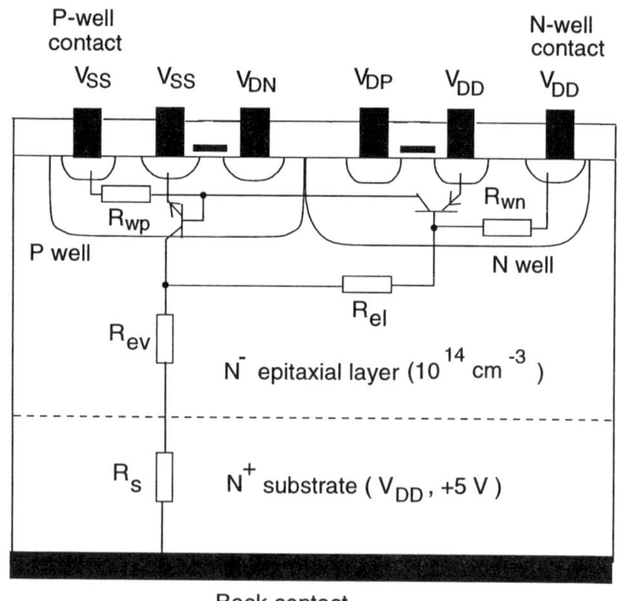

Fig. 3.38. Schematic of the parasitic bipolar transistors being responsible for latch-up in a CMOS circuit [87]

The N^+ source-island of the N-channel MOSFET forms the emitter of the parasitic NPN transistor, the P well forms the base, and the epitaxial substrate forms the collector of the parasitic NPN transistor. Analogously, the emitter of the parasitic PNP transistor consists of the P^+ source island, the base consists of the N well, and the collector of the PNP transistor consists of the P well. The collector of one transistor simultaneously forms the base of the other transistor and vice versa. The NPN and PNP transistors, therefore, form a thyristor structure. The thyristor can turn on, for instance, when the base-emitter-potential of one of the transistors exceeds a value of 0.6 V, approximately, due to a lateral current in one of the wells across R_{wp} or R_{wn} as a result of voltage spikes on the well-interconnect lines. The turn-on of the thyristor leads to a shorting of V_{DD} and V_{SS} and the transistors lose their function in the circuit at least until the next power-off and power-on. They may even be destroyed due to large currents resulting in strong heating.

A reduction of the doping concentration in the epitaxial layer for PIN photodiode integration reduces the value of the base Gummel integral of the PNP transistor and increases the PNP current gain accordingly. The value of the NPN transistor's collector series resistance R_{ev} increases when the doping concentration of the epitaxial layer is reduced. These two aspects may lead to a reduced latch-up immunity. The lateral resistor R_{el} (see Fig. 3.38), however, is also increased. This aspect increases the latch-up immunity again, because R_{el} forms a voltage divider together with R_{wn}, which is independent of the doping concentration of the epitaxial layer.

The latch-up immunity can be enhanced by appropriate design measures. When the source of the NMOS transistor is closer to the N well than the drain of the NMOS transistor, the parasitic NPN transistor may partially inject laterally into the N well as a collector instead of injecting only vertically as shown in Fig. 3.38. Therefore, the distance of neighboring N- and P-channel transistors (more accurately between N^+ and P^+ islands in wells of different type which are in contact with each other) must not be smaller than a mimimum N^+–P^+ distance defined in the design rules in order to keep the so-called holding voltage above the V_{DD} voltage value. Using a larger distance increases the latch-up immunity due to a reduction of the bipolar transistor current gain with a wider base. Paying attention to this design rule results in an extinguishing current flow in the thyristor structure after the end of the firing impulse. The minimum N^+–P^+ distance depends on the substrate doping concentration. In order to minimize the N^+–P^+ distance, epitaxial wafers with highly doped substrates are used. Thus the minimum distance depends on the doping concentration of the epitaxial layer. Chapman et al. [88] from Texas Instruments reported a thickness of the epitaxial layer of 5 µm for a 0.8 µm CMOS technology. In order to keep the holding voltage above a value of 5 V, the N^+–P^+ distance had to be larger than 4 µm. For a constant distance of the well contacts, the holding voltage reduced to 1.5 V when the thickness of the epitaxial layer was increased to 7 µm. The holding voltage reduced to approximately 1 V for an epitaxial layer thickness of 9 μm. In order to obtain holding voltage values larger than 5 V for the thicker epitaxial layers, the distance to the well contacts had to be reduced [88].

The influence of the substrate doping concentration on the latch-up behavior of a twin-well CMOS process was investigated in [89]. The result of this work was that with a highly doped substrate (0.01–0.02 Ωcm) and with a 10 μm thick epitaxial layer (1.7–2.5 Ωcm), the emitter firing currents of the NPN and PNP transistors could be increased from 30 mA to 250 mA by implementing an N^+ guard ring around the P emitter in the N well. For a low substrate doping (1.7–2.5 Ωcm), an N^+ guard ring around the P emitter increases the emitter firing current of the NPN transistor from about 7 mA to 235 mA. An additional P^+ guard ring around the N emitter increases the emitter firing current of the PNP transistor from about 9 mA to 60 mA. From the results of [89] we can draw the conclusion that in OEICs with reduced doping concentration in the epitaxial layer or/and increased epitaxial layer thickness, the positive influence of the highly doped substrate on the latch-up immunity is reduced. A second conclusion, however, is that the latch-up immunity can almost be regained with the implementation of guard rings.

Another very effective design measure for the suppression of latch-up is the implementation of guard rings around the wells. These design measures applied as a precaution for OEICs with a reduced doping concentration in the epitaxial layer, however, would require die area and they might result in an inacceptable increase of the die area of highly integrated digital OEICs. For analog OEICs, which contain far fewer transistors, the additional die area for latch-up precaution would be easier to accept. Because of the large economical importance of digital ICs and because of the potential high-volume market for OEICs in optical interconnects, however, the latch-up immunity of PIN-CMOS-OEICs was examined experimentally.

In order to investigate the influence of the doping concentration in the epitaxial layer on latch-up immunity a test structure (see Fig. 3.39) was fabricated together with the CMOS-integrated PIN photodiodes.

The distance of the emitters from the well boundary was 2 μm each. A V_{DD} voltage of 5.5 V was used in order to account for tolerances of power supplies. The hole current injected into the P^+ emitter of the parasitic PNP transistor necessary for triggering the thyristor structure was measured [87]. The injected hole current necessary for triggering latch-up is shown in Fig. 3.40 for several doping concentrations in the epitaxial layer. The injected hole current necessary for triggering latch-up is constant when the doping concentration of the epitaxial layer is reduced from $1 \times 10^{15}\,\mathrm{cm}^{-3}$ to $2 \times 10^{13}\,\mathrm{cm}^{-3}$ for room temperature and for an elevated temperature of 85 °C [87]. Latch-up immunity, therefore, is not degraded when PIN photodiodes are integrated. Accordingly, the latch-up aspect does not require a modification of the design rules of the CMOS process used for PIN photodiode integration and die-

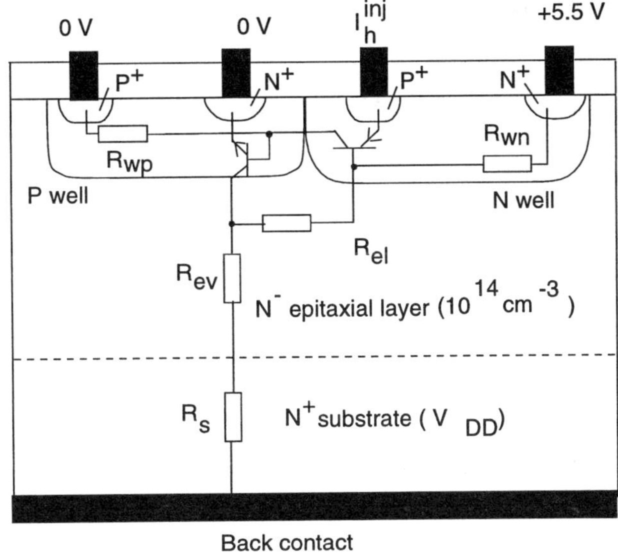

Fig. 3.39. Schematic cross section of the latch-up test structure [87]

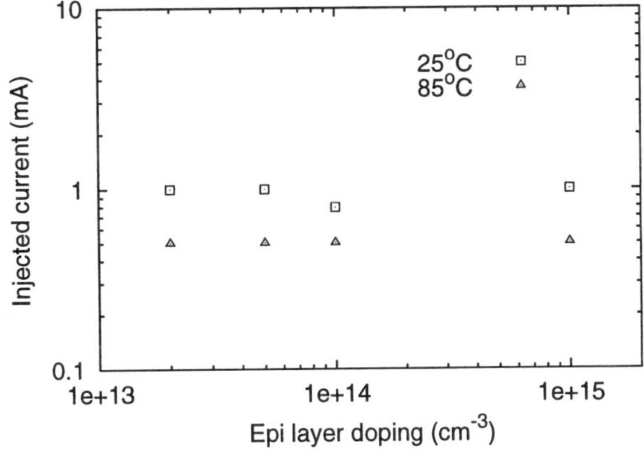

Fig. 3.40. Injected hole current $I_{\mathrm{h}}^{\mathrm{inj}}$ necessary for triggering latch-up versus doping concentration in the epitaxial layer [87]

area-consuming guard rings are not necessary for analog and digital CMOS OEICs.

Reach-through Effect. The reach-through or punch-through effect is important between neighboring analog wells which have a doping type different from that of the substrate. Analog wells are not at the power supply voltage rails like digital wells. Let us assume for simplicity that one of the wells is at the substrate potential (Fig. 3.41).

Then the potential difference between the two wells can be maximum. The consequence is that the space-charge region extends most widely from one well towards the other. When the space-charge region extends to the

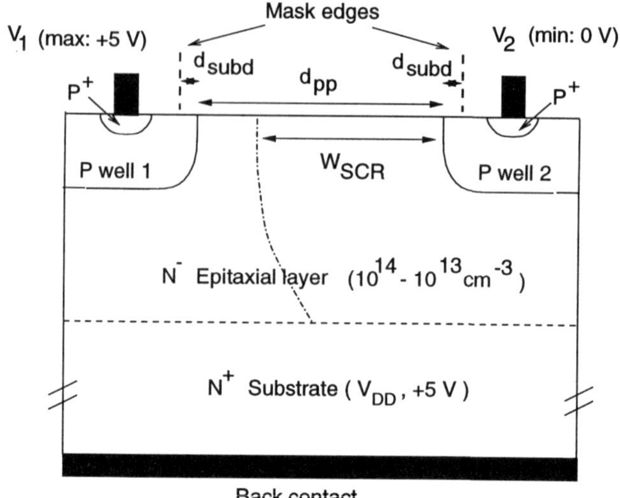

Fig. 3.41. Schematic to illustrate the reach-through effect between analog wells of the same type [90]

other well for a large potential difference of the two wells, a current begins to flow between the two wells (Fig. 3.42), which may change the operating point in an analog circuit. Therefore, there is a design rule which sets the minimum distance of analog wells.

When we reduce the doping concentration of the epitaxial layer, the space-charge region extends further for a constant difference voltage. The design rule for the minimum distance of analog wells, therefore, has to be modified [87]. According to (2.12) with the difference voltage of the two wells U_{12} and considering the lateral subdiffusion of the two wells d_{subd} plus a security distance d_{sec}, the new minimum distance of analog wells $d_{\mathrm{pp}}^{\mathrm{min}}$ [90] results

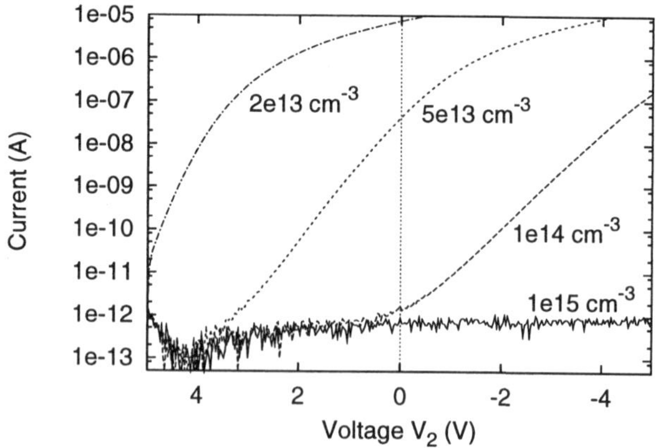

Fig. 3.42. Measured reach-through current for a P-island distance of 10 μm ($V_1 = +5\,\mathrm{V}$; V_2 is varied from +5 to −5 V) [90]

$$d_{\mathrm{pp}}^{\mathrm{min}} = \sqrt{\frac{2\epsilon\epsilon_0}{qN_{\mathrm{I}}}(U_{\mathrm{D}} + |U_{12}| - \frac{2k_{\mathrm{B}}T}{q})} + 2d_{\mathrm{subd}} + d_{\mathrm{sec}}. \tag{3.3}$$

The difference in this equation compared to the original expression for the design rule is that N_{I} instead of the standard doping concentration in the epitaxial layer is used. For N_{I}, the low tolerance limit of the epitaxial doping concentration has to be used. $|U_{12}|$ is the maximum potential difference between two neighboring wells which have a doping type opposite to that of the substrate. This maximum potential difference depends on the actual circuit.

The worst case for the maximum potential difference would be, for instance, $V_1 = 5\,\mathrm{V}$ and $V_2 = 0\,\mathrm{V}$. Figure 3.42 shows reach-through currents between a P well and a P^+ island measured in a test structure actually designed for investigation of diffusion of photogenerated carriers. For $N_{\mathrm{I}} = 10^{15}\,\mathrm{cm}^{-3}$, there is no reach-through current. For $N_{\mathrm{I}} = 10^{14}\,\mathrm{cm}^{-3}$, there is a very low reach-through current of about $1.5 \times 10^{-12}\,\mathrm{A}$ at $V_2 = 0\,\mathrm{V}$. For $N_{\mathrm{I}} = 5 \times 10^{13}\,\mathrm{cm}^{-3}$, the reach-through current for $V_2 = 0\,\mathrm{V}$ is about $5 \times 10^{-8}\,\mathrm{A}$. This reach-through current is tolerable for the $\mathrm{N}^-\mathrm{N}^+$-CMOS-OEIC (see Fig. 12.19) fabricated on epitaxial material with $N_{\mathrm{I}} = 5 \times 10^{13}\,\mathrm{cm}^{-3}$ [85]. Actually the potential difference between P wells is only about 1 V in this circuit, and measurements confirmed that the $\mathrm{N}^-\mathrm{N}^+$-CMOS-OEIC also works well with a doping level of $N_{\mathrm{I}} = 2 \times 10^{13}\,\mathrm{cm}^{-3}$. The distance of two analog wells of 10 μm is sufficient for the circuit shown in Fig. 12.19. For many analog circuits this value, therefore, does not increase dramatically above the original minimum analog well distance of 6.6 μm.

3.5.3 Finger Photodiodes

Future digital-video-disk or digital-versatile-disk (DVD) systems will use wavelengths shorter than red in order to increase the storage capacity and the play time [91]. The announcement of the Japanese company Nichia that the first blue diode laser [92, 93] will be available commercially at the end of 1998 made it very likely that optical storage systems like DVD, CD-ROM and audio CD will use blue laser light. Finger photodiodes, therefore, were suggested for a high quantum efficiency in the blue spectral range [83]. Now, engineering samples of blue laser diodes are available making optical storage systems with blue lasers soon possible and creating a need for finger photodiodes. Figure 3.43 shows the cross section of such interdigitated photodiodes, which can be fabricated in a CMOS process not using self-adjusting well formation with little process modification. The photodiode protection mask was the only additional mask necessary to block out an originally unmasked P-type threshold implantation from the photodiode area.

The internal quantum efficiency for PIN and finger photodiodes was determined by device simulations including photogeneration. The results are listed in Table 3.3. The internal quantum efficiency η_{i} of PIN photodiodes for the red spectral range is very good. In the blue spectral range, however,

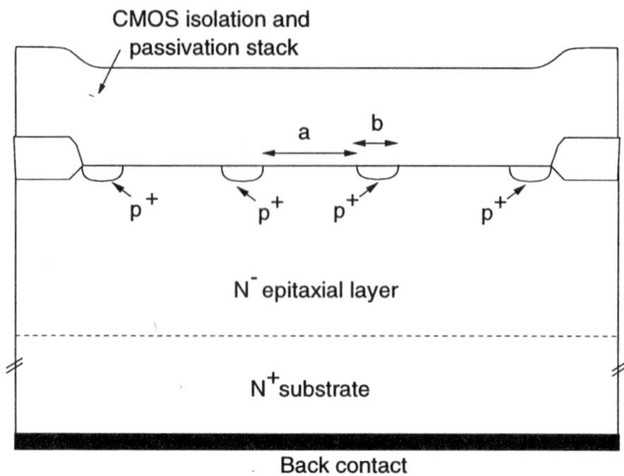

Fig. 3.43. Cross section of a finger photodiode [83]

Table 3.3. Internal quantum efficiency of PIN-photodiodes with a standard P^+-type source/drain anode

Wavelength (nm)	Quantum efficiency (%)
635	98
452	53
430	40

η_i is small due to the low penetration depth of light into silicon or due to the large absorption coefficients, $\alpha = 5.7\,\mu m^{-1}$ for 430 nm and $\alpha = 3.7\,\mu m^{-1}$ for 452 nm. Many photogenerated carriers recombine therefore in the P^+-anode island of a PIN photodiode before they reach the drift region or the anode contact at the boundary of the photodiode. One way to increase the quantum efficiency would be to use a shallower P^+ island for the anodes. This solution, however, requires a lower implantation energy and the implementation of the anode formation after the source/drain annealing and after the BPSG-flow, in order to reduce the PN-junction depth. In order to avoid this complex process modification, finger photodiodes (Fig. 3.43) were investigated. The finger distance and the finger width were varied (Table 3.4) keeping the photodiode area constant. The distances a and b were chosen in such a way that η_i is expected to be larger than 80 %.

The rise and fall times of these finger photodiodes were determined for $C_e = 10^{15}\,cm^{-3}$ and $d_e = 10\,\mu m$ [83]. The rise and fall times for a finger number m of 3, i.e. for a finger distance of about 18 µm, are approximately 75 and 65 ns, respectively, for illumination with red light. There is a large contribution of slow carrier diffusion. When the finger distance is reduced

Table 3.4. Dimensions of the finger photodiodes

Finger number	Finger distance a (μm)	Finger width b (μm)
3	18.4	4.4
4	12.6	3.0
6	7.8	1.8
9	4.6	1.4

and the finger number is increased, the rise and fall times reduce considerably to values below 30 ns. In the blue spectral range, rise and fall times below 0.25 ns are obtained for $m = 9$, when the space-charge regions of neighboring fingers meet [83].

Transmission of Isolation and Passivation Layers. The total quantum efficiency (η_e) of a photodiode is the product of the internal (η_i) and the optical (η_o) quantum efficiencies. The internal quantum efficiency was dealt with above. The optical quantum efficiency is determined by the transmission of the isolation and passivation layers covering the integrated photodiodes. The thicknesses of these layers are of the order of the optical wavelength. Destructive interference deteriorates the quantum efficiency as a consequence. Optical simulations using the program MEDICI-ODAAM were performed assuming a CMOS process with only one metal layer for simplicity in order to study this effect [83]. The results for the red and blue spectral ranges are shown in Figs. 3.44 and 3.45, respectively. The optical transmission strongly depends on the wavelength and varies between approximately 0.4 and 0.95. For the nominal oxide and nitride layer thicknesses, the transmission at 635 nm and 430 nm is close to a minimum (Figs. 3.44 and 3.45).

Then, the thickness of the passivating nitride layer (plasma nitride, PN) was assumed to have a tolerance of 10 %, keeping the thicknesses of the TEOS, of the BPSG, and of the passivation oxide (POX) layers fixed at their nominal values. As a result, minima of the transmission are converted to maxima and vice versa.

These results lead to very severe conclusions: (i) If only one layer of the CMOS isolation and passivation stack is allowed to have a tolerance of 10 %, the transmission of a stack, which is optimized for one wavelength, can be reduced by approximately a factor of 2. Very small tolerances of the layer thicknesses of the order of 1 % would be necessary to guarantee a constantly high transmission. These small tolerances are not possible in a CMOS process. (ii) It is not possible to optimize the stack for the complete red or blue spectral range by small changes of the layer thicknesses. (iii) Special antireflection coating layers have to be implemented in the CMOS process in order to obtain a reproducibly high photodiode quantum efficiency. It should be mentioned

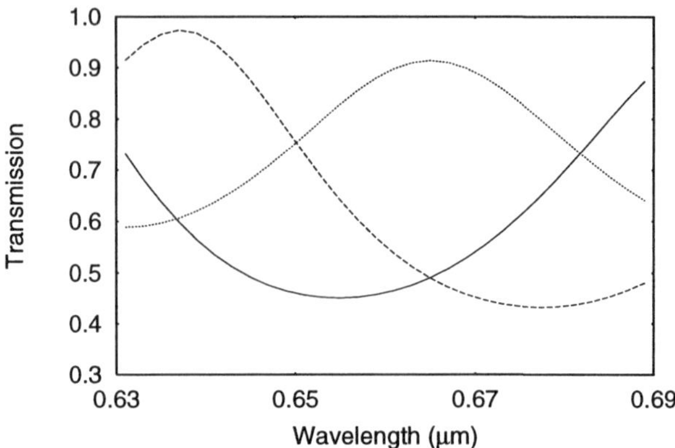

Fig. 3.44. Simulated transmission of CMOS TEOS, BPSG, POX, and PN isolation and passivation layers for various nitride layer thicknesses (*broken line*: +10 %, *solid line*: ± 0 %, *dotted line*: −10 %) in the red spectral range [83]

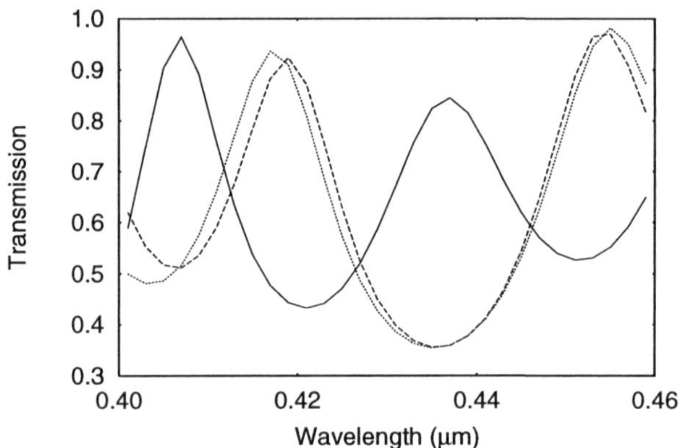

Fig. 3.45. Simulated transmission of CMOS TEOS, BPSG, POX, and PN isolation and passivation layers for various nitride layer thicknesses (*broken line*: +10 %, *solid line*: ± 0 %, *dotted line*: −10 %) in the blue spectral range [83]

that these statements are also valid for two- or more-metal-layer processes as well as for bipolar and BiCMOS processes.

A special antireflection coating consisting of a thin SiO_2 and a thin Si_3N_4 layer was integrated on the photodiodes [87]. Figure 3.46 shows the calculated transmission together with two measured results.

The measured external quantum efficiency η_e for 638 nm is included in Fig. 3.46. A mean value of 94 % was obtained for η_e of a PIN photodiode on an N-type substrate with the SiO_2/Si_3N_4 antireflection coating. Therefore, the internal quantum efficiency η_i of the PIN photodiodes is almost equal to 1 for $\lambda = 638$ nm.

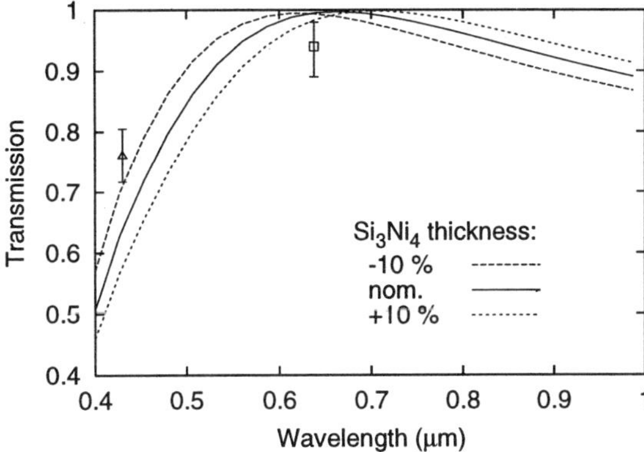

Fig. 3.46. Measured total quantum efficiency η_e and calculated spectral transmission η_o with SiO_2 and Si_3N_4 antireflection coating optimized for λ=638 nm

For 430 nm a measured value for η_e is also included in Fig. 3.46 for finger photodiodes. The measured $\eta_e = 76\%$ value for finger photodiodes with the ARC layer shown in Fig. 3.46 for 430 nm indicates that η_i of the finger photodiodes is much larger than 0.9 and that surface recombination of photogenerated carriers is small. Summarizing these results, it can be stated that the finger photodiodes with small finger distances are very well suited for the integration in future blue-sensitive OEICs for optical storage systems, because of their fast transient response, their large quantum efficiency and their simple integratability.

3.5.4 Particle Detector

Silicon PIN detectors for infrared illumination, x-ray, γ-ray, i. e. high energetic photons up to approximately 100 keV, and for high-energy ionizing particles can have an intrinsic layer thickness of 300 μm (see the following example in Fig. 3.47). This figure shows a schematic three-dimensional view of a position-sensitive pixel matrix for such a detector. The PIN diode is formed by the N^+ diffusion from the backside, by the 300 μm thick P-type substrate with a resistivity of 10 kΩcm, which corresponds to an acceptor concentration of $1.2 \times 10^{12}\,cm^{-3}$, and by the P^+ contacts of the pixels. The complete intrinsic layer could be depleted with voltages in the range of +60 to +80 V at the backside cathode [94]. The electric field guides the electrons of generated electron–hole pairs to the backside cathode and the holes to the nearest P^+ pixel contact. In order to guarantee the complete collection of the charge on the P^+ contacts, i. e. not losing charge to the N well between the P^+ contacts, the N well has to be reverse-biased with respect to the P^+ contacts. Voltages less than +12 V at the N well were sufficient for cathode biases of +55 to +80 V. For a cathode bias of +65 V, a voltage of +2 V at the N well was sufficient to collect all the charge on the P^+ contacts. From these contacts

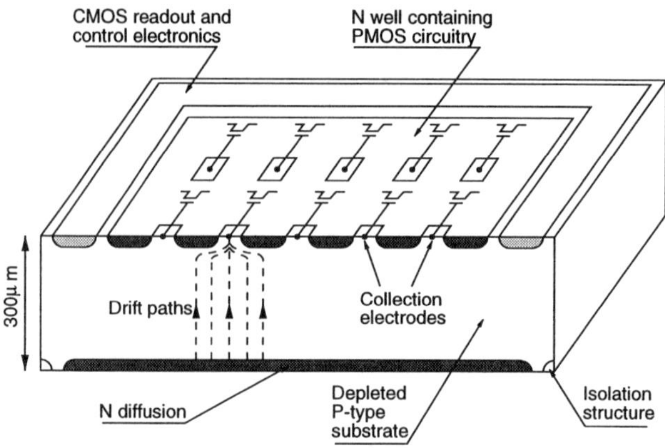

Fig. 3.47. Schematic 3D view of the complete particle detector [94]

the charge is put onto a small capacitance transistor gate, where it controls the readout current. Each collection electrode collects the charge generated in a certain fraction of the substrate allowing position sensitive radiation or particle detection.

The array size was 30 by 10 pixels, each 34 µm by 125 µm. The collection electrode had a size of 14 µm by 14 µm. The integrated detector was fabricated in a 2 µm CMOS process with 13 masks on the front side of the wafer and three masks at the backside. A gettering step was also implemented to achieve a minority carrier lifetime of 500 µs, which is much larger than the drift time of the generated carriers in the thick I-layer.

The readout of the pixels is done sequentially. The active chip area was 1.8 by 2.2 mm^2 without bondpads. A γ-quantum with an energy of 60 keV from an ^{241}Am source resulted in an output voltage of 100 mV with a signal-to-noise ratio of 150 to 1.

3.5.5 Charge-Coupled-Device Image Sensors

Charge-Coupled-Device (CCD) image sensors are an economically very important sub-division of integrated optical receivers. CCD image sensors were invented at Bell Labs [95]. Since then, there have been numerous publications on CCD image sensors (for instance [96–108]), because these sensors are key devices for high-volume video cameras and still-photography cameras. There is also a huge number of CCD imaging devices available on the market [109]. Record minimum pixel size was reduced from 40 µm in 1972 to 5 µm in 1995 [110]. The reported record maximum pixel number has increased from less than 2000 to 26 million in the same period [110].

Usually CCD image sensors are simply called CCDs. An advanced CCD image sensor, however, consists of photodetectors, CCDs for readout, and MOSFET source follower output drivers. For the photodetectors, PN photodiodes can be used. The CCD is a much more complicated device than the photodiode. Therefore, the CCD will be explained in some detail. The CCD is basically an array of closely spaced MOS capacitors (see Fig. 3.48). A quantity of charge can be stored below one gate when an appropriate biasing of this and the neighboring gates is chosen (see Fig. 3.48a). The electrons representing the stored charge are confined in the potential well below the electrode with the higher voltage. When the third electrode to the right of this electrode is pulsed to a higher voltage the electrons begin to transfer to its deeper potential well (see Fig. 3.48b). The voltages 5, 10, and 15 V, are chosen arbitrarily; meanwhile they have lower values (e. g. 5, 8, and 10 V [111]; 2.0 and −8.0 V [112]; 1.8 V [113]; 1.1 V [105]) in advanced CCDs.

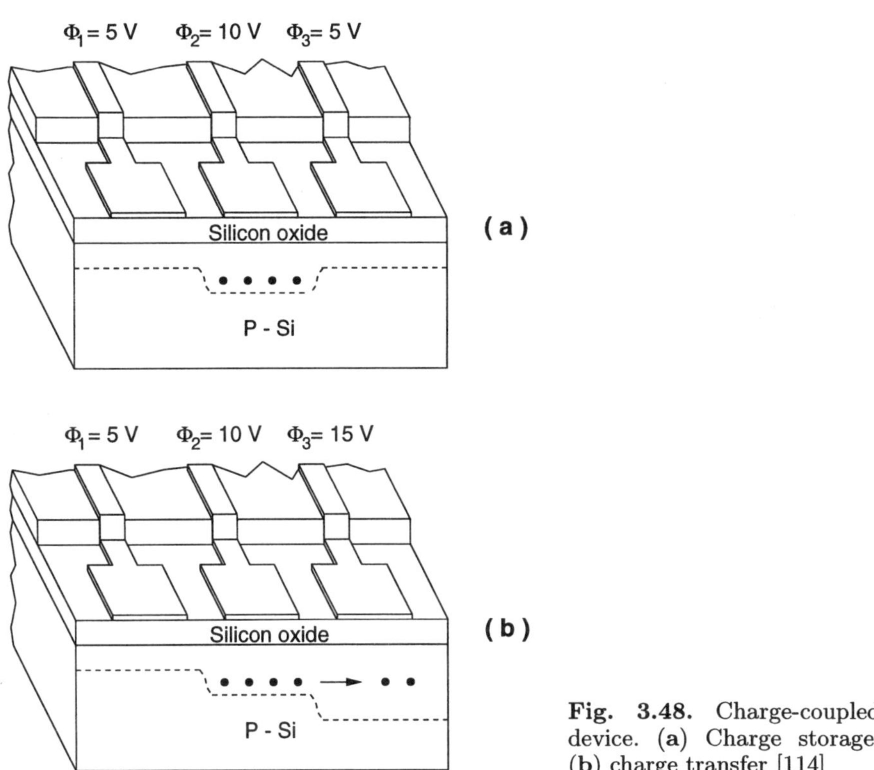

Fig. 3.48. Charge-coupled device. (**a**) Charge storage; (**b**) charge transfer [114]

Figure 3.49 shows the schematic cross section of a three phase N-channel CCD together with its basic input and output structures. The six (in real devices of course many more) MOS capacitors with their gates connected to the clock lines Φ_1, Φ_2, and Φ_3 form the main body of the CCD. The input diode (ID) and input gate (IG) inject charge packets into the main CCD body. The output gate (OG) and output diode (OD) extract the transferred charge packets from the main CCD body. Before the charge transfer, the input and output diodes are at high positive voltage in order to deeply deplete the surfaces under IG and OG. ID and IG, therefore, cannot supply electrons into the main CCD array. The potential wells under the CCD array, therefore, can be assumed to be empty. Then voltages $\Phi_1 > \Phi_2$, Φ_3 are applied. The potential well below Φ_1 is deeper than the others. When the voltage of ID is lowered electrons flow to the potential well under the first Φ_1 electrode through IG which is at a potential higher than 5 V but lower than 10 V. At the end of the injection, the surface potentials under the IG and Φ_1 electrodes are the same as the input diode potential and the electrons are now stored under the IG and first Φ_1 electrode. Now ID can be returned to a higher voltage. Then, the voltage at the Φ_2 electrode is increased and the voltage at the Φ_1 electrode is reduced in order to transfer the charge to the second electrode. Now, the voltage at the Φ_3 electrode is increased and the voltage at the Φ_2 electrode is reduced in order to transfer the charge to the third electrode. This pulse sequence is repeated until the charge packet is stored under the second Φ_3 (or in a real device the last Φ_3) electrode. OG and OD are at high potentials and when the voltage at the Φ_3 electrode is reduced, the charge is pushed to the output diode, thereby giving an output signal proportional to the size of the charge packet at the output terminal.

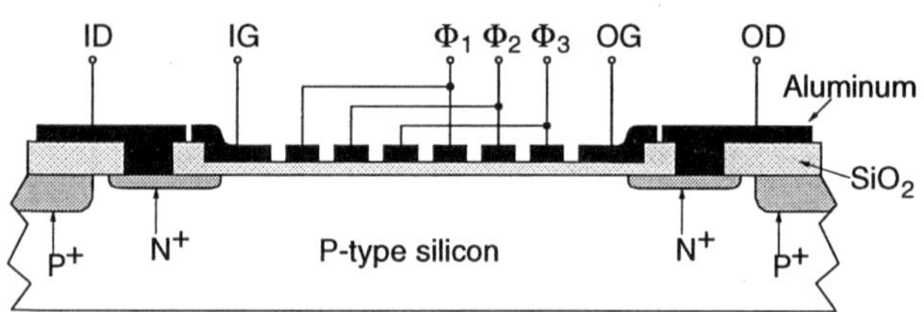

Fig. 3.49. Schematic cross section of a CCD [40]

In a CCD image sensor, the input diode is the photodiode. Therefore, ID is larger in a CCD image sensor than shown in Fig. 3.49 and not covered by metal, of course. During the image integration time i. e. when the ID voltage is high, the electrons photogenerated in the space-charge region are collected and stored under the N^+ island of ID. The photogenerated holes drift to the substrate. Electrons photogenerated in the substrate below the space-charge region diffuse slowly and may recombine with holes or diffuse to neighboring pixels giving rise to cross-talk. The collected charge is proportional to the integration time and to the light intensity as long as the well is not filled. During the charge transfer in the CCD, almost no charge is lost and the output signal is proportional to the light intensity detected by the ID photodiode.

The minimum of the potential well is at the Si/SiO_2 interface, which, therefore, is a critical part of the described MOS structure. The Si/SiO_2 interface may capture electrons in localized states, preventing their proper readout. To improve the situation, buried-channel structures are used, where a thin top layer of the substrate is N-type doped [115]. The resulting potential distribution has a minimum in the silicon somewhat below the Si/SiO_2 interface, which represents a 'cleaner' place for storing electrons. A transfer inefficiency, defined as the relative amount of electrons left behind after transfer from one stage to the next, as low as 10^{-5} per stage is typical for buried-channel CCD structures [116].

In an advanced CCD image sensor, there are arrays of photodiodes and arrays of input gates (Fig. 3.50). The advantage of this interline transfer CCD (IT-CCD) arrangement [117–119] is that the charge transfer from the detector to the storage cells can be made very quickly since only one transfer stage is involved. In older CCD image sensors the CCD main bodies in image registers themselves were used as the photodetectors and the complete image was transferred to a storage register requiring 3N charge transfers [120]. The electrodes were made from polysilicon resulting in a low sensitivity for the blue spectral range. These two disadvantages led to the IT-CCD structure. The larger noise of the IT-CCD with photodiodes [110, 121] was less important. Each third MOS capacitor can have its own photodiode and input gate. One line of a picture with N pixels can be read out through the CCD output register in a serial mode to a single video output (Fig. 3.50). A complete CCD image sensor consists of a two-dimensional matrix ($N \times M$) of photodiodes and input gates as well as M CCD main bodies. The active areas are separated by channel stops consisting of narrow regions with heavy doping in order to prevent blooming, i. e. the spilling over of electrons from full wells into neighboring channels.

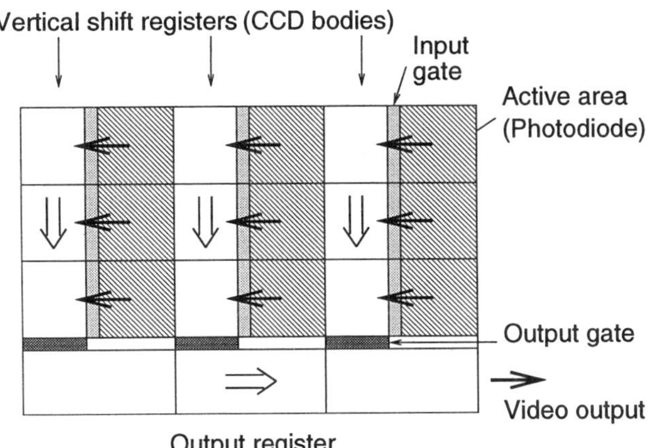

Fig. 3.50. Schematic of a interline-transfer CCD (IT-CCD)

For the sake of completeness it should be mentioned that two-phase CCDs are also possible. These two-phase CCDs require an additional doping structure for generating an internal lateral electric field [105, 113, 120].

Single-chip color image sensors are obtained using three photodiodes in each pixel. These three photodiodes carry a red, a green, or a blue color filter, which are printed on the insulator on the CCD image sensor surface. The color filters are arranged in a mosaic pattern for achieving a better isotropy in resolution than using stripes of the three color filters [120].

A 2/3-inch 1280×1600 CCD image sensor for still photography was fabricated with a minimum feature size of 0.5 μm [107]. Its chip size was 90 mm^2 with a pixel size of 5.1×5.1 μm^2. Supply voltages of 3.3 and 12 V were necessary. The maximum charge storage capacity per pixel was 40 000 electrons. A sensitivity of 390 mV/(Lux s) was reported for green light. The output driver consisted of a triple source follower. The output sensitivity was larger than 19 μV/electron with a saturation output voltage of 0.9 V and an output noise of 10 electrons root-mean-square (RMS). This CCD image sensor was capable of taking six images per second in full resolution [107].

The fill factor of CCD imagers is considerably less than 1. Therefore microlenses were integrated in order to increase the sensitivity [122, 123]. A clock frequency of 37.125 MHz was used in a 2 million pixel CCD image sensor for a HDTV camera [124]. A similar clock frequency of 40 MHz was reported in [112]. Maximum output data rates of 250 Mb/s for CCD arrays [125] were reported. The clock rate of CCD arrays finally is limited as the transfer inefficiency increases at higher clock frequencies, depending on the actual device structure [110]. A clock rate of 325 MHz has been reached [126]. The larger the pixel number the smaller the frame rate and vice versa. A high frame rate of 10^6 frames/s was obtained for a rather small 360×360 pixel CCD imager. The readout frequency of CCDs, therefore, is rather limited. It is, however, sufficient for video applications. CCD image sensors are rather well developed and the reduction in cell size is coming to an end. The limit is set by diffraction occurring at the aperture above the photodiode in IT-CCD image sensors [127]. When the width of the aperture in the tungsten photoshield becomes smaller than about 1 μm, the response of the photodiode begins to drop [127].

CCD image sensors were integrated in CMOS technology. For instance a CMOS/buried-N-channel CCD compatible process has been described [128]. Another CCD/CMOS process was suggested in [129]. The efforts to integrate CMOS with high-pixel-count CCD imagers to increase CCD functionality, however, has not been fruitful due to cost [130]. The aspect which makes it difficult to integrate CCD devices in CMOS technology is the need for closely spaced, thin polysilicon electrodes for charge transfer being incompatible with the self-adjusting gate-source/drain processing scheme. Furthermore, the well doping profiles and threshold implants are not optimized for CCD devices. Silicided gates in advanced CMOS processes increase the difficulties.

A low-pixel-count 128×128 pixel CCD image sensor in a 2 μm CMOS technology with 128 charge-sensitive amplifiers and 128 analog-to-digital converters has been reported [131]. Only one 5 V power supply was necessary. The maximum picture rate was 25 pictures per second.

3.5.6 Active-Pixel Sensors

Nowadays, there is a trend towards a camera-on-a-chip, which consists of optical sensor elements and highly integrated signal-processing circuitry with a low power consumption. CCDs, therefore, have to compete with a severe rival, the active-pixel sensor (APS), for many imaging applications. It has a performance competitive with CCDs with respect to output noise, dynamic range, and sensitivity but with vastly increased functionality due to its CMOS compatibility and with the potential for lower system cost as well as for a camera-on-a-chip [130].

The APS was described first by Noble in 1968 [132]. Advanced APSs are now emerging from the most advanced image-sensor research and development laboratories in Japan [133–136]. Laboratories in the United States [137–143] and in Europe [144, 145] are trying to catch up. The charge is shifted from pixel to pixel in a CCD in order to read out the image. An APS (see Fig. 3.51), by contrast, acts similarly to a random access memory (RAM), wherein each pixel contains its own selection and readout transistors, which serve as amplifiers or buffers. The steadily decreasing feature sizes of VLSI/ULSI technologies make it possible to integrate source follower MOSFETs or amplifiers together with each photodetector for all pixels in a large two-dimensional array.

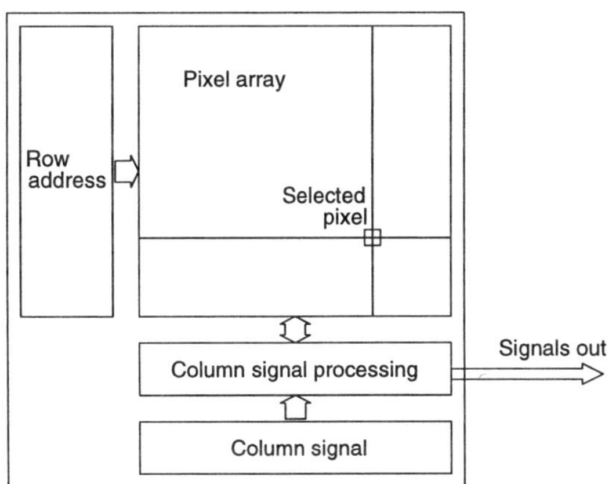

Fig. 3.51. Block diagram of an active pixel sensor imager with random access to each pixel [146, 147]

The advantages of an APS compared to a CCD sensor accordingly are:
(i) random access
(ii) nondestructive readout
(iii) easy integratability with on-chip electronics.

Although CCDs can be integrated in CMOS processes, undesirable compromises in the performance of CMOS or CCD devices have to be made [130]. Important weaknesses of the CCDs are that (i) the charge-transfer efficiency

(CTE) has to come very close to 100 % for large pixel numbers, i. e. that the charge-transfer inefficiency has to come close to 0 %, that (ii) large pixel sizes at high data rates are difficult to implement, and that (iii) readout rates are limited.

There are two types of active-pixel sensors: (i) photo-gate APS and (ii) photodiode-type APS. A photo-gate APS requires one charge transfer from the photo-gate into the storage region FD (Fig. 3.52). The photo-gate APS in its function, therefore, resembles a sample&hold circuit. The pixel is selected using transistor S in order to read out the signal. The storage region or output node is reset using transistor R. After reset, the signal charge is transferred from under the photo-gate into the storage node FD. The change in the source-follower voltage between the reset level and the final level is the output signal from the pixel. In more detail, the operation of the sensor is as follows: The photogenerated charge is integrated under the pixel photo-gate PG, which is biased at +5 V. The transfer gate TX and the gate of the reset transistor R are biased at +2.5 V and the selection transistor S is also biased off with 0 V at its gate for charge integration. After the signal integration, all pixels or only the selected ones in the row can be read out simultaneously onto column lines. For this purpose, the pixels are first addressed by the row selection transistor S biased at +5 V, which activates the source-follower output transistor in each pixel. Now, the reset gate R can be pulsed briefly to +5 V in order to reset the floating diffusion output node FD. Then, the photo-gate PG is pulsed low to 0 V, whereby TX is held at +2.5 V, to transfer the integrated signal charge under the photo-gate to the floating diffusion output node FD. The gate potential of the source follower changes and the output voltage of the pixel changes depending on the amount of integrated charge. It should be mentioned explicitly that without resetting the storage

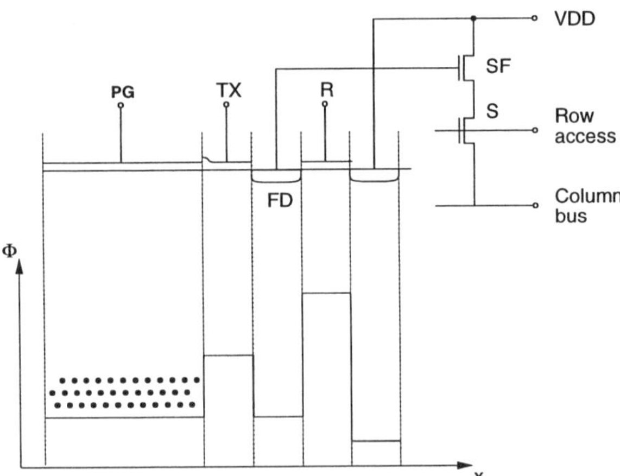

Fig. 3.52. Pixel circuit of a photo-gate active-pixel image sensor [148]

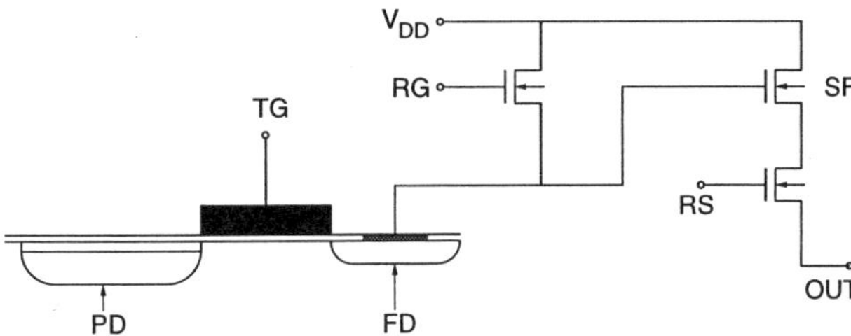

Fig. 3.53. Cross section of photodiode and transfer gate with the schematic of the remaining transistors in the 4-transistor pixel cell of a photodiode-type APS [147]

node FD before pixel selection, the readout is nondestructive, i. e. the pixel can be read out much more often than once.

A 128 × 128 array with the pixel element shown in Fig. 3.52 was integrated in a 2 μm P-well CMOS process. A 100 % compatibility of the image sensor with the CMOS process was reported [148]. A total area of $6.8 \times 6.8\,\text{mm}^2$ was necessary for this array of pixels with the size of $40 \times 40\,\mu\text{m}^2$. The larger part of this pixel size was occupied by the transistors and the active, i. e. light sensitive, area of each pixel was 26 %. The output conversion factor was 4.0 μV/electron. A noise floor of 42 electrons-rms and a dynamic range of 71 dB were reported [148].

An example of a photodiode-type APS pixel unit with a 4-transistor pixel cell architecture is shown in Fig. 3.53. Here the charge is integrated with the photodiode PD. For readout, this charge can be transferred by the transfer gate TG to the storage node FD. The rest of the circuit is similar to the schematic in Fig. 3.52. A larger conversion gain of 13 μV/electron was measured for a 256 × 256 pixel imager array with a pixel size of 7.8 μm and a fill factor of 35 % in a 0.6 μm CMOS technology [147]. A dynamic range of 66 dB was obtained. The above mentioned inferiority of photodiode-type CCD imagers to MOS-detector-type CCDs with respect to noise is avoided here by providing each photodiode with its own amplifier and reset transistor [110].

A photodiode-type APS actually does not require a transfer gate. The photogenerated charge can be stored in the photodiode itself [149]. Such an APS schematic is shown in Fig. 3.54.

It is worthwhile to compare the photodiode-type APS and the photo-gate APS. The photo-gate APS has a lower quantum efficiency due to photons being absorbed in the polysilicon gate. It also needs a larger pixel size due to an additional transfer switch and it is less compatible with standard CMOS processes because it requires a thin polysilicon layer with a thickness of only about 50 nm. The advantages of the photo-gate APS are a lower noise and a higher conversion gain due to a smaller floating node capacitance [149]. A 256 × 256 CMOS active pixel sensor was previously combined with circuitry for motion detection [146]. The area of each pixel was $20 \times 20\,\mu\text{m}^2$ with a fill factor (photo-gate area/total pixel area) of 25 %. This APS was realized in a 0.9 μm CMOS process. A conversion gain of 7 μV/electron was measured. The pixel output transistor was found to be the primary APS noise source.

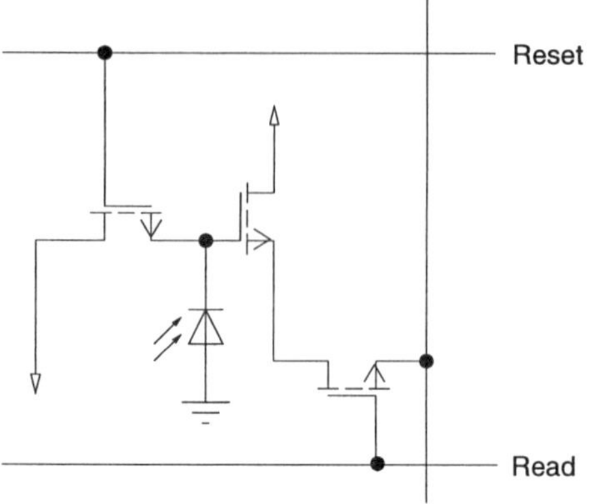

Fig. 3.54. Active pixel with three MOSFETs [133]

A typical output referred noise level of 29 electrons rms and a dynamic range of 74 dB was reported.

As a step towards a camera-on-a-chip, an APS and an embedded dynamic-random-access-memory (DRAM) have been combined [149]. A triple-well CMOS technology (Fig. 3.55), ideally, should be chosen in order to isolate the DRAM cell and the sensor array from noisy logic circuits like I/O buffers and (refresh) clock drivers as is done for DRAMs embedded in logic chips.

A trench DRAM was chosen because of its low area needs compared to DRAM cells with planar capacitors [150]. Compared to stacked capacitors, the trench capacitor is fully planarized and the trench capacitor is formed before the MOS transistors [151–153], which is advantageous for APS inte-

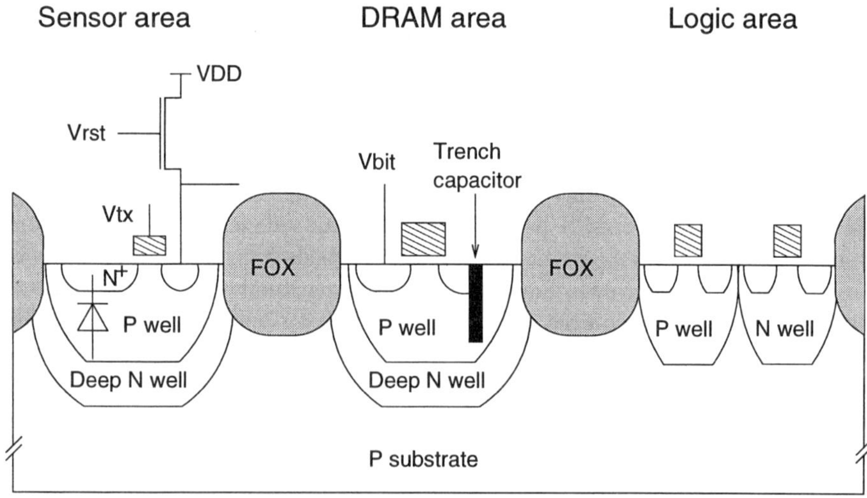

Fig. 3.55. Schematic cross section of an active pixel image sensor integrated with CMOS logic and with an embedded DRAM [149]

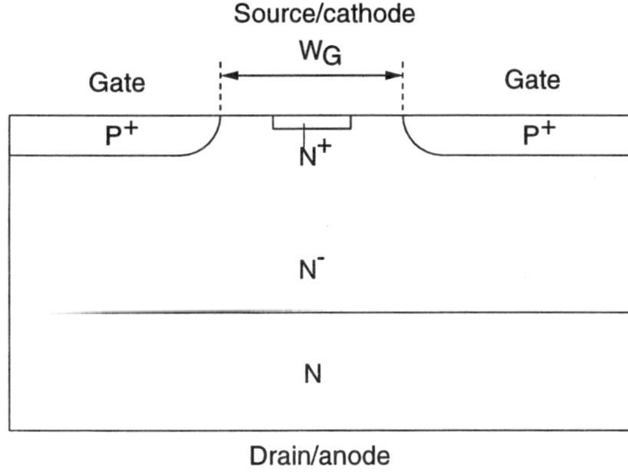

Fig. 3.56. Cross section of a static induction transistor [154]

gration. Nevertheless, one measure has to be taken to guarantee low sensor junction leakage currents, because it is known that the junction leakage current for self-aligned silicide (salicide) logic processes is higher. Thus, one extra mask is necessary to exclude the silicide from the photodiode and sensor area [149].

A special kind of APS using a static induction transistor (SIT) known from power device applications as a photodetector (Fig. 3.56) and for the amplification in the pixel was investigated [154]. Only two transistors were needed in each pixel (Fig. 3.57). During the reset period, the gate voltage is set so that the SIT is pinched off. After reset, in the integration period the SIT gate floats while maintaining the reverse-bias condition. The photogenerated holes are collected by the P^+ gate, while the photogenerated electrons are swept away to the substrate. The holes accumulated on the gate, raise the gate voltage and decrease the potential barrier beneath the source region. The charge is integrated on C_T. During readout, the SIT is turned on via C_T and the SIT acts as a voltage follower with a load capacitor consisting of the vertical signal line parasitic capacitance C_S. The voltage corresponding to the SIT gate voltage, i. e. to the integrated charge on C_T, appears in the vertical

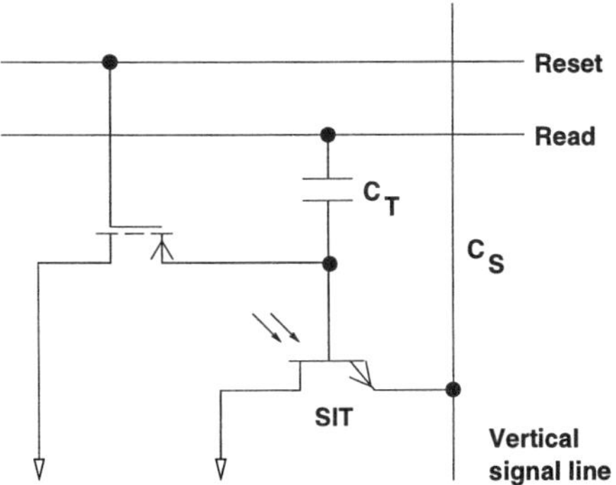

Fig. 3.57. Active pixel with a static induction transistor [154]

signal line (the SIT source line). The SIT, therefore, provides a charge gain of about C_S/C_T.

The pixel size in a test chip was $100 \times 100\,\mu m^2$ and the gate width W_G of the SIT was 3.8 µm. A dynamic range of 81.6 dB and a random noise floor of 34 µV rms was reported. The charge-to-voltage conversion factor, however, was only 0.4 µV/hole. A broad spectral response range of 400–1000 nm [155] was obtained with a relatively high responsivity in the blue spectral region due to the shallow P^+/N^- junction depth of 0.34 µm in the photo-conversion area.

3.5.7 Schottky Photodiodes

Schottky diodes are metal-semiconductor devices. Figure 3.58 shows the cross section of a Schottky photodiode with an N-type guard ring. Like PN diodes, Schottky diodes possess a forward and a reverse direction. The theory of Schottky diodes was reviewed in [156–158].

Schottky photodiodes open up the way to extend the spectral sensitivity range beyond the cutoff wavelength of approximately 1.1 µm of silicon into the infrared spectral range. Furthermore, Schottky photodiodes are a possible route to increase the quantum efficiency of silicon-based photodetectors in the blue and ultraviolet spectral range.

Schottky photodiodes consist of a metal/semiconductor contact. The metal, however, can also be replaced by a silicide. Silicide formation temperatures are known to be quite low [160]. The resistivity of many silicides is much lower than the resistivity of polysilicon gates. Due to these two properties, silicides have been used for many years in VLSI and ULSI technology in order to reduce gate and source/drain series resistances [161–164]. These silicides available in many advanced processes can be used for the integration of infrared detectors or UV-enhanced photodetectors. For such a purpose, an additional mask becomes necessary to block out the originally unmasked P-type threshold implantation of a CMOS process (compare Fig. 3.19g). The resist masks shown in Figs. 3.19l and m have to be applied in the Schottky-photodiode area, to prevent the detector area from source/drain N^+ or P^+

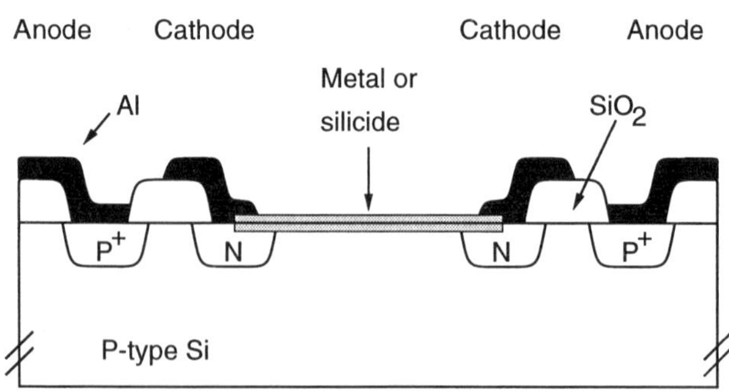

Fig. 3.58. Cross section of a Schottky photodiode [159]

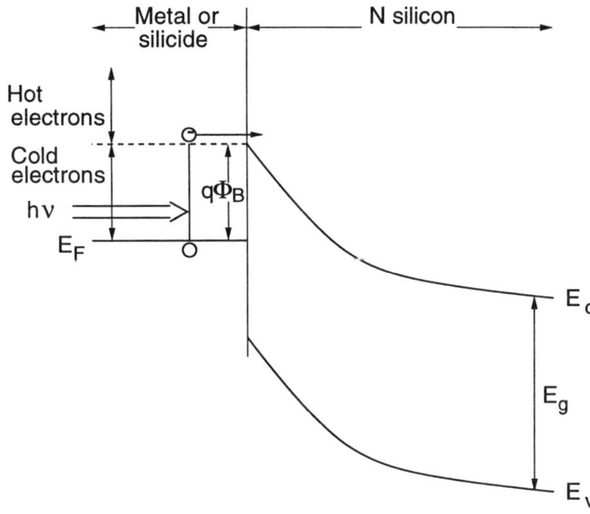

Fig. 3.59. Band diagram of a Schottky photodiode on N-type silicon

implantation doping, because metals and silicides form ohmic contacts on highly doped silicon instead of Schottky contacts. Usually, good Schottky diodes with a low reverse current are obtained for silicon doping levels below approximately $10^{16}\,\mathrm{cm}^{-3}$. For some metals and silicides, the doping type of the silicon is important, i. e. on N-type silicon a Schottky contact is formed whereas on P-type silicon an ohmic contact is formed or vice versa.

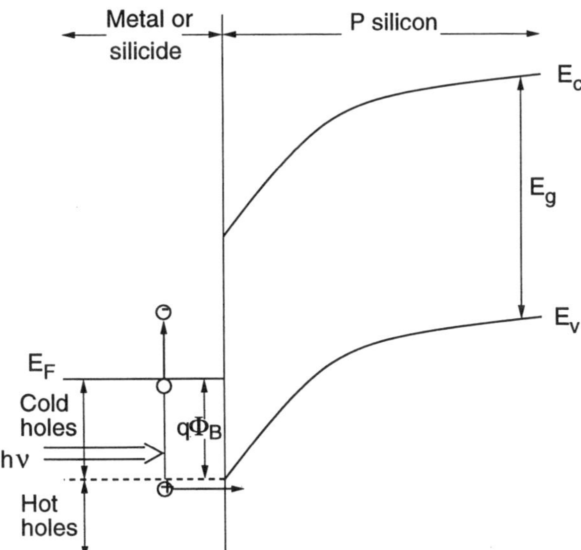

Fig. 3.60. Band diagram of a Schottky photodiode on P-type silicon

The detectable energy range in the infrared spectral range due to the absorption of photons in the metal or in the silicide and the generation of hot carriers, which can overcome the energy barrier Φ_B and enter into the silicon (see Figs. 3.59 and 3.60), is:

$$\Phi_B < h\nu < E_g. \tag{3.4}$$

Figure 3.61 shows the extension of the sensitivity range of a Schottky photodiode on P-type silicon into the infrared spectral range due to this internal photoemission. For $\lambda > 1.1\,\mu m$, the values for the quantum efficiency η are systematically lower for the front-illumination mode than for the back-illumination mode. The reduction in η for front illumination occurs because photon absorption is decreased by reflection losses resulting from the high reflectivity of PtSi. For back-illumination, these losses are lower because the substrate serves as an index-matching layer for the PtSi [159].

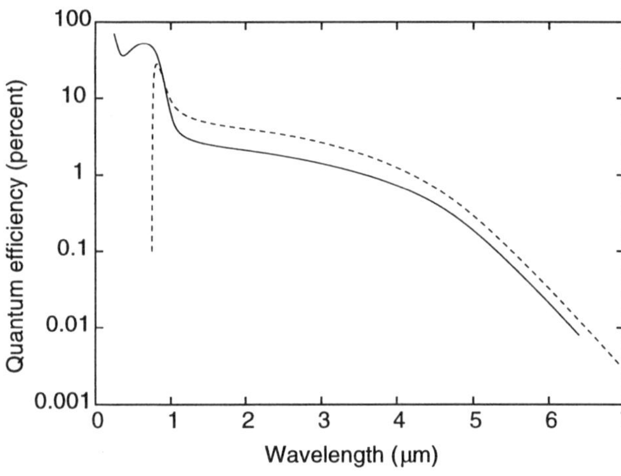

Fig. 3.61. Quantum efficiency of a PtSi Schottky photodiode on P-type silicon (*solid line*: front illumination, *dashed line*: back illumination [159])

Photons with $h\nu > E_g$ penetrate through a semitransparent metal or silicide layer with a thickness of the order of 1–10 nm and generate electron–hole pairs in the silicon. The space-charge region in a Schottky diode begins immediately at the metal(silicide)-silicon interface (see Fig. 3.62). The region I in a PN or PIN photodiode (see Fig. 2.7), where diffusion and recombination of photogenerated carriers occur, is not present in a Schottky photodiode. Schottky photodiodes, therefore, enable a higher speed than PN and PIN photodiodes and their quantum efficiency does not decrease much in the blue and ultraviolet (UV) spectral range (see Fig. 3.63). The PtSi Schottky-barrier detector was characterized in [159].

Several silicides were investigated for application in infrared image sensors for night vision and thermal imaging [165]. Such silicides were Pd_2Si for short wavelength infrared (SWIR, 1 to 3 μm) [166], PtSi for medium wavelength

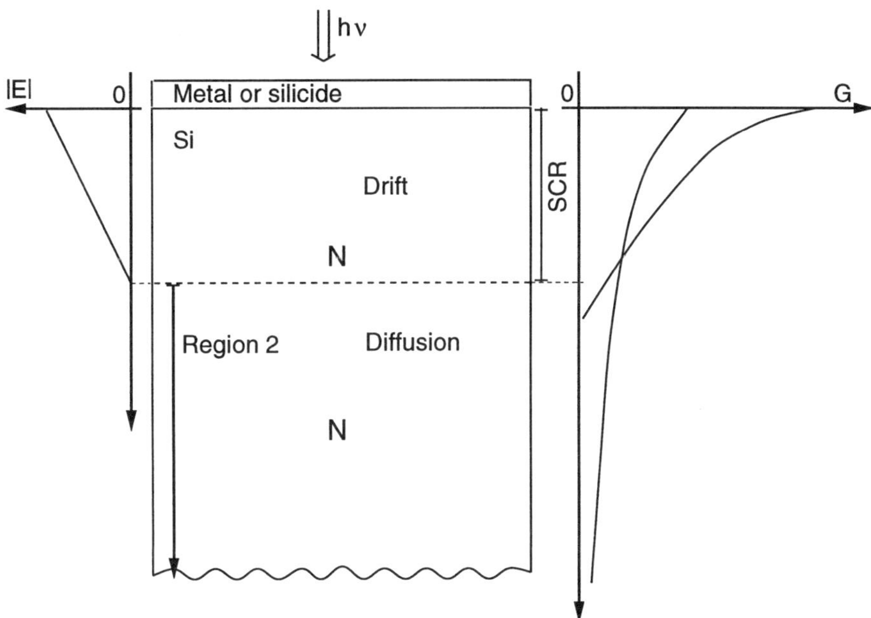

Fig. 3.62. Drift and diffusion regions in a blue and UV-sensitive Schottky photodiode with front illumination and electron–hole pair generation in the silicon

infrared (MWIR, 3 to 5 μm) [165, 167–170], and IrSi for long wavelength infrared (LWIR, 8 to 10 μm) [170, 171] sensitive photodetectors. A review of Schottky-barrier IR-CCD vision sensors reported from 1978 to 1990 is given in [172]. The best results were obtained for PtSi. PtSi has a barrier height of approximately 0.22 eV on P-type silicon. PtSi detectors, therefore, are appropriate for wavelengths up to 5.6 μm. A dark current density in the range

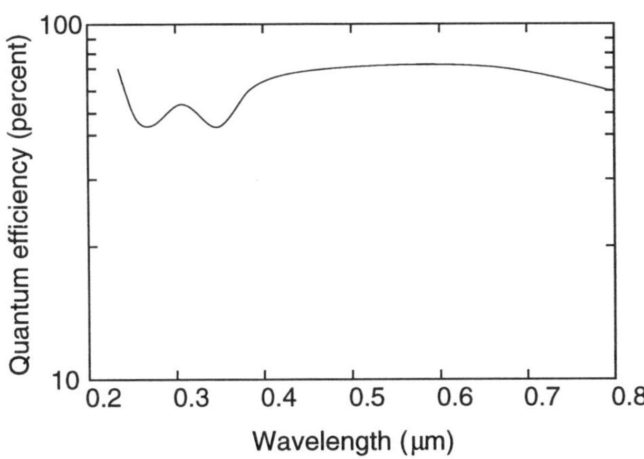

Fig. 3.63. Quantum efficiency of a PtSi Schottky photodiode in the visible and ultraviolet spectral range for front illumination [159]

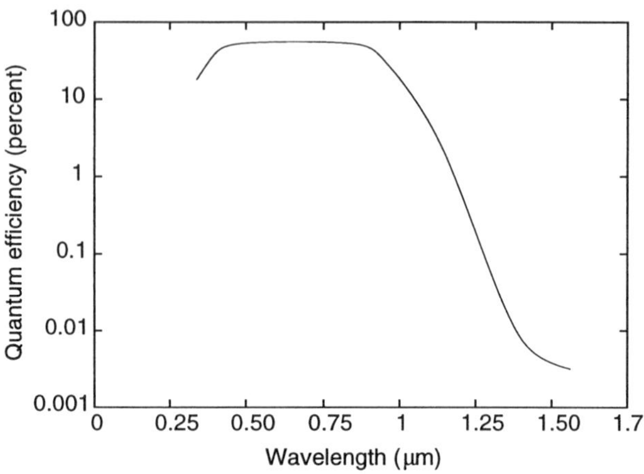

Fig. 3.64. Quantum efficiency of a PtSi Schottky photodiode on N-type substrate in the visible and SWIR spectral range for front illumination [159]

of 1 to 4 nA/cm^2 was obtained for PtSi Schottky-barrier detectors (SBDs) at 77 K. PtSi SBDs have quantum efficiencies between 0.5 and 1 % at 4.0 μm. On N-type silicon, the barrier height of PtSi is approximately 0.9 eV. The quantum efficiency coefficient of PtSi SBDs on N-type silicon for $h\nu < E_g$, therefore, is much lower than on P-type silicon (compare Figs. 3.64 and 3.61).

The quantum efficiency for the internal photoemission was derived in [173–176]:

$$\eta = C_1 \frac{(h\nu - \Phi_B)^2}{h\nu}. \tag{3.5}$$

The quantum efficiency coefficient C_1 depends on the thickness of the metal or silicide film and on the barrier height Φ_B. The thickness of the metal or silicide film determines the absorption of photons in the film. The thicker the film the more photons can be absorbed. An the other hand, the hot carriers are scattered at phonons in the film and the escape probability of the hot carriers for leaving the film and entering into the silicon decreases when the film thickness exceeds the mean free path of the hot carriers in the metal or silicide film. Actually, the quantum efficiency can be increased by a gain factor due to multiple reflections of the hot carriers at the film boundaries, when the film thickness is much smaller than the mean free path of the hot carriers [176]. Diffuse wall scattering of hot carriers at the interfaces also can be exploited for the enhancement of the quantum efficiency coefficient C_1 [177]. In addition, an optical cavity can be used in order to increase C_1 [176]. The optimization of the quantum efficiency coefficient C_1, therefore, is a complex task.

The quantum efficiency coefficient C_1 was found to be 0.542 eV^{-1} for PtSi with a barrier height $\Phi_B = 0.208$ eV resulting in a responsivity R of about 0.3 A/W for $\lambda = 1.5$ μm, $R = 0.1$ A/W for $\lambda = 3.5$ μm, and $R = 0.003$ A/W for $\lambda = 5.5$ μm [176]. Back illumination and an optical cavity (see Fig. 3.65)

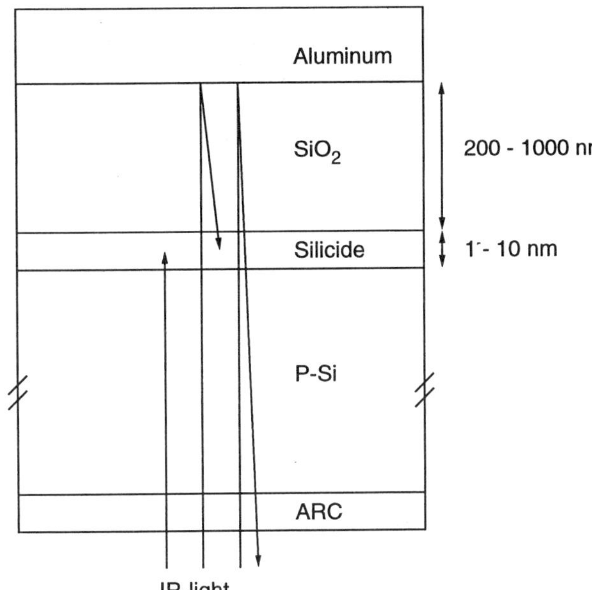

Fig. 3.65. Simplified structure of a PtSi Schottky photodiode with an aluminum reflector and an optical cavity for back illumination [176]

were necessary for these large values of the quantum efficiency coefficient C_1 [176]. Due to the aluminum mirror the IR light passes the silicide twice and the optimum oxide thickness in the optical cavity enhances the quantum efficiency of the Schottky-barrier detector. The structure of the IR sensitive PtSi SBD array combined with a CCD for readout is shown in Fig. 3.66. Due to the low barrier height, the PtSi detector has to be cooled.

Although PtSi gives the best results for SBDs, PtSi is not used in modern VLSI and ULSI processes. Pt is known to be a recombination center [178,179] and a carrier-lifetime killer and is, therefore, avoided in the fabrication process of integrated circuits. $CoSi_2$ [162,180] and $TiSi_2$ [163,164] are used for low-resistance source/drain contacts in modern submicrometer processes. We, therefore, will review the properties of $CoSi_2$ and $TiSi_2$ SBDs.

First, the results of [181] are summarized. In this work, Co was sputtered onto the bare Si surface in the field oxide opening immediately after an HF-dip. The size of the field oxide openings, i.e. the diode area, was $100 \times 130\,\mu m^2$. The silicidation was performed with rapid thermal annealing (RTA) at 575 °C for 30 s and a second RTA step at 750 °C for 30 s was performed in order to improve the silicide quality and to reduce its sheet resistance. Both RTA steps were performed in an N_2 ambient. Silicide thicknesses of about 25 nm and 60 nm were obtained. The oxide thickness on top of the silicide for passivation was 485 nm, which was not optimized as an antireflection coating layer. The light of diode lasers with $\lambda = 1.31\,\mu m$ or $\lambda = 1.54\,\mu m$ was coupled into the integrated Schottky photodiodes on a wafer level via an optical fiber. The photocurrent of the SBDs was compared to that of a

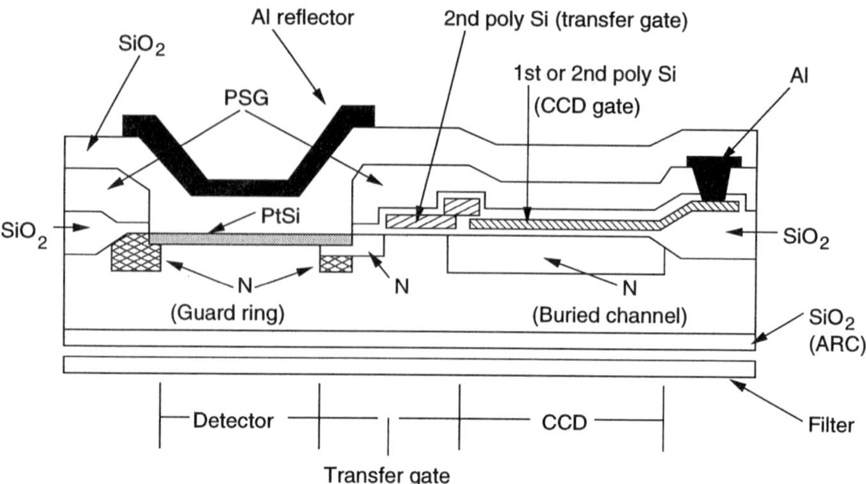

Fig. 3.66. Structure of a PtSi Schottky barrier detector for readout with a CCD [169]

calibrated InGaAs photodiode and the quantum efficiencies as well as the responsivities were determined. Tables 3.5 and 3.6 contain the measured results for $CoSi_2$/Si Schottky photodiodes [181].

Another investigation compared the infrared response of epitaxial and polycrystalline $CoSi_2$/Si Schottky diodes [182]. The epitaxial, i. e. crystalline, $CoSi_2$ film was formed by sequential sputtering of 6 nm Ti and 10 nm Co layers. Silicidation was carried out by a step at 900 °C for 10 s followed by a second step at 1100 °C for 10 s. During the first anneal, the Ti atoms re-

Table 3.5. Responsivity and external quantum efficiency of $CoSi_2$/Si IR photodiodes on N-type silicon

	25 nm $CoSi_2$	60 nm $CoSi_2$
Responsivity R_{1310nm} (A/W)	1.45×10^{-3}	0.68×10^{-3}
Quantum efficiency η_{1310nm} (%)	0.137	0.064
Responsivity R_{1550nm} (A/W)	0.87×10^{-3}	0.29×10^{-3}
Quantum efficiency η_{1550nm} (%)	0.070	0.023

Table 3.6. Responsivity and external quantum efficiency of $CoSi_2$/Si IR photodiodes on P-type silicon

	25 nm $CoSi_2$	60 nm $CoSi_2$
Responsivity R_{1310nm} (A/W)	1.85×10^{-3}	0.77×10^{-3}
Quantum efficiency η_{1310nm} (%)	0.175	0.073
Responsivity R_{1550nm} (A/W)	1.56×10^{-3}	0.55×10^{-3}
Quantum efficiency η_{1550nm} (%)	0.125	0.044

act with the oxygen atoms remaining on the silicon surface after cleaning and prior to Ti sputtering. Meanwhile, the Co atoms diffuse through the Ti compound layer to form epitaxial $CoSi_2$ by reacting with the silicon surface underneath. The Ti compound remaining on top of the epitaxial $CoSi_2$ is removed by selective wet etching. The second anneal step is done to improve the quality of the epitaxial $CoSi_2$ layer by annealing residual defects [182]. The thickness of the epitaxial $CoSi_2$ film was 36 nm.

The polycrystalline $CoSi_2$ layer with a final thickness of 30 nm was formed by magnetron sputtering of Co onto lightly doped P-type silicon. Subsequent rapid thermal annealing in an N_2 ambient produced the $CoSi_2$ layer. The average grain size of the $CoSi_2$ layer was determined by transmission electron microscopy (TEM) to be about 150 nm. An oxide layer of 200 nm thickness for passivation and antireflection coating purposes was present on top of the epitaxial and polycrystalline silicides [182].

The quantum efficiency was measured at 100 K with a 1000 °C black-body light source and interference narrow-bandpass filters. The hot-carrier emission coefficient for the epitaxial Schottky diodes was 0.61 %/eV and for the polycrystalline Schottky diodes 1.56 %/eV. The barrier heights were 0.447 eV and 0.454 eV, respectively. It was concluded that the scattering of the photogenerated carriers at the grain boundaries of the polycrystalline silicide (see Fig. 3.67) increases their emission probability into the silicon.

Table 3.7 contains the measured results for epitaxial and polycrystalline $CoSi_2$/Si Schottky photodiodes [182].

$TiSi_2$/Si SBDs are another possibility for VLSI compatibility. $TiSi_2$/Si SBDs were investigated in [181]. In this work, Ti was sputtered onto the bare Si surface in the field oxide opening immediately after an HF-dip. The size of the $TiSi_2$ Schottky diodes was $400 \times 250\ \mu m^2$. The sputtered Ti thickness was

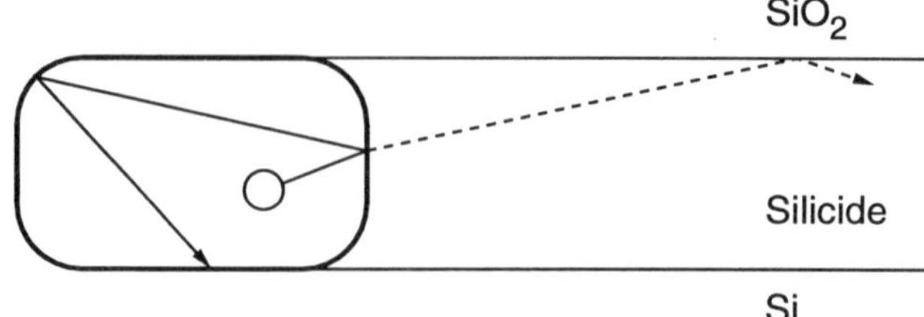

Fig. 3.67. Scattering of hot carriers at the grain boundaries of the polycrystalline silicide and scattering at the silicide-oxide interface for the epitaxial silicide

Table 3.7. Quantum efficiencies (QE) at different wavelengths for epitaxial and polycrystalline $CoSi_2$/Si Schottky photodiodes

	QE (%)	QE (%)	QE (%)	QE (%)	QE (%)
Wavelength (μm)	1.35	1.57	1.7	2.2	2.4
Epitaxial	0.15	0.074	0.064	0.014	0.006
Polycrystalline	0.37	0.23	0.18	0.025	0.011

approximately 20 nm. The first silicidation step was performed with rapid thermal annealing at 600 °C for 15 s in an N_2 ambient resulting in an incomplete silicidation. The selective etch step after this RTA treatment removed unreacted Ti. The second RTA step at 850 °C for 30 s in an Ar ambient formed the low-resistivity C54-phase of $TiSi_2$. A silicide thickness of about 20 nm was obtained. The silicide layer was polycrystalline with rather large grain sizes of up to roughly 1 μm. An oxide on top of the silicide with a thickness of approximately 110 nm was deposited for passivation. The oxide thickness was not optimized as an antireflection coating layer. Table 3.8 contains the measured results for $TiSi_2$/Si Schottky photodiodes [181].

The fabrication processes of $CoSi_2$ and $TiSi_2$ SBDs in [181] were comparable. $TiSi_2$/Si SBDs show a larger quantum efficiency than the $CoSi_2$/Si SBDs [181]. With the fabrication process in [182], however, the largest quantum efficiency was obtained with $CoSi_2$/Si SBDs. It should be mentioned that

Table 3.8. Responsivity and external quantum efficiency of $TiSi_2$/Si IR photodiodes with a $TiSi_2$ thickness of 20 nm

	N-type silicon	P-type silicon
R_{1310nm} (A/W)	2.85×10^{-3}	1.01×10^{-3}
η_{1310nm} (%)	0.270	0.096
R_{1550nm} (A/W)	1.58×10^{-3}	0.60×10^{-3}
η_{1550nm} (%)	0.127	0.048

an optimized ARC layer thickness of approximately 200 nm would increase the quantum efficiency of $TiSi_2$ Schottky photodiodes on N-type silicon to similar values as for $CoSi_2$/Si Schottky photodiodes listed in Table 3.7.

3.5.8 MSM-Photodetectors

Metal-semiconductor-metal (MSM) photodetectors consist of two metal/semiconductor junctions. One of these junctions is forward-biased and the other is reverse-biased. The light is coupled into the MSM detector between the two metal/semiconductor junctions (see Fig. 3.68). For a low dark current a large enough barrier height of the metal/semiconductor junctions is necessary.

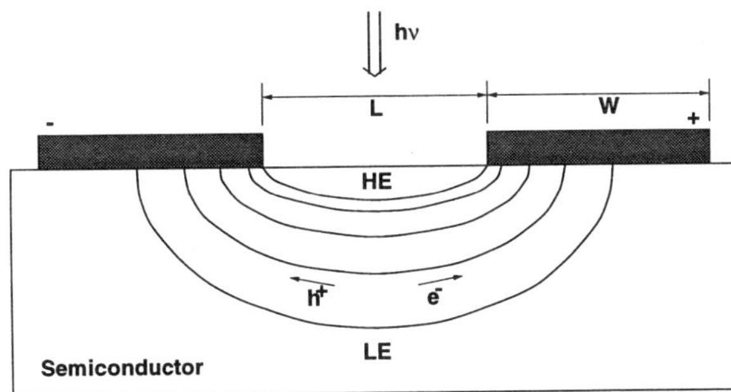

Fig. 3.68. Cross section of a MSM photodetector with electric field lines (HE: high electric field, LE: low electric field)

The planar or horizontal contact arrangement of metal-semiconductor-metal (MSM) photodetectors has speed advantages compared to vertical PN and PIN photodiodes mainly for short wavelengths. Advanced MSM devices can be designed in such a way that they do not have field-free regions with slow carrier diffusion. The process temperatures and the thermal budget for the fabrication of MSM detectors are low. MSM detectors, therefore, are thought to be compatible with silicon ULSI and gigascale integrated (GSI) technologies.

We will first consider the planar MSM structure with an interdigitated finger structure (Figs. 3.68 and 3.69) or with just one gap (Fig. 3.70). The capacitance per unit area for a MSM photodetector is [183]:

$$C_{\text{MSM}}(L, W) = \frac{K(k)}{K\sqrt{1-k^2}} \frac{\epsilon_0(\epsilon_s + \epsilon_d)}{L + W}, \tag{3.6}$$

where $K(k)$ is the complete elliptic integral of the first kind with

$$k = \tan^2\left(\frac{\pi W}{4(L + W)}\right). \tag{3.7}$$

The parameter ϵ_s is the dielectric constant of the semiconductor and ϵ_d is the dielectric constant of the overlaying passivation layer. The capacitance

Fig. 3.69. Top view of a MSM photodetector

of the MSM photodetector was shown to be less than half that of a PIN photodiode [183].

Usually the light penetrates into the semiconductor only between the metal fingers. For a large responsivity it is, therefore, important to have metal fingers with a small width W and a large finger spacing L. The quantum efficiency of a MSM photodetector is

$$\eta_{\mathrm{MSM}} = \eta_{\mathrm{o}} \frac{L}{L+W}(1 - \exp(-\alpha d)), \tag{3.8}$$

where d is the thickness of the MSM absorbing region.

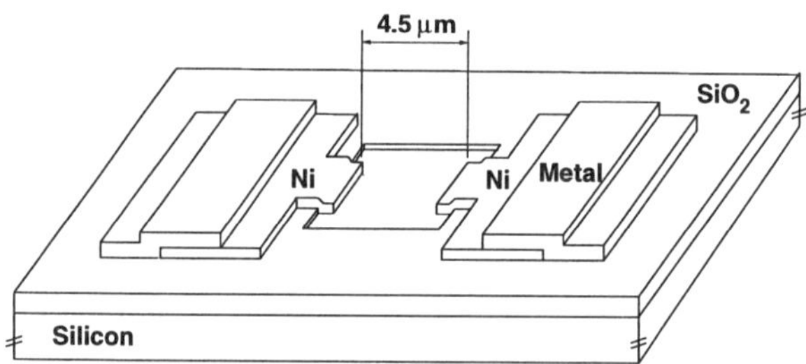

Fig. 3.70. MSM photodetector with only one light-sensitive gap [184]

A Ni-Si-Ni MSM photodetector fabricated with a simple three-level lithography process on bulk N-type Si with $N_{\mathrm{D}} \approx 8 \times 10^{14}\,\mathrm{cm}^{-3}$ has been reported [184]. Figure 3.70 shows this MSM detector with a light sensitive gap with a relatively large width of $L = 4.5\,\mu\mathrm{m}$. A −3dB bandwidth of 16 GHz was determined at a detector bias of 10 V. The dark current of the detector was less than 2 nA per micrometer detector length for this bias. A responsivity of 32 mA/W ($\eta_e = 12\,\%$), which is a relatively large value for a detector with such a high −3dB bandwidth, for $\lambda = 335$ nm was measured for the detector without an antireflection coating.

For MSM detectors with higher speed, the finger spacing, however, has to be smaller, because it determines the maximum electric field $E = U_{MSM}/L$ and the maximum carrier drift velocity. From Fig. 2.5 we obtain an electron drift velocity of $\approx 8 \times 10^6$ cm/s and a hole drift velocity of $\approx 3 \times 10^6$ cm/s for $E = 1V/\mu m$. These values result in drift times of 12.5 ps/μm for electrons and 33.3 ps/μm for holes. The electrons and holes reach their saturation velocities of 1×10^7 cm/s, i. e. 0.1 μm/ps for $E > 10^5$ V/cm. Very short rise and fall times of the photocurrent, therefore, can be expected for sub-μm finger spacings, especially because the average carrier drift length is only $L/2$ for a finger spacing L [185]. The drift time and consequently the rise/fall time are

$$\tau_{r/f} \approx \tau_{drift} = L/(2v_s). \tag{3.9}$$

Equation (2.20) for a PIN photodiode is not applicable for the −3dB frequency f_{3dB} of MSM detectors, because of the spatially different photogeneration. The −3dB frequency f_{3dB} of MSM detectors, however, can be estimated [186]:

$$f_{3dB} \approx \frac{1}{2\pi\tau_{r/f}}. \tag{3.10}$$

A finger spacing of 1 μm, therefore, enables a drift time of 5 ps and a transit-limited bandwidth of 32 GHz [186].

When the speed of a detector is limited by the time constant $R_{imp}C$ of the impedance R_{imp} of the connecting microstrip line and by the capacitance C_{MSM}, the −3dB bandwidth [185] is

$$f_{3dB} = \frac{1}{2\pi\sqrt{\tau_{r/f}^2 + R_{imp}^2 C_{MSM}^2}}. \tag{3.11}$$

The structure in Fig. 3.68 was shown to allow pulse response times of 3.7 ps full width at half maximum (FWHM) with finger widths and spacings of 200 nm each for 0.15 ps optical pulses with $\lambda = 465$ nm [187, 188]. This short pulse response time is due to the low penetration depth for violet light with $\lambda = 465$ nm of 0.28 μm (compare with Table 1.1). For red light with a penetration depth of 2.5–4 μm, the pulse response time is larger due to low electric field and drift velocity values in the depth of the Si [189]. MSM photodetectors on thin-film silicon-on-insulator (SOI) (see Chap. 4) eliminate the slow contributions from the depth to the photocurrent for red and IR light.

Structures with feature sizes of down to 200 nm were fabricated using electron-beam lithography. With the reduction of CMOS gate lengths down to 0.18 and 0.10 μm with optical deep-UV lithography using an ArF light source with $\lambda = 193$ nm [190], electron-beam lithography will not be required for the integration of MSM photodetectors with $L, W < 200$ nm in monolithic OEICs.

MSM photodetectors on thin-film SOI [187, 188, 191] eliminate the slow contributions from the depth to the photocurrent for red and IR light. A pulse response time of 3.2 ps FWHM and a −3dB frequency of 140 GHz for $\lambda = 780$ nm were reported for a SOI-MSM detector with a SOI layer thickness of 0.1 μm [191]. This large −3dB frequency is at the expense of the responsivity, of course, which was only 12 mA/W for $\lambda = 633$ nm in this case [191]. Continuous-wave (CW) operation of lasers for ultrashort pulse generation of down to 0.15 ps is not possible and the responsivity for $\lambda = 780$ nm was not determined in [191]. It is, however, much smaller than that measured for $\lambda = 633$ nm with a CW laser diode. A backside reflector was shown to enhance the responsivity [192].

Another way to improve the field distribution in a MSM detector without decreasing the responsivity was chosen in [193]. 9 μm deep trenches with a spacing of 1 μm were etched into the P-type Si substrate using reactive ion etching. The trench sidewalls were metallized. A constant horizontal electric field is obtained in the 1 μm wide Si ridges avoiding the low electric field region shown in Fig. 3.68. A pulse width (FWHM) of 28.2 ps and a −3dB bandwidth of 2.2 GHz were achieved at 5 V. The responsivity measured for 790 nm was 0.14 A/W, corresponding to an external quantum efficiency of 22.1 %. A bias of 2 V was sufficient without reducing the responsivity and still achieving a FWHM of 29.4 ps [193].

A further very sophisticated approach used a vertical MSM detector in order to achieve a homogenous field distribution (Fig. 3.71) [185, 194] and to avoid slow contributions to the photocurrent from low-field regions in the depth of the Si (compare with Fig. 3.68).

The epitaxial silicon-on-metal (SOM) structure [195] was fabricated in the following way: Co was implanted to form a buried metal-like $CoSi_2$ layer during subsequent annealing. The $CoSi_2$ forms a Schottky contact with Si. Then silicon was grown epitaxially to a total Si thickness of 330 nm. An 8 nm thick semitransparent Cr layer was used on top of the Si layer as the second Schottky contact of the vertical MSM detector. Standard UV lithography was sufficient for this approach, because no deep-sub-μm fingers are needed for the vertical MSM detector with an area of $7 \times 10\,\mu m^2$. A pulse response time of 3.5 ps FWHM for $\lambda = 800$ nm at a bias of $U_{MSM} = 5$ V was measured at 300 K [185] for this vertical MSM detector. Its responsivity was 36 mA/W (η_e = 4.6 %) for this wavelength. The pulse response, however, was not ideal because measurements at 40 K showed a much faster decay of the photocurrent. The nonideal response at 300 K was attributed to carrier trapping and detrapping at defects within the Si layer between the buried silicide and the top metal layers. The dark current was below 10 nA for the 70 μm^2 device, despite these

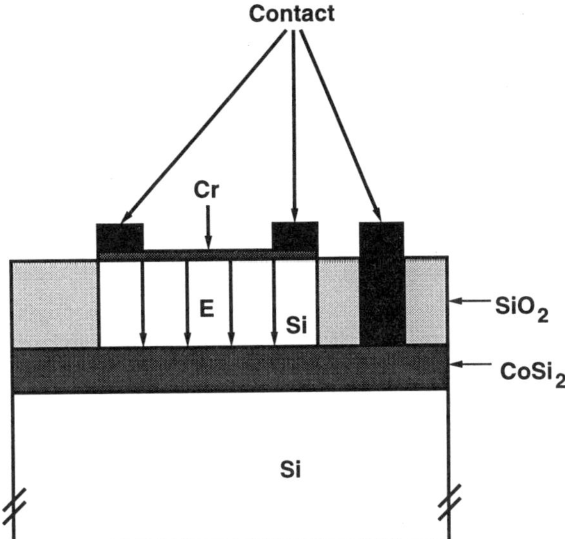

Fig. 3.71. Electric field distribution in a vertical MSM photodetector [194]

defects [196]. These good results, however, were achieved on (111)-oriented silicon. On (100)-oriented silicon, which is used and needed for VLSI, ULSI, and GSI MOS-compatibility, a pulse response time of 6.7 ps was obtained [196]. This slower response of the (100)-device was attributed to a higher dislocation density in the top Si layer giving rise to electron hopping [196]. The higher dislocation density in the SOM structure on a (100)-oriented Si substrate is due to a lower-quality epitaxial $CoSi_2$ layer than on a (111)-oriented Si substrate. The dark current of the vertical MSM photodetector on (100)-oriented Si, therefore, is larger than on (111)-oriented Si.

Another advantage in addition to the homogeneous field distribution of a vertical MSM photodetector is the application as an infrared detector for $\Phi_B < E_{photon} < E_g^{Si}$. The barrier height Φ_B for $CoSi_2$/Si is 0.64 eV for N-type Si and 0.46 eV for P-type Si. The corresponding values for Cr/Si Schottky diodes are 0.61 eV and 0.50 eV, respectively. Light with wavelengths of 1.3 and 1.55 μm, therefore, can be detected with the vertical MSM structure, however, with a relatively low quantum efficiency (compare with Sect. 3.5.7).

3.5.9 Amorphous-Silicon Detectors

Hydrogenated amorphous silicon (a-Si:H) has been employed for a variety of imaging devices for low cost applications in large area sensor technology [197–201]. Page-sized two-dimensional a-Si:H photodetector arrays have been realized [201]. Usually, glass substrates are used for the low-temperature fabrication of a-Si:H imaging devices. A huge number of dangling bonds is present in amorphous silicon. In order to obtain stable devices, the a-Si is

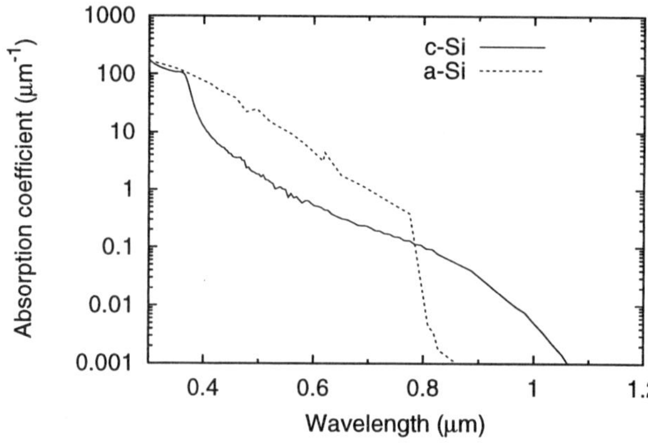

Fig. 3.72. Absorption coefficient of a-Si:H [202]

hydrogenated, i. e. the dangling bonds are saturated and in such a way passivated with hydrogen atoms.

The bandgap of a-Si:H is approximately 1.72–1.78 eV slightly depending on the deposition process. The bandgap of a-Si:H is much larger than that of crystalline silicon and the spectral response of a-Si:H suits the sensitivity of the human eye much better than that of crystalline silicon. The sensitivity of a-Si:H ranges from blue to red. Infrared light with $\lambda > 780\,\text{nm}$ cannot be detected with a-Si:H devices. The absorption coefficient of a-Si:H depends on the hydrogen content and on the deposition process. A typical curve for the absorption coefficient is shown in Fig. 3.72 [202]. The absorption coefficient of a-Si:H is much larger than that of crystalline Si in the visible spectrum. An a-Si:H layer with a thickness of 1 μm is sufficient for completely absorbing light with $\lambda < 0.7\,\mu\text{m}$ [203].

Amorphous Silicon Image Sensors. Amorphous silicon image sensors can be integrated vertically on microelectronic circuits due to the low-temperature deposition of a-Si:H (below 300 °C). In contrast to the low fill factor of active-pixel sensors, vertically integrated a-Si:H image sensors provide a fill factor close to 100 %. The crystalline integrated circuit typically consists of identical sub-circuits underneath each pixel detector and peripheral circuitry at the boundary of the light sensitive area. An insulating layer separates the optical detector from the crystalline chip.

Another advantage of a-Si:H image sensors is their large dynamic range of 60 dB, which can be further extended considerably by electronic measures. The concept of *Thin Film on ASIC* (TFA) [204, 205] adds much flexibility to a-Si:H image sensors. The TFA concept allows to design and optimize both the thin-film detector and the application specific integrated circuit (ASIC) independently. Figure 3.73 shows the layer sequence of a TFA sensor.

The a-Si:H thin film system of a TFA sensor does not need to be patterned and can be fabricated in a plasma-enhanced chemical vapor deposition (PECVD) cluster tool without temporarily being taken out of the vacuum

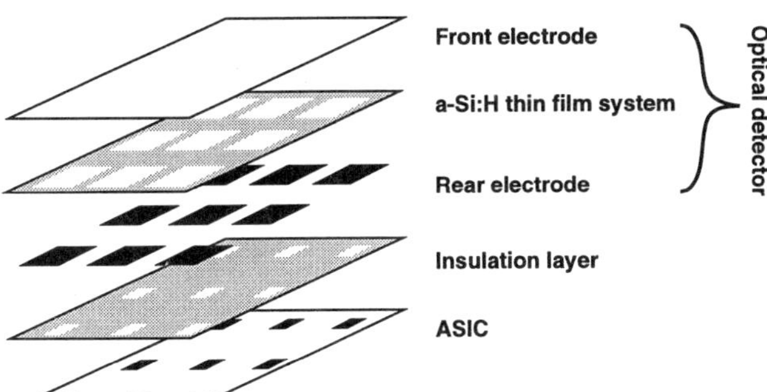

Fig. 3.73. Layer sequence of a TFA sensor [204]

for lithography. A very high yield of almost 100 %, therefore, is possible for the thin film system. Furthermore, fabrication costs are very low and barely exceed those of an ordinary ASIC. Without lithography, the pixel is simply defined by the size of its rear electrode. The continuous thin-film layer permits lateral balance currents between adjacent pixel detectors, resulting in cross-talk and in a reduced local contrast. A self-structuring technology for the suppression of this effect was suggested in [206].

A pixel of the least complex a-Si:H image sensor, which provides minimum pixel sizes and maximum resolution, consists of an a-Si:H photodiode and a PMOS transfer transistor. The function principle is based on a simple charge transfer from the pixel to the readout line [207]. The photocurrent discharges the capacitance of the reverse-biased photodiode during the integration time, resulting in a signal charge proportional to the illumination intensity and to the integration time. This operation mode is called charge storage mode. The amount of storable charge can be increased by an additional MOS capacitance, which can be integrated into each pixel on the ASIC in order to achieve a higher signal-to-noise ratio. In [205] this type of image sensor was called a varactor analog image detector (VALID). VALID is addressed linewise and read out columnwise. A random access sensor is also feasible employing a modified peripheral circuitry. 128×128 pixels with an area of $16 \times 16\,\mu m^2$ each were integrated in VALID using a standard $0.7\,\mu m$ CMOS ASIC process. A saturation illumination of 2500 Lux after an integration time of 20 ms was achieved with an integration capacitance of 250 fF. The dynamic range amounted to 60 dB. A good linearity and no visible image lag was observed. The circuits of more complex TFA image sensors for even higher dynamic ranges, which are essential in the field of artificial vision, will be discussed in Sect. 12.4.2.

Color Detectors. Here we will review color image sensors, which can easily be realized with a-Si:H technology. These color sensor arrays can be realized without color filters [208–210]. In contrast to CCD-color sensors, which employ RGB (red, green, blue) color filters, with three photodetectors per pixel for the generation of an RGB signal, only one vertical photodetector within one pixel is needed in an a-Si:H color sensor. This vertical a-Si:H color de-

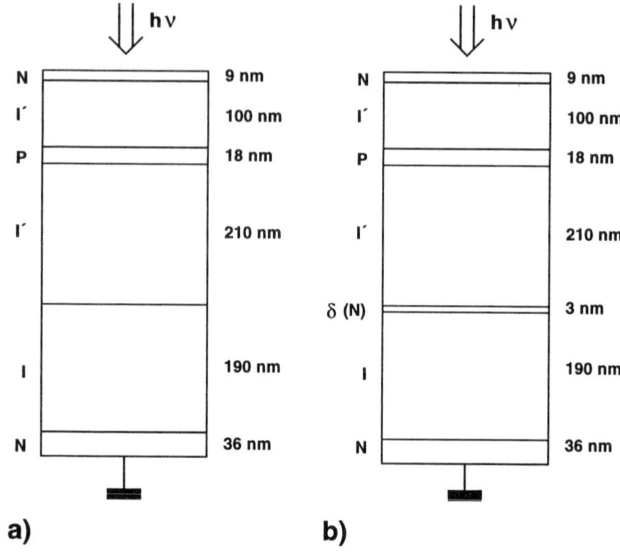

Fig. 3.74. Structures of a-Si:H based color sensors. (**a**) NI′PI′IN structure. (**b**) NI′PI′δ(N)IN structure [211]

tector has only a bottom and a top electrode. For an easier understanding of a-Si:H color sensors, we will begin with the explanation of a two-color sensor. The structure of a two-color sensor is shown in Fig. 3.74a. Carbonized 'intrinsic' (I′-type) a-Si:H (a-Si(C):H) layers with a bandgap of 1.91 eV are used in order to increase the blue-sensitivity of the top NI′P-detector and to increase its red-transmittance. The schematic band diagram of the NI′PI′IN detector is shown in Fig. 3.75.

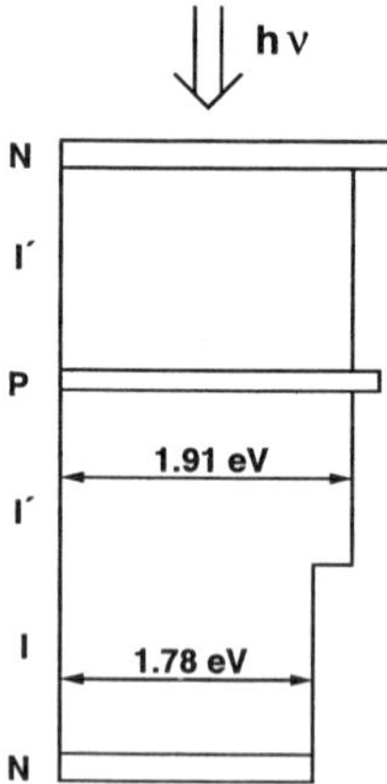

Fig. 3.75. Schematic band diagram of a NI′PI′IN color detector [211]

In a NIPIN-structure, the top diode of the color detector is reverse biased, when the potential of the top contact is positive compared to the bottom contact. A strong electric field then extends across the top diode. Most of the photons with short wavelengths are absorbed there. The detector is highly sensitive to blue light for a positive bias voltage at the top electrode. Under long wavelength illumination most carriers are generated in the bottom diode and for this positive bias voltage the photocurrent is low, because the bottom diode is forward-biased and the electric field is low in the bottom diode. The collection efficiency of photogenerated carriers, which can be characterized by the drift length ($\mu\tau\bar{E}$) where μ is carrier mobility, τ is carrier lifetime, and $\bar{E}$ is average electric field [212], is low in the bottom diode and most of the photogenerated carriers recombine.

When the bias voltage at the top electrode is negative, the top diode is forward-biased and the bottom diode is reverse-biased resulting in a low electric field in the top diode and a high electric field in the bottom diode. For this negative bias voltage, the detector is highly sensitive to red light.

It should be mentioned that such color detectors fabricated in crystalline Si (c-Si) would not work, because the carrier lifetimes in c-Si are much longer than in a-Si:H and in a-Si(C):H. The low electric field in the forward-biased diodes would be sufficient for efficient carrier collection. Due to the short carrier lifetimes and low carrier mobilities in a-Si:H, however, efficient color detectors are possible.

Two-color detectors can easily be realized with a-Si:H. The three colors red, green, and blue, however, have to be detected with color television and color video cameras. Three-color detectors also can be fabricated in a-Si:H thin-film technology, introducing a thin N-doped δ(N) layer between the bottom a-Si:H I-layer and the bottom a-Si(C):H I′-layer.

Table 3.9. Process parameters for a-Si(C):H color detectors

Feature	Process parameter
Substrate	Corning glass coated with TCO
RF-power	0.02 to 0.04 W/cm^2
Deposition facility	RF-PECVD system
Deposition rate	$\approx$0.35 nm/s
Deposition parameters	Substrate temperature & reaction gas
N-type layers	300 °C, $SiH_4 + PH_3$
P-type layers	250 °C, $SiH_4 + B_2H_6$
I-type layers	300 °C, Silane
I′-type carbonized layers	300 °C, Silane + CH_4
Top contact	Aluminum or chromium, 200 nm thick

This δ(N) layer increases the electric field in the bottom I′-layer and reduces it in the I-layer. In such a way, the detector is sensitive to blue light for a positive bias between 1 and 4.5 V at the top electrode, to green light between −1 and −2 V, and to red light for a bias below −6 V [211]. Response times of the order of 1 ms were observed for the color detectors [213]. For the sake of completeness, Table 3.9 lists the process parameters for the fabrication of a-Si:H and a-Si(C):H color detectors [211]. The glass substrate would be on top of the detectors shown in Fig. 3.74 and the top contact in Table 3.9 is on the bottom of the detectors in Fig. 3.74. The light has to transmit through the glass. SnO_2 is used on the glass substrate as a transparent conductive oxide (TCO). The deposition temperatures for the a-Si layers are 300 °C or less. Therefore, instead of a TCO-coated glass substrate, silicon wafers with already metallized integrated circuits can be used and TFA color sensors with TCO top electrodes are feasible.

X-ray Detector. The capability for large area electronics and detectors as well as the resistance to radiation damage extend the application field for the a-Si:H material to a medical x-ray detector (Fig. 3.76). One pixel element of such an x-ray detector is shown in Fig. 3.76. The pixel element consists of an a-Si:H thin film transistor (TFT) for addressing and readout as well of a molybdenum/a-Si:H Schottky diode. This detector is based on the photoelectric interaction of x-ray photons in the Mo layer and on the ejection of energetic electrons from the Mo layer into the a-Si:H layer. These energetic electrons undergo various scattering events inside the a-Si:H layer by which they transfer energy to the generation of electron–hole pairs. The electron–hole pairs are separated by the electric field in the depletion region of the reverse biased Schottky diode.

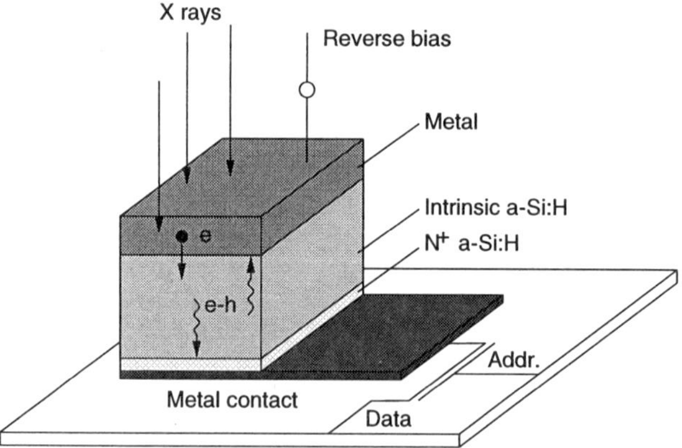

Fig. 3.76. X-ray detector using the external photoeffect of a metal/a-Si Schottky diode

Details of the fabrication process, which was stated to be fully compatible with that of the a-Si:H thin film transistor [214], are given in [215]. The a-Si:H layer had a thickness of approximately 400 nm. The Mo layer should have a thickness of 500 nm for optimal overall performance of the system, because not all photoelectrons generated in the Mo layer reach the a-Si:H layer when the Mo thickness increases. A thin Mo layer, however, does not absorb the x-ray photons effectively. Therefore, an intermediate Mo layer thickness gives the optimum conversion efficiency.

A ratio of 10^6 was observed for the forward and reverse currents at a bias of ± 1 V. The barrier height Φ_B was 0.77 eV and a low reverse current density of less than 10 nA/cm^2 at a reverse bias of 1 V was found. For x-ray energies in the range 20–100 keV, the number of electrons in a $200 \times 200\,\mu\text{m}^2$ detector ranged from 10^7 to 10^8. The measured photocurrent response was found to be linear with the x-ray intensity.

3.5.10 Bipolar Phototransistors

Photodetectors with a built-in amplification are most desirable for weak optical signals, especially in a comparatively simple CMOS process because of the high degree of standardization, its low cost, and the widespread interest in CMOS circuits.

The simplest way to integrate a bipolar phototransistor in a CMOS process without any process modifications is to use the N well as the floating base of a PNP transistor, the P$^+$ source/drain diffusion as the emitter and the P$^-$P$^+$ substrate as the collector (Fig. 3.77). The phototransistor then can be used as an emitter follower only, because the collector is at substrate potential, i. e. at 0 V. The complementary NPN phototransistor on an N$^-$N$^+$ substrate material is also possible, of course.

The simple structure shown in Fig. 3.77, however, has several disadvantages:

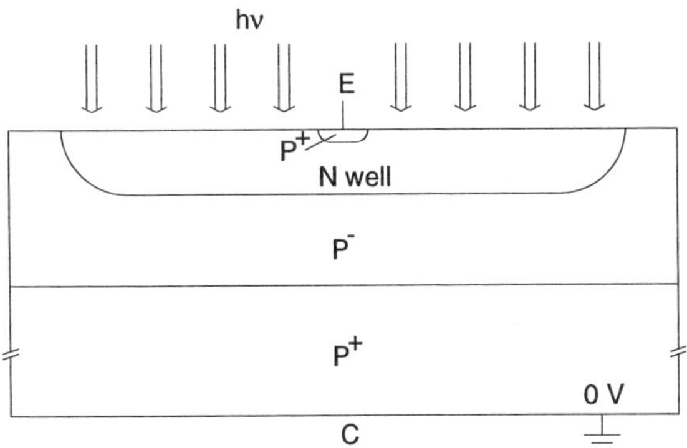

Fig. 3.77. Cross section of the simplest CMOS-integrated PNP phototransistor [216]

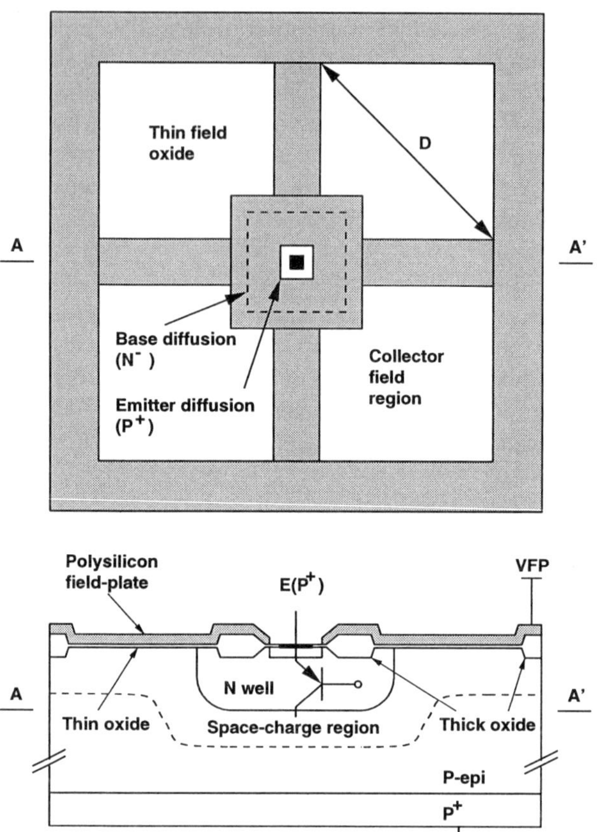

Fig. 3.78. Layout and cross section of a CMOS-integrated PNP phototransistor with reduced base-collector capacitance [216]

(i) low sensitivity for short and medium wavelengths due to the large N-well/P-substrate junction depth.

(ii) large base-collector capacitance due to the large base area.

In order to overcome these disadvantages, a polysilicon field-plate structure was proposed (Fig. 3.78), which allows a reduction of the N-well base area, whereby the space-charge region below the field plate and carrier diffusion in the areas without the field plate are used for the collection of photogenerated electrons [216]. This special layout reduces the base-collector capacitance and improves the sensitivity in the short to medium wavelengths spectral range.

The distance for all carriers being generated in the regions of thin field oxide without a polysilicon field-plate to diffuse to the space-charge regions near the field-plates should not (Fig. 3.78) exceed the carrier diffusion length L_{D}, in order to collect a maximum of photogenerated carriers. This is why

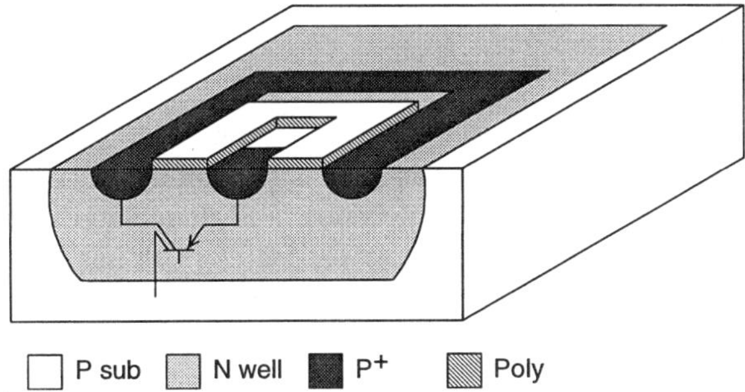

Fig. 3.79. Structure of a CMOS-integrated lateral PNP phototransistor with increased current gain [218]

this area was given a squared layout with a diagonal D less than $2\,L_D$ [216]. The bipolar phototransistor with the polysilicon field-plate structure was fabricated in a 2 µm one-polysilicon-two-Al-metal N-well CMOS technology with an N-well doping of $1.9 \times 10^{16}\,\text{cm}^{-3}$, a P^- epitaxial layer doping of $5.3 \times 10^{15}\,\text{cm}^{-3}$, an epitaxial layer thickness of 10 µm, an N-well penetration depth of 4.5 µm, a P^+ source/drain diffusion depth of 0.5 µm, and a gate oxide thickness of 40 nm. The light sensitive area of the special bipolar phototransistor was 8000 µm^2. This transistor showed a current gain of 120, a photoresponse time of 4 to 5 µs, and a conversion efficiency of 0.75 electrons per photon at $\lambda = 550$ nm. The dynamic range was examined for light intensities between 1 and 4000 Lux and the linearity of the emitter photocurrent with the light intensity was confirmed.

As already mentioned, the vertical bipolar phototransistor (VBPT), being available in a CMOS technology without process modification, is limited to the use in a common-collector configuration. The lateral bipolar transistor [217] being also available without process modification, however, can be freely connected. A typical lateral bipolar transistor suffers from a lower current gain and lower collector efficiency than the vertical bipolar device. In order to achieve a higher responsivity as an identically sized vertical bipolar phototransistor, modifications have been made to the typical lateral bipolar transistor [218]. The structure of this improved lateral bipolar phototransistor (LBPT) is shown in Fig. 3.79.

The lateral PNP phototransistor in a rectangular structure is formed with a P^+ emitter surrounded by a P^+ collector situated in an N-well base. The rectangular structure with the emitter in the center allows the optimum exploitation of the emitter efficiency. Because the base region is more lightly doped than the collector, the collector-base depletion region extends almost completely into the base. This space-charge region must not reach the emitter in order to avoid reach-through between emitter and collector. The base, therefore, has to be made wide enough. The current gain β, therefore, is rather low since β decreases when the base width W_B increases [218]:

$$\beta \approx \frac{1}{(W_{\mathrm{B}}^2/2\tau_{\mathrm{B}}D_{\mathrm{N}}) + (D_{\mathrm{P}}W_{\mathrm{B}}N_{\mathrm{A}}/D_{\mathrm{N}}L_{\mathrm{P}}N_{\mathrm{D}})}. \tag{3.12}$$

In this equation, τ_{B} is the minority carrier lifetime in the base, D_{N} and D_{P} are the carrier diffusion coefficients in the base, L_{P} is the diffusion length of holes in the emitter, and N_{A} and N_{D} are the acceptor and donor densities in the emitter and base, respectively. In [218], therefore, a trick [219] has been used in order to increase the current gain β. For this purpose, a polysilicon ring around the emitter is implemented (see Fig. 3.79) in order to set the base width. By connecting the polysilicon gate to the most positive potential in the circuit, majority carriers in the base accumulate directly under the gate, retarding the widening of the collector-base depletion region and setting the base width to that of the minimum polysilicon gate width.

It should be mentioned that the lateral phototransistor of Fig. 3.79 also contains a parasitic vertical PNP transistor. The effect is a reduction of the collector efficiency of up to 40 % [217, 219], i. e. up to 40 % of the collector current is lost to the substrate. In order to achieve a 100 % exploitation of the amplification of the LBPT, the photodetector current, therefore, has to be drawn from the emitter.

The base of the LBPT is left floating. Base bias current is provided when incident photons create electron–hole pairs. The photodetector current taken from the emitter is $(\beta + 1)$ times the base current, which would correspond to the photocurrent of a similar sized photodiode.

LBPT and VBPT devices with a size of $60 \times 60\,\mu\mathrm{m}^2$ were fabricated in a standard 2 µm digital CMOS process. An LED light source with a peak wavelength of 660 nm was used for the illumination of the devices and its illumination intensity was varied over four orders of magnitude. LBPT and VBPT output currents were approximately proportional to the light intensity. The LBPT output current was constantly about nine times greater than the VBPT output current. The output current of the LBPT was about 20 µA for an illumination intensity of $1\,\mathrm{mW/cm^2}$. For these values we can calculate a photoamplification of about 1040 compared to a photodiode of the same size. For an illumination intensity of $P' = 1\,\mathrm{mW/cm^2}$ at 660 nm we obtain a photon flux density $\Phi = P'\lambda/(hc) = 3.33 \times 10^{15}\,\mathrm{cm^{-2}s^{-1}}$. The photocurrent density of a photodiode $j_{\mathrm{ph}}^{\mathrm{PD}} = \Phi q$, therefore, is $5.33 \times 10^{-4}\,\mathrm{A/cm^2}$. For the area of $60 \times 60\,\mu\mathrm{m}^2$, $I_{\mathrm{ph}}^{\mathrm{PD}} = 19.2\,\mathrm{nA}$ results. If we divide the measured LBPT photocurrent of 20 µA by this value, we obtain a photoamplification of 1040 for the LBPT.

LBPT devices also were fabricated in 1.2 and 0.8 µm CMOS processes. The active area of these devices scaled directly with the processes, i. e. the LBPT cell dimensions in the 1.2 µm process were 36 µm per side and 24 µm in the 0.8 µm process. The highest LBPT output current was observed for the CMOS process with the 1.2 µm design rules. However, no explanation was given for this finding.

The transient behavior of a photodetector determines how fast an array of photodetectors can be scanned in an image sensor. Due to charge storage, the fall time of a phototransistor is usually greater than the rise time [220]. The rise time, however, fortunately limits the scan speed of an image sensor. The fall time does not pose a problem as long as it is shorter than the time needed to scan the array. The worst-case scenario for the rise time is when a brightly illuminated cell is next to a dark cell. Step response of the LBPT with $W_B = 2\,\mu m$ was measured by switching between dark and brightly illuminated cells in an array. In such a way, a rise time of 3.1 µs was determined for the LBPT. In order to be complete, the dark current of the LBPT of less than 10 pA has to be mentioned.

3.5.11 Field-Effect Phototransistors

For certain applications, photodetectors with an internal amplification of the photocurrent are desirable or necessary. These amplifying photodetectors do not need separate amplifiers or at least simpler amplifiers are sufficient. The amplifying photodetectors, therefore, help to save chip area and reduce system costs.

The photo-MOSFET shown in Fig. 3.80 is an amplifying photodetector being readily available in CMOS technology without any modifications. The P-channel MOS transistor is placed in an N well, which can be made floating by a MOSFET switch. The space-charge region at the N-well/P-substrate junction separates the photogenerated electron–hole pairs. The N well is charged negatively, because the electrons drift to the N well. The

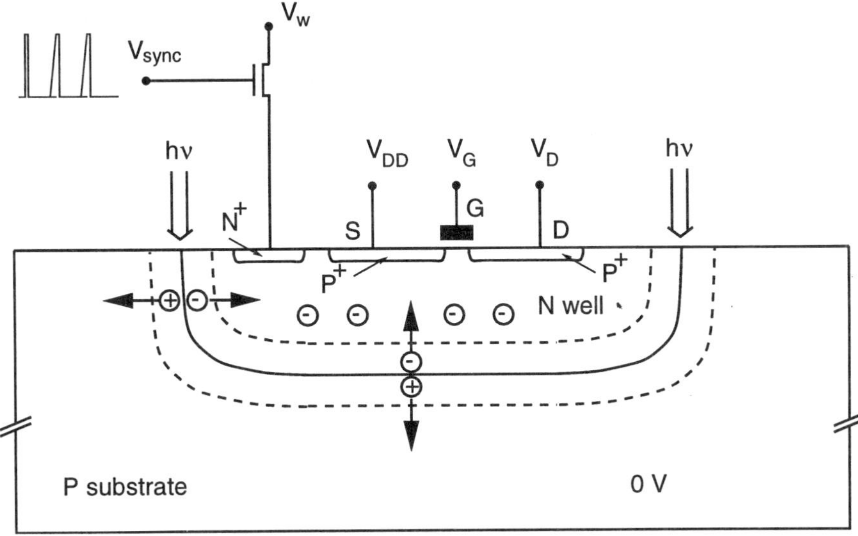

Fig. 3.80. Cross section of a photo-MOSFET

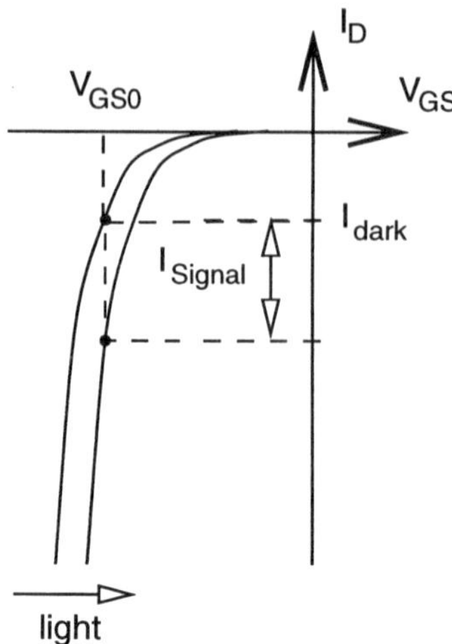

Fig. 3.81. Current–voltage characteristics of a photo-MOSFET

photo-MOS-transistor exhibits a high sensitivity since the photogenerated charge affects the threshold voltage and, hence, modulates the output drain current (Fig. 3.81).

The dark current, unfortunately, is also amplified and must be cancelled to ensure proper operation of image sensors [221]. The sensitivity depends on the operating point defined by the quiescent well potential V_w, which determines the drain current. This quiescent potential can be applied periodically to the well by turning on a MOSFET switch, which is connected to the desired bias voltage V_w, for a short time.

The space-charge region extends only approximately 1–2 μm laterally from the N well and slow carrier diffusion limits the speed of the photo-MOS-transistor for usual sizes of light sensitive areas. Another disadvantage of the photo-MOS-transistor is a rather high fixed-pattern noise (FPN). The FPN depends mainly on the gate area ($W \times L$) of the transistor and it decreases for large areas, as expected from matching properties of MOS devices [222]. The photo-MOS-transistor exhibits a very high dynamic range of 140 dB and a wide temperature range from at least −40°C to 120°C. With temperature compensation, a temperature drift of 8 mV/K for the output voltage of an image sensor was reported [221]. The minimum detectable illumination was 1 mLux.

The response time of the photo-MOS-transistor depends on the operating point and capacitive loading. It is of the order of tens of microseconds, when the sensor is illuminated up to its saturation [221]. The maximum pixel clock frequency with a 128-array one-dimensional active-pixel sensor in a 2 μm CMOS technology was reported to be 50 kHz [221]. Due to the dependence of the sensitivity on the operating point, i. e. on the well potential, the sensitivity of the photo-MOS-transistor is programmable.

A CMOS linear photosensor array for edge extraction, containing a 64 pixel one-dimensional photo-MOS-transistor array in a CMOS technology with 2 µm design rules, occupied a silicon area of 7.7 mm^2 and consumed 80 mW at a 10 V power supply voltage. The edge extraction required 8 µs with a 16 MHz clock frequency. The size of the photo-MOS-transistor was 80 µm × 3.2 µm and the pitch of the photo-MOS-transistors in the one-dimensional array was 52 µm. The fixed-pattern-noise was found to be 0.6 % [221].

Low power CMOS imaging systems are currently the subject of many investigations. A high responsivity and especially a photodetector with internal amplification allow a simpler amplifier or no amplifier is necessary at all. Therefore, a photodetector with internal amplification helps to save power. In [223], a highly responsive photodetector in a standard 0.8 µm N-well CMOS technology was reported. The photodetector is obtained by connecting the N well with the N^+ polysilicon gate of a PMOSFET. Figure 3.82 shows the cross section and the layout of the amplifying photodetector. The PMOS

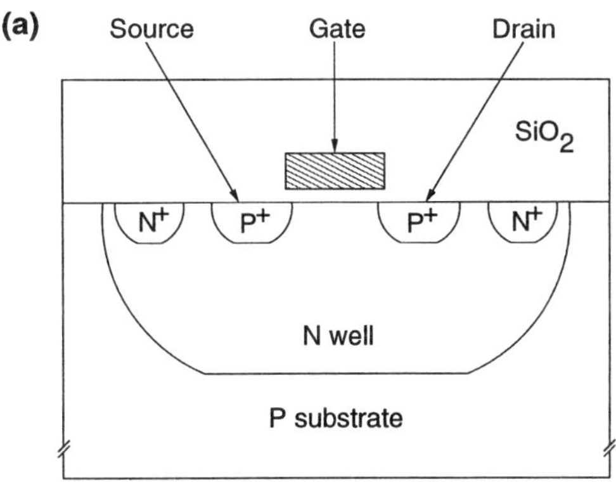

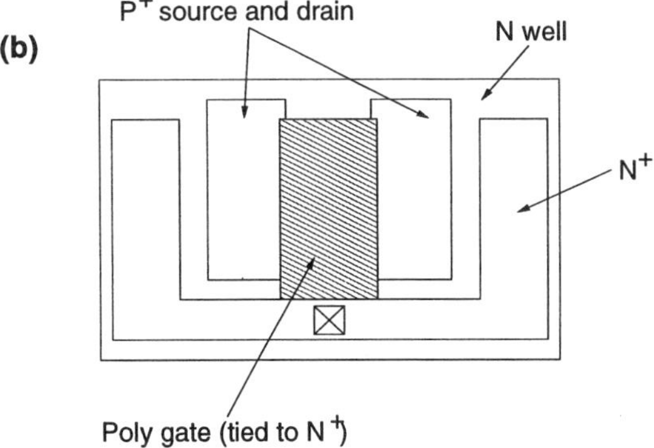

Fig. 3.82. Cross section (**a**) and top view (**b**) of a PMOS-photo-FET [223]

transistor is situated in the N well. The N well is connected to the polysilicon gate by the first metal layer. The N well and the polysilicon gate are left floating.

It should be mentioned that due to the N^+ polysilicon gate used, the PMOSFET is a buried channel device [223]. Illumination generates electron–hole pairs in the channel of the PMOS-photo-FET and in the N well. The electron–hole pairs are separated vertically due to the vertical electric field resulting from the vertical band structure. The potential has a minimum at the location of the buried channel, where holes can flow between source and drain, when the gate potential is sufficiently negative. The holes drift upwards from the N well towards the potential minimum at the channel while the electrons drift to the bottom of the N well. Therefore, a negative bias is induced in the N well and the magnitude of the threshold voltage of the PMOSFET is reduced. In turn, the subthreshold current increases. The negative voltage is fed back to the gate as the well is shorted to the polysilicon gate turning the PMOSFET further on. This positive feedback mechanism results in a high current gain.

Compared to an N-well/P-substrate diode, when the substrate and therefore the anode is usually grounded, the PMOS-photo-FET can be operated in two different modes: the source of the PMOS-photo-FET can be connected to V_{DD} or the drain can be tied to V_{SS}. In both cases, a voltage V_{DS} of only 0.3 V is sufficient to operate the device, which enables low power operation [223]. For a $W/L = 8.2/0.8\,\mu m$ device, a current gain of 80 to 290 compared to a photodiode of the same size as the N well was reported. The current gain was larger for lower light intensity [223]. Therefore, the drain current does not increase linearly with light intensity. A drain current of $20\,\mu A$ for instance was reported for a light intensity of $122\,mW/cm^2$ for the $W/L = 8.2/0.8\,\mu m$ device.

Another advantageous property of the PMOS-photo-FET is that its current gain increases with decreasing gate length L. The dark current was below 70 pA for $W = 8.2\,\mu m$ and $L = 0.8 - 4\,\mu m$. Unfortunately, no results were reported for the transient behavior of the PMOS-photo-FET. A signal-to-noise ratio of 65 dB was determined for the PMOS-photo-FET up to an operating frequency of 100 kHz. The performance of the PMOS-photo-FET makes it very attractive for low power image sensors.

3.6 BiCMOS Processes

BiCMOS stands for Bipolar&CMOS. It combines both bipolar transistors and NMOS as well as PMOS transistors in the same chip. In digital BiCMOS circuits it is possible to exploit the advantages of bipolar transistors and of MOS transistors. The capabilities of BiCMOS circuits, therefore, exceed those that are obtained when only one of the device types is used. For example, CMOS allows low-power, high-density digital ICs, which, however, are slower than bipolar ECL-based ICs due to the poor driver capability of MOS transistors, especially when large capacitive loads are encountered. Such loads occur at long interconnects on chip, at high fan-out circuits, as well as at chip outputs. The driver capability of bipolar transistors can be used to advantage in BiCMOS circuits to charge such heavy loads rapidly. Bipolar digital circuits implemented with ECL gates can also operate with small logic swings and high noise immunity. ECL gates, however, exhibit high power consumption, poor density, and limited circuit options. BiCMOS offers the benefits of both bipolar and CMOS circuits by appropriately trading off the characteristics of each technology.

BiCMOS can also offer analog and digital functions on the same chip. For instance, better device matching, lower offset voltages of operational amplifiers, and enhanced bandwidths compared to analog CMOS circuits can be obtained with analog bipolar sub-circuits in BiCMOS ICs or with BiCMOS circuits. From CMOS, the high input resistance, i. e. the zero input bias current, can be used advantageously in BiCMOS circuits. BiCMOS circuits, therefore, pushed mixed analog/digital (mixed signal) applications tremendously.

The interested reader is recommended to read [224] to study the advantages of BiCMOS in more detail. The benefits of BiCMOS, however, are attained at the expense of a more difficult technology development and at the expense of more complex chip-manufacturing tasks leading to longer chip-fabrication times and higher costs.

BiCMOS processes can be derived from bipolar and CMOS processes. The simplest low-cost BiCMOS process is obtained adding one mask for the P base of the NPN device to a well established CMOS process (Fig. 3.83). The added mask is used to selectively produce lightly doped P regions with a doping level of about $10^{17}\,\mathrm{cm}^{-3}$. The N-well fabrication step also forms the collector regions of the NPN transistors. The heavy NMOS source/drain implant is employed to produce the emitter regions and collector contacts of the bipolar transistors. The PMOS source/drain implant is used to create the base contact, since this serves to lower the base series resistance R_B. The starting substrate is either a P-type wafer or a P-type epitaxial layer on a P^+ wafer. Such simple processes are called *triple-diffused (3-D)* BiCMOS processes, because they produce triple-diffused (C, B, E) bipolar transistor structures [224,226]. In order to optimize both the CMOS and bipolar device characteristics independently, the doping profiles in the CMOS N well and in

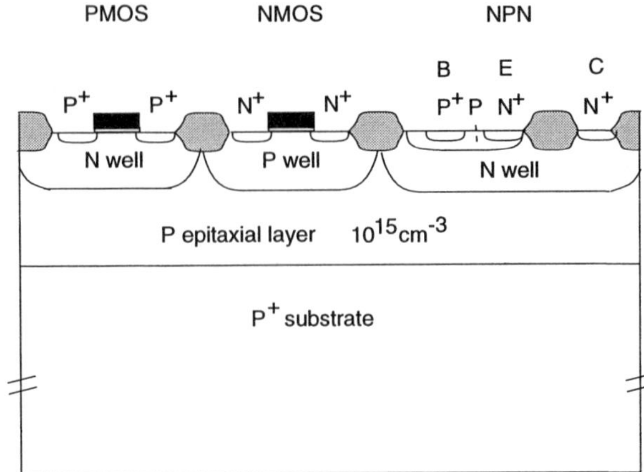

Fig. 3.83. Cross section of a triple-diffused BiCMOS structure [88, 225]

the bipolar N well have to be made different. The base-collector breakdown voltage and therefore the collector-emitter breakdown voltage of the NPN transistor may be rather small with the CMOS N well as the collector. In order to increase these breakdown voltages the doping concentration has to be reduced making a separate collector doping, i. e. a bipolar N well, necessary. The resulting cross section is shown in Fig. 3.84.

The main drawback of 3-D bipolar transistors is a large collector series resistance R_C. In order to improve the properties of the NPN transistors (e. g. current gain β and transit frequency f_T), polysilicon emitters allowing a thinner base were added to a 3-D BiCMOS process in [227], of course increasing the process complexity.

The most widely used high-performance BiCMOS technology is based on a twin-well CMOS process. Figure 3.85 shows the cross section of such a high-performance BiCMOS structure [225, 228]. N^+ buried layers are used

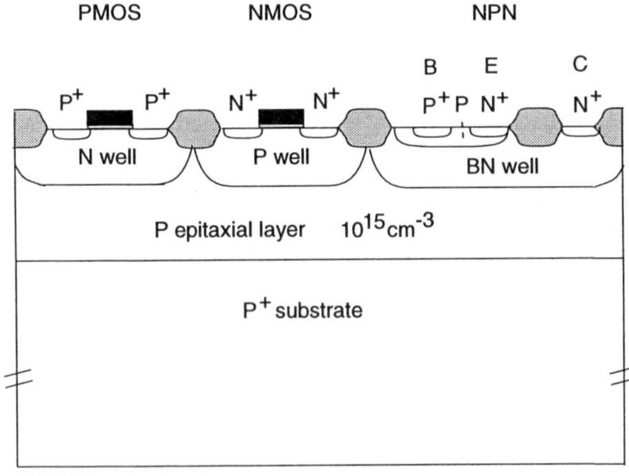

Fig. 3.84. Cross section of a 3-D BiCMOS structure with separately optimized CMOS and bipolar N wells

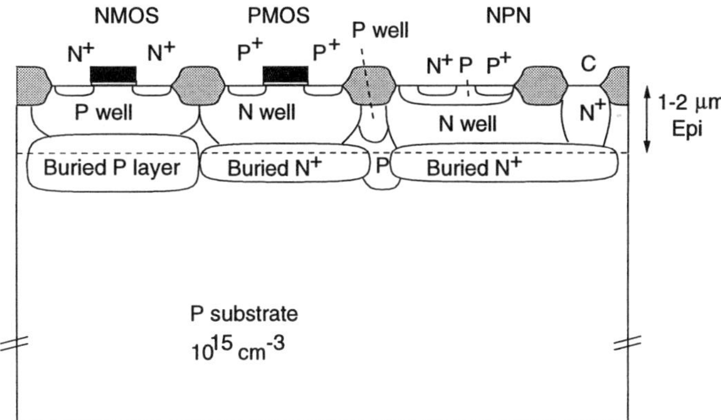

Fig. 3.85. Optimized high-performance BiCMOS structure

to reduce the collector series resistance. These N^+ buried layers and the P buried layers also reduce the susceptibility to latch-up in the CMOS part of the BiCMOS ICs.

Four masking steps were added to an advanced 0.8 μm 12-mask CMOS process described in [88] to integrate high-performance bipolar transistors. The starting material is a lightly doped ($\approx 10\,\Omega$cm) silicon wafer with a <100> orientation. At the beginning of the process, a mask has to be added to pattern the N^+ buried layers (see Fig. 3.86a). Next, a selective boron implant is performed, which places an additional P-type dopant between the N^+ buried layers (see Fig. 3.86b), to reduce the minimum buried layer distance. This boron implant also creates the P-type buried layers. This self-aligned approach requiring only one mask to create both the N^+- and P-buried layers is widely used [224]. After the removal of all oxides, a thin (1.0–1.5 μm), near-intrinsic (less than $10^{15}\,\mathrm{cm}^{-3}$) epitaxial layer is then deposited (see Fig. 3.86c). The steps for twin-well formation are conducted next (compare Fig. 3.19a and b). A single-mask process can be used to form both the N well and the P well. A dual-phosphorus implant (deep and shallow) is used to form the N well (Fig. 3.86d). A boron implant is used to form the P well (Fig. 3.86e). Following the implantations that introduce the well dopants, the well drive-in is performed. Then the active regions are defined (see Fig. 3.86f and compare with Fig. 3.19d). The boron channel stop is implanted next and the field oxide is grown.

At this stage, the second mask is added for the formation of the deep N^+-collector contact (Fig. 3.86g). A deep phosphorus implant and a subsequent drive-in/anneal step are applied to form the deep-collector-contact structure. The third added mask (Fig. 3.86h) is used for the patterning of the P base, which is implanted by boron and annealed subsequently.

The fourth added mask is needed to open the areas for the polysilicon emitters (Fig. 3.86i). This process is rather complex and the interested reader

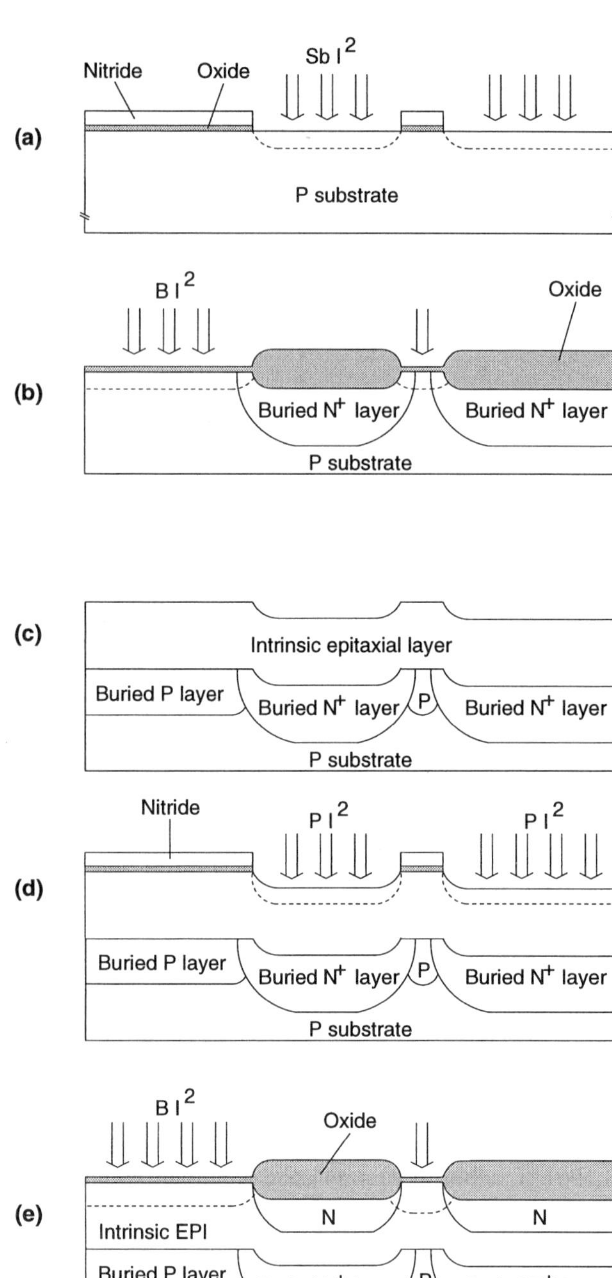

Fig. 3.86a–e. For caption see the page after next page

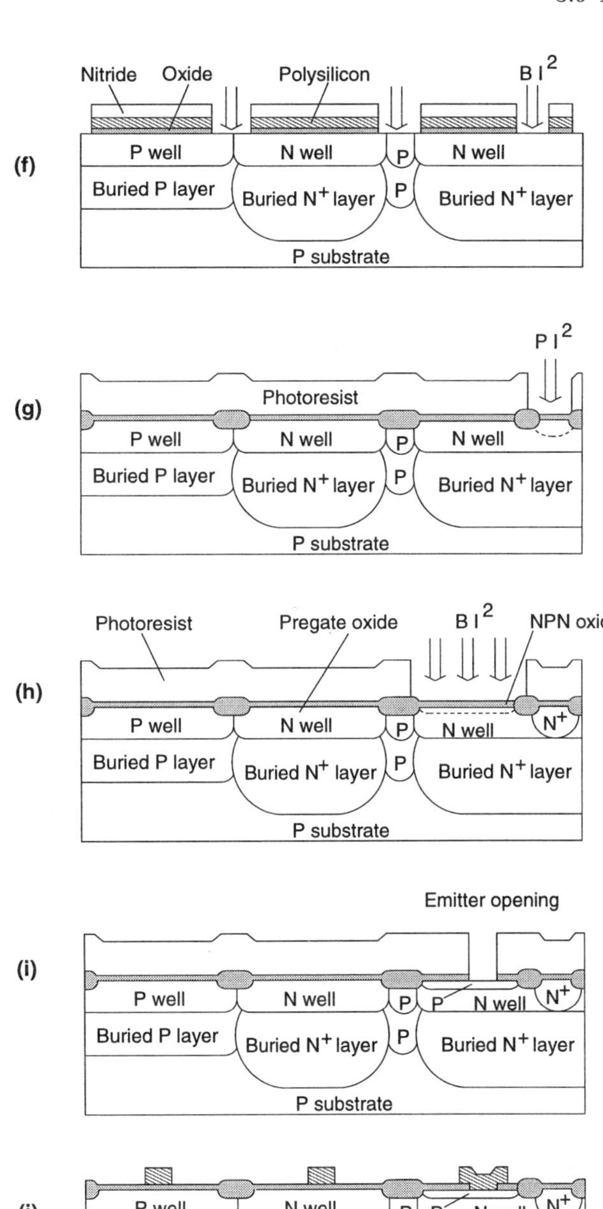

Fig. 3.86f–j. For caption see next page

(k)

(l)

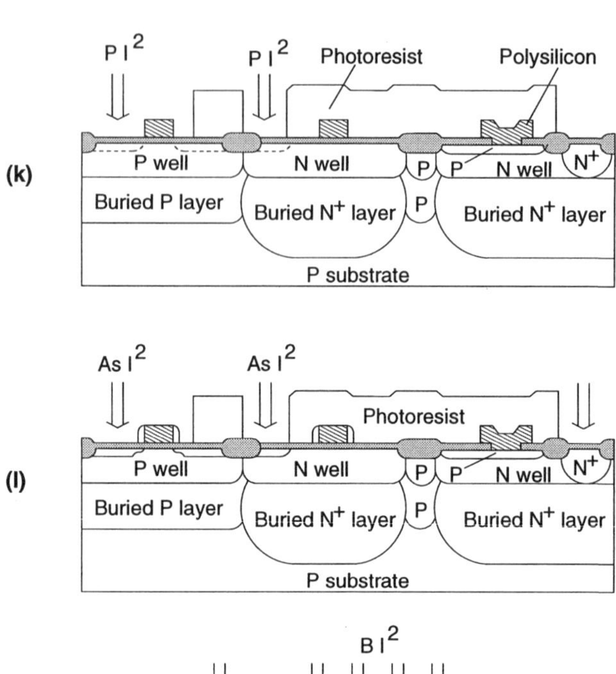

(m)

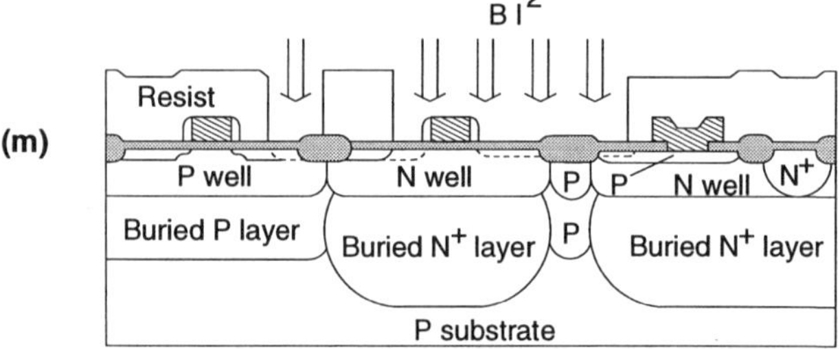

(n)

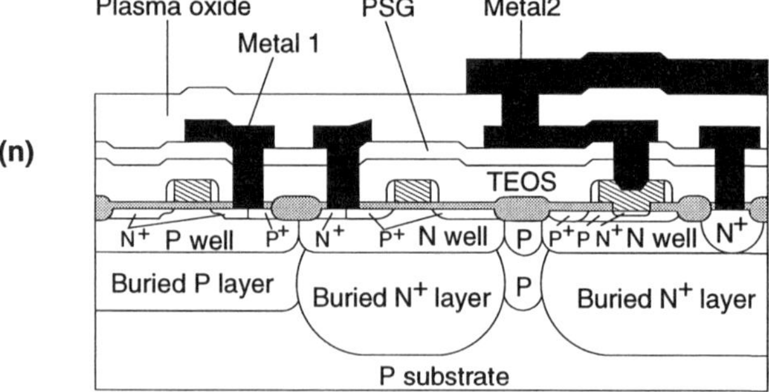

Fig. 3.86a–n. Process flow of an optimized high-performance BiCMOS process [224]

will find a thorough description in [224]. The polysilicon on top of the emitters is patterned together with the polysilicon gate definition (Fig. 3.86j). From this stage on, processing (Figs. 3.86k–n) is identical to the CMOS process, from which the BiCMOS process was derived.

In order to achieve transit frequencies of the order of 10 GHz for the NPN transistors, the epitaxial N collector has a thickness of only about 1–2 μm [228–233]. In a 0.8 μm BiCMOS process, an epitaxial layer with a thickness of only 0.8 μm was used [234]. In this process, different from Fig. 3.86a, As instead of Sb was used for the buried collector and due to the thin epitaxial layer, autodoping caused by the N^+ As buried collectors during epitaxy had to be avoided. For this purpose, the pressure in the epitaxial dichlorosilane process was reduced to 25 Torr and a special temperature ramp starting at 850 °C and increasing the temperature to 1150 °C within 10 min resulted after 1 min at 1150 °C in an intrinsic — i. e. intentionally undoped — epitaxial layer thickness of 0.15 μm. The main epitaxial cycle was carried out at 860 °C for approximately 18 min where the epitaxial layer was grown to a thickness of 0.8 μm. The transit frequency of the NPN transistor was 11.5 GHz. A 0.5 μm BiCMOS process, for instance, is described in [235].

3.7 BiCMOS-Integrated Detectors

BiCMOS circuits enable the use of photodetectors described in Sects. 3.2 and 3.5. We, therefore, have to discuss only a few photodetectors here. BiCMOS-OEICs have already been introduced [236, 237]. The OEICs in both cases consisted of a PIN photodiode and an amplifier, which exploited only MOSFETs and no bipolar transistors. The BiCMOS technology, therefore, was chosen merely for the integration of the PIN photodiode. A standard BiCMOS technology with a minimum effective channel length of 0.45 μm without any modifications was used [236, 237]. This effective channel length corresponds to a drawn or nominal channel length of about 0.6 μm. The buried N^+-collector in Fig. 3.87 was used for the cathode of the PIN photodiode. The P^+–source/drain island served for the anode. The intrinsic zone of the PIN photodiode was formed by the N well and, therefore, had only a thickness of 0.7 μm. Consequently, the responsivity of the photodiode was only 0.07 A/W for a wavelength of 850 nm [236]. For a bias of 2.5 V across the $75 \times 75\,\mu m^2$ PIN photodiode with a capacitance of 1.8 pF, a −3dB bandwidth of 700 MHz was reported. The OEIC reached a bit rate of 531 Mb/s with a bit error rate of 10^{-9} and a sensitivity of −14.8 dBm. This bit rate was limited by the capacitance of the photodiode and the feedback resistor of 1.4 kΩ in the amplifier transimpedance input stage. The dark current of the photodiode was 10 nA for a reverse voltage of 2.5 V at room temperature.

In [237] a laser with a wavelength of 670 nm was used for the characterization of the same OEIC as in [236]. The data rate was increased to 622 Mb/s for this wavelength. No measured result for the responsivity was given in [237].

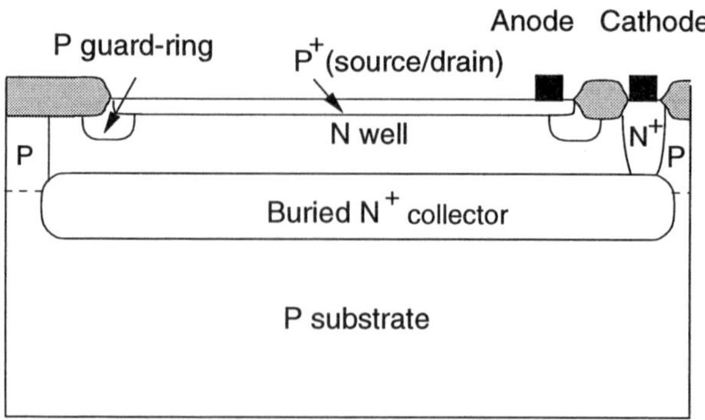

Fig. 3.87. PIN photodiode in a SBC-based BiCMOS technology [236]

Instead, a three times larger responsivity value of approximately 0.2 A/W for $\lambda = 670$ nm was estimated due to the lower penetration depth than for $\lambda = 850$ nm. This estimation, however, may be doubtful, because possible destructive interference in the isolation and passivation stack is neglected. The main drawback for the photodiode used in [236, 237] was its low responsivity due to the thin epitaxial layer in the SBC-based BiCMOS process.

In the following we will investigate an N^+/P-substrate photodiode in a CMOS-based BiCMOS technology. Figure 3.88 shows the cross section of this photodiode.

The process used for the fabrication of this photodiode is optimized with respect to a minimum number of masks and implements self-aligned wells. Therefore, there is a P well incorporated in the N^+/P-substrate photodiode. This leads to a thin N^+/P-substrate space-charge region and the effect of slow carrier diffusion is very pronounced (see Fig. 3.89). Rise and fall times of 26 and 28 ns, respectively, are determined due to the pronounced diffusion tail. The measured −3dB bandwidth is 6.7 MHz.

Another example for a photodetector in a standard BiCMOS technology will be added here. In a CMOS based (3-D) BiCMOS process, an N well can be used as the collector of a bipolar NPN transistor. This N well can also be used in order to form the cathode of a so-called double photodiode (DPD). Figure 3.90 shows the cross section of such a DPD. The two anodes are connected to ground. The cathode is connected to the amplifier in the OEIC

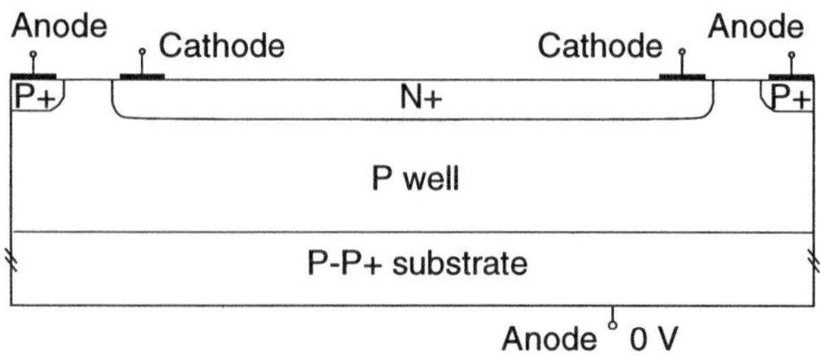

Fig. 3.88. Cross section of a N^+/P-substrate photodiode in self-adjusting well BiCMOS technology

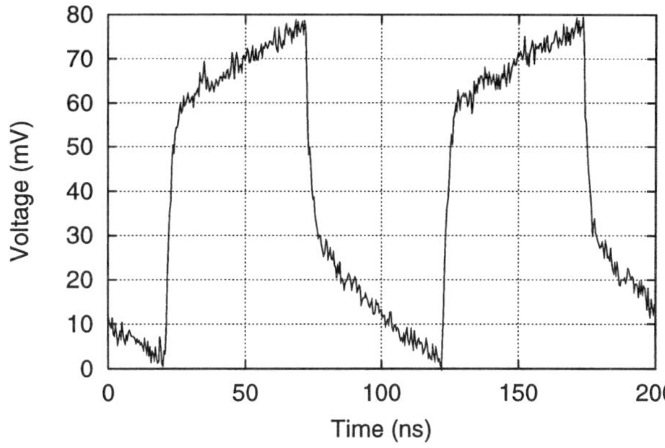

Fig. 3.89. Transient response of a N^+/P-substrate photodiode in a self-adjusting well BiCMOS technology

(see Fig. 12.25). Two PN junctions are vertically arranged. The N-well/P-substrate junction minimizes the effect of slow diffusion of photogenerated carriers from the substrate.

In addition to the two space-charge regions at the two vertically arranged PN junctions, an electric field is also present between the two space-charge regions (see Fig. 3.91) due to the doping gradient of the N well. Therefore, there is no contribution of slow carrier diffusion from the region between the two space-charge regions.

With an integrated polysilicon resistor of 1 kΩ, rise and fall times of 3.2 and 2.8 ns, respectively, were measured for the DPD with $\lambda = 638$ nm and $U_r = 2.5$ V [239]. There is no indication of a so-called diffusion tail in the photocurrent (Fig. 3.92).

The transient response of such a DPD is compared to that of a PIN photodiode with a P^+ anode at the surface from Sect. 3.5.2 in Fig. 3.93. The PIN photodiode possesses shorter rise and fall times. Furthermore, the behavior of a DPD with a CMOS N well is compared to that of a DPD with a bipolar N well. The DPD with the bipolar N well shows shorter rise and fall times because the doping concentration in this well is lower than in the CMOS N well and the P^+/N-well space-charge region is thicker in the DPD with the bipolar N well.

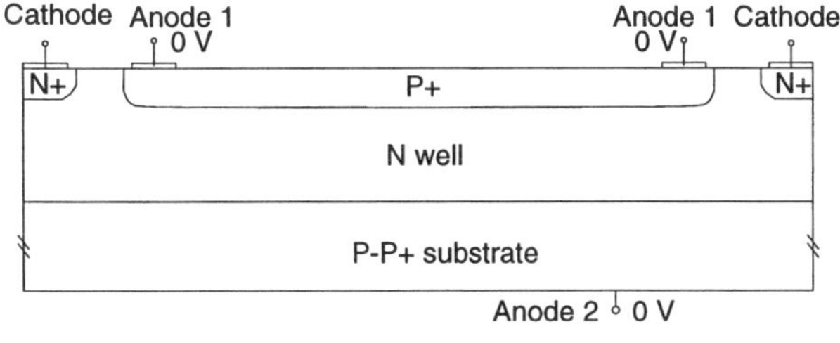

Fig. 3.90. Cross section of a double photodiode in BiCMOS technology [238]

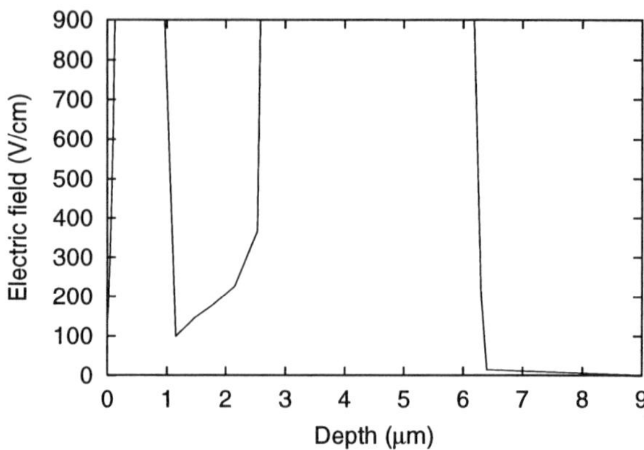

Fig. 3.91. Electric field in a double photodiode [72]

With an integrated polysilicon resistor of 500 Ω, rise and fall times of 1.8 and 1.9 ns, respectively, and a −3dB bandwidth of 156 MHz were measured for the DPD with $\lambda = 638$ nm and $U_r = 2.5$ V [238]. Later a bandwidth of 220 MHz was measured for an OEIC with a DPD of the size $50 \times 50\,\mu m^2$ and with a simple transimpedance bipolar amplifier representing a low input impedance for the photocurrent of the DPD [240], leading to the conclusion that the rise and fall times of 1.8 and 1.9 ns as well as the bandwidth of 156 MHz were limited by the RC constant of load resistor and DPD capacitance.

For a DPD with the area of $23 \times 23\,\mu m^2$, finally rise and fall times of less than 0.37 ns and 0.70 ns, respectively, and a bandwidth of 367 MHz were determined in [72], where an OEIC with this smaller DPD and with a transimpedance bipolar amplifier has been investigated. The transient response

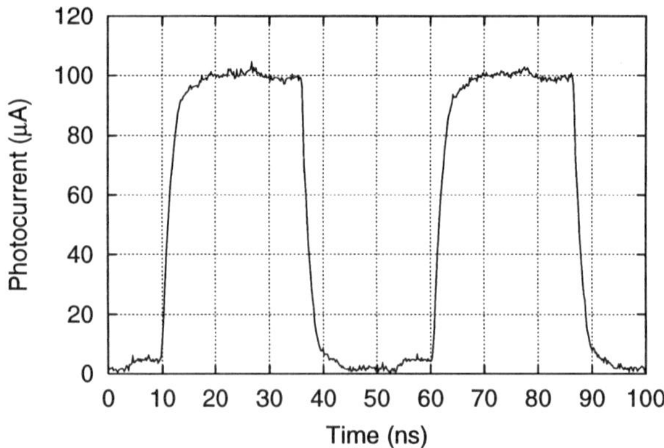

Fig. 3.92. Transient response of the double photodiode in BiCMOS technology shown in Fig. 3.90 [239]

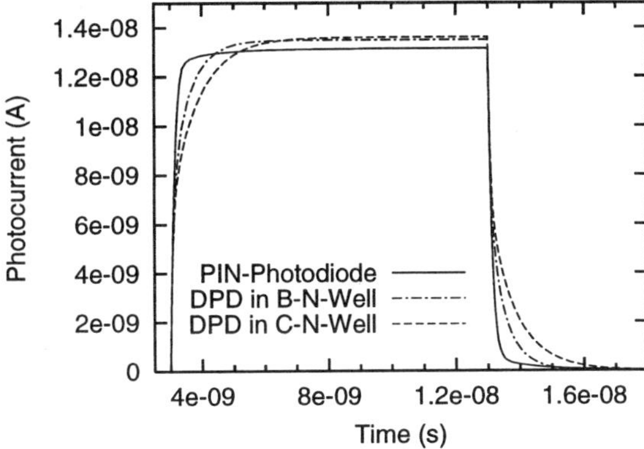

Fig. 3.93. Comparison of the transient response of double photodiodes with a bipolar N well and with a CMOS N well to that of a PIN-photodiode with $C_e = 5 \times 10^{13}\,\mathrm{cm}^{-3}$ ($\lambda = 638\,\mathrm{nm}$, $U_r = 2.5\,\mathrm{V}$)

of this OEIC with the small DPD is shown in Fig. 3.94. The faster speed of the smaller DPD can be explained with a lower series resistance in the N well of the DPD and with a lower space-charge capacitance due to its smaller size in addition to the low input impedance of the transimpedance amplifier. The values of 0.37 ns and 0.70 ns for the rise and fall times of the DPD determined in [72], therefore, better characterize the intrinsic speed of the DPD. With the conservative estimate a non-return-to-zero data rate $\mathrm{DR} = 2/(3(t_r + t_f)) \geq 620\,\mathrm{Mb/s}$ could be derived for the DPD [72].

The integration of photodiodes in BiCMOS technology, for which process modifications are necessary, is similar to the integration of photodiodes in bipolar or CMOS technology. The integration of PIN photodiodes in optimized high-performance BiCMOS technologies (see Fig. 3.85) requires a similar additional process complexity as the integration of PIN photodiodes

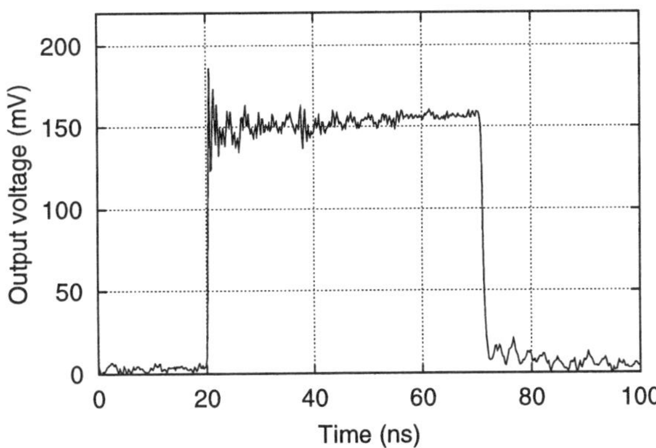

Fig. 3.94. Transient response of the double photodiode in BiCMOS technology with an area of 530 μm^2 [72]

in bipolar technology (see Fig. 3.8). The N well is used for the collector in the BiCMOS process. The third mask for the doping of the collector in Fig. 3.8, therefore, is not necessary for the PIN-BiCMOS integration in a process with a structure as in Fig. 3.85. Consequently, two additional masks are necessary for PIN-BiCMOS integration in such a structure. The integration of PIN photodiodes in 3-D BiCMOS technology (see Fig. 3.83), however, is similar to the PIN photodiode integration in twin-well CMOS technology (see Fig. 3.20).

The integration of a PIN photodiode in a high-performance BiCMOS technology was investigated in [79]. The cross section of such a modified BiCMOS OEIC is shown in Fig. 3.95. A three-step N^{--} epitaxial process was assumed in this figure. A two-step N^{--} epitaxial process similar to that shown in Fig. 3.8, however, may be sufficient also for a comparable photodiode speed. In order to avoid the shorting of the buried N^+ collectors and N wells, the P-isolation layers are needed. These P-isolation layers, furthermore, restrict the PIN cathode to the photodiode region. For the fabrication of the BiCMOS OEIC according to Fig. 3.95, the process complexity is increased considerably: three masks are necessary for the N^+ cathode, cathode contact implantations, and P-isolation layer formation. These masks have to be applied twice in a three-step epitaxial process and only once in a two-step epitaxial process. The P^+ source/drain implant can be used for the formation of the PIN anode. The PIN photodiode obtained is appropriate for the wavelengths 780 and 850 nm.

An advantage of the structure in Fig. 3.95 is that the voltage across the PIN photodiode is not limited by the maximum operating voltage of the circuit of about 5 V as is the case for the PIN CMOS integration shown in Fig. 3.20. For the PIN BiCMOS integration the carrier drift velocities can be

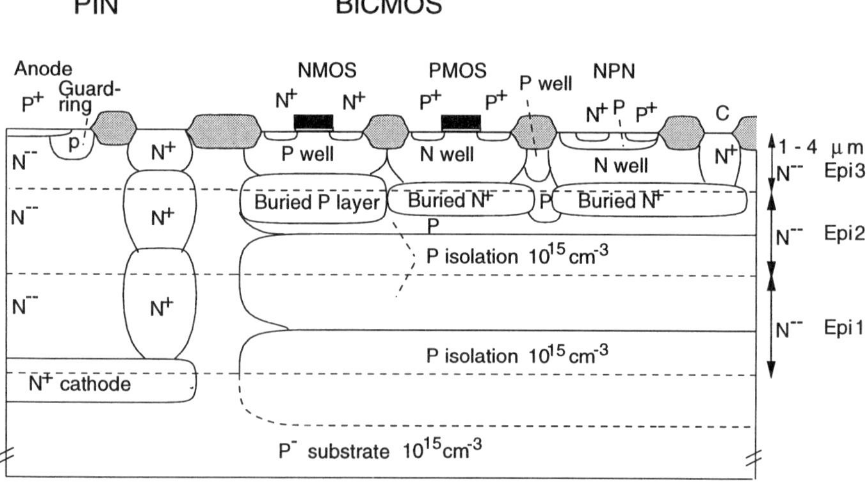

Fig. 3.95. Structure of a BiCMOS-OEIC requiring a modified BiCMOS process [79]

Table 3.10. Rise and fall times of the photocurrent for different doping levels in the 'intrinsic' layer of a BiCMOS-integrated PIN photodiode for $\lambda = 780\,\text{nm}$ and $U_r = 20\,\text{V}$

I-Doping (cm^{-3})	t_r (ns)	t_f (ns)
1×10^{15}	4.73	9.06
5×10^{14}	2.95	3.42
2×10^{14}	0.30	0.26
1×10^{14}	0.27	0.22
5×10^{13}	0.26	0.21

increased by a higher photodiode voltage and the rise and fall times of the PIN photocurrent can be reduced.

Lateral autodoping during epitaxy may cause problems. These problems, however, were solved even for arsenic [234], where the autodoping effect is most pronounced. For phosphorus the autodoping problem was solved in [38]. For the operation of the photodiodes at a higher reverse bias, the doping concentration in the 'intrinsic' region of the PIN photodiode does not have to be reduced as much as for a low reverse bias. This aspect also reduces the importance of autodoping.

In the following, some results of simulations for the structure in Fig. 3.95 are brought together. Table 3.10 contains the rise and fall times of the BiCMOS-integrated PIN photodiode for a total epitaxial growth thickness of 15 μm assuming a phosphorus cathode [79].

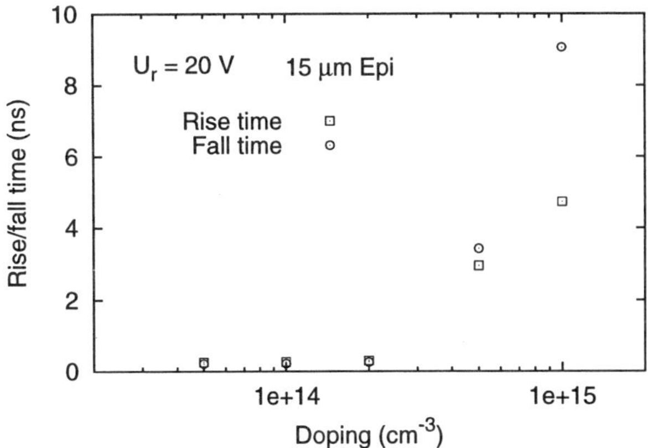

Fig. 3.96. Rise and fall times of the photocurrent for different doping concentrations in the 'intrinsic' layer of the BiCMOS structure shown in Fig. 3.95 for $\lambda = 780\,\text{nm}$

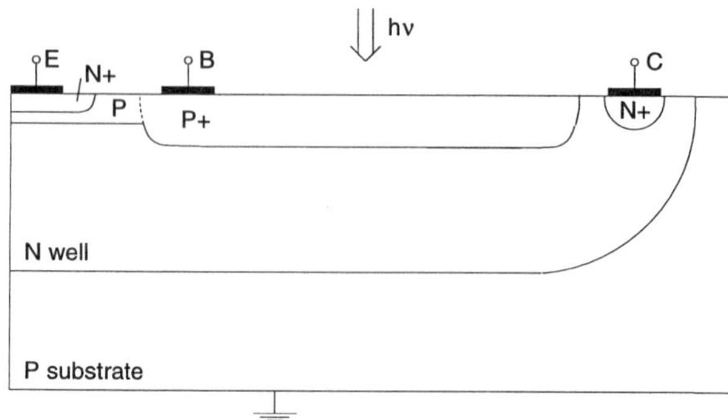

Fig. 3.97. Schematic cross section of a phototransistor in BiCMOS technology [240]

Compared to the results for CMOS-integrated PIN photodiodes in Table 3.1, the rise and fall time values are reduced by about a factor of 2.5 due to the larger photodiode bias and a higher doping concentration of the I-layer of $5 \times 10^{13}\,\text{cm}^{-3}$ is sufficient to obtain the minimum rise and fall time values. The rise and fall times of the photocurrent for the BiCMOS-integrated PIN photodiode are plotted in Fig. 3.96.

As the last example of BiCMOS-integrated photodetectors, the NPN transistor allowing an amplification of the photocurrent will be investigated. In order to obtain a phototransistor, the P^+-base contact doping area is made larger (Fig. 3.97). This increased P^+-base island serves as the light-sensitive area. The photogenerated carriers are separated by the electric field at the P^+/N-well base/collector junction. The P^+-base island is used instead of the P base in order to achieve a low series resistance in the base.

The frequency response of the phototransistor shown in Fig. 3.97 can be seen in Fig. 3.98 for a common-emitter circuit with open base and three differ-

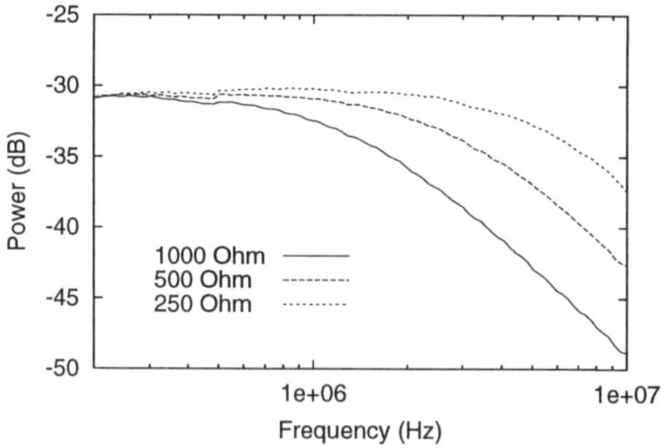

Fig. 3.98. Frequency responses of a phototransistor in BiCMOS technology with a supply voltage of 5.0 V for three different load resistors R_C [240]

ent load resistors R_C at the collector. The phototransistor had a size of about 2700 μm^2. The −3dB bandwidths of 1.4, 3.9, and 7.8 MHz for $R_C = 1\,k\Omega$, 500 Ω, and 250 Ω, respectively, are approximately inversely proportional to the load resistor [240]. This is clear evidence that an RC time constant limits the frequency response. The base-collector and the collector-substrate capacitances can be made responsible for the rather low bandwidths.

The rise and fall times measured for the phototransistor in Fig. 3.97 were 189 and 266 ns, respectively, with a load resistor R_C of 1 kΩ. For $R_C = 500\,\Omega$, they reduced to 107 and 160 ns and for $R_C = 250\,\Omega$, 56 and 83 ns were obtained, respectively.

4. Optoelectronic Devices in Silicon-on-Insulator

Silicon-on-Insulator (SOI) has been investigated mainly as a material for CMOS technology as a substitute for bulk silicon [241]. The main advantages of thin-film SOI for CMOS are: (i) The die area consumed by device isolation can be less than for bulk material. (ii) The number of process steps can be reduced. (iii) The substrate current is suppressed. (iv) There is no latch-up effect. (v) Lower parasitic capacitances enable faster circuits and a reduction of the power consumption.

In addition to CMOS technology, SOI can be interesting for the integration of photodetectors. SOI was considered as a possibility to avoid the slow diffusion current, which is caused by carriers being generated by light with long wavelengths (in the region of 850 nm), from the highly doped substrate of integrated photodiodes. For photodetectors in a thin SOI layer, this light generates carriers in the silicon substrate wafer, which is isolated from the device SOI layer by the buried oxide. Therefore, SOI avoids the slow diffusion current known from photodiodes in bulk silicon.

Furthermore, the SOI material allows the exploitation of a resonant optical cavity for an increased quantum efficiency. The optical reflection and absorption in such a cavity can be controlled by electron injection resulting in optical modulation. In addition to photodetectors in SOI films, therefore, an optical modulator will be described.

4.1 Photodiodes in SOI

A lateral PIN photodiode in a thin SOI layer was investigated first by Colinge [242]. The structure of this lateral PIN photodiode is shown in Fig. 4.1. The SOI layer was made by laser-recrystallization. The thickness of the silicon film was (440 ± 50) nm at the end of the process. The buried oxide layer had a thickness of 1 µm and the oxide on top of the silicon thin-film layer was 0.5 µm thick. The width of the intrinsic region was designed to be 7 µm. The N^+ and P^+ regions were each 8 µm wide. The carrier lifetime in the laser-recrystallized SOI film was found to be approximately 8 µs and the dark current of the lateral PIN photodiode was reported to be 0.1 pA per 1 µm junction length for a reverse voltage of 5 V. This value for the dark current is large and is due to the fabrication of the SOI layer by the laser-recrystallization technique,

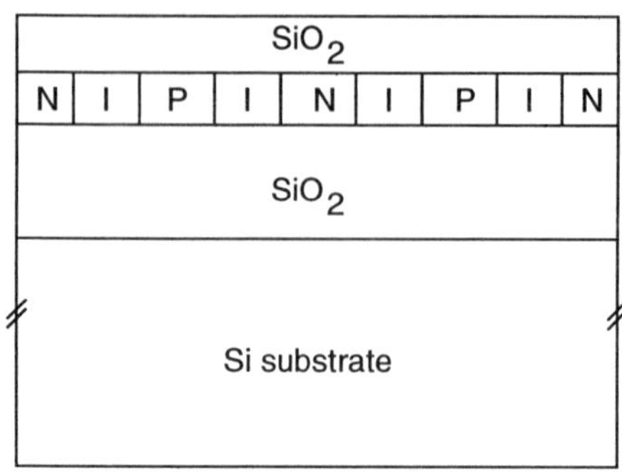

Fig. 4.1. Cross section of a lateral PIN photodetector in a thin-film SOI layer [242]

which results in a lower thin-film quality than bonded and etched-back SOI (BESOI). Due to the low thickness of the Si film and the buried and top oxide layers, interference effects were present in the spectral photoresponse. The maximum quantum efficiency of approximately 80 % was found at an interference maximum for green light. Due to the low thickness of the Si film, the quantum efficiency decreased strongly for longer wavelengths with smaller absorption coefficients. The quantum efficiency for $\lambda = 780$ nm for instance was approximately 9 % due to an interference maximum, whereas for $\lambda = 850$ nm the quantum efficiency was reduced to only approximately 2 %. No results for the transient response of the lateral PIN photodiode were reported.

Ghioni et al. [243] described a lateral PIN photodetector, which is compatible with trench isolation VLSI technology. A larger thickness of (2.8 ± 0.5) μm was chosen for the SOI layer in order to obtain a larger quantum efficiency than mentioned above. The resistivity of the N-type BESOI layer was in the range of 10 to 15 Ωcm, which corresponds to a doping concentration in the range of 4.5×10^{14} cm^{-3} to 3.0×10^{14} cm^{-3}. The thickness of the buried oxide was 1 μm. Due to the thickness of the SOI layer (2.8 μm) the trench isolation technique as shown in Fig. 4.2 was used. The interdigitated PIN detector biased at 3.5 V attained −3dB bandwidths in excess of 1 GHz with $\lambda = 840$ nm for a finger width of 2 μm and for finger distances of less than 4 μm. Photodetectors located in different positions on the wafer exhibited slightly different bandwidths (± 0.2 GHz), due to the thickness variations of the SOI layer. The dark current was less than 10 pA at a bias of 5 V for a photodetector area of 75×75 μm^2. The responsivity was 0.048 A/W ($\eta = 0.071$, quantum efficiency = 7.1 %) for a finger spacing of 1.5 μm and 0.091 A/W ($\eta = 0.135$) for a finger spacing of 7.0 μm. A ± 15 % fluctuation was observed for these values due to the thickness variation of the SOI film over the wafer area. For the smaller finger spacing a larger amount of the incident light is reflected by the aluminum contacts and the quantum efficiency is lower than for the larger finger spacing.

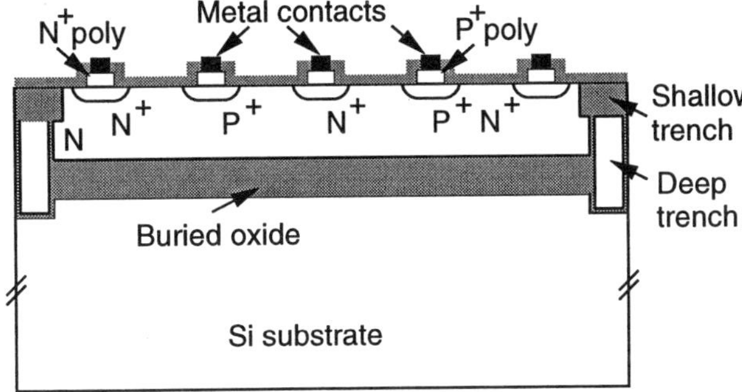

Fig. 4.2. Cross section of a trench isolated lateral PIN photodetector in a SOI layer [243]

Ghioni et al. [243] reported a calculated sensitivity of −22.5 dBm for a photoreceiver comprising a MOS transimpedance preamplifier and the lateral PIN photodetector in SOI with $R = 0.09$ A/W at a bit-error rate of 10^{-12}.

In order to increase the responsivity, interference effects were investigated. The stack consisting of the top passivating oxide, the SOI layer, and the buried oxide layer forms a Fabry–Perot cavity, with the passivating oxide as the top reflector and the buried oxide as the bottom reflector. In such a way the lateral PIN photodetector can be considered as a resonant photodetector that can be tuned to the desired wavelength in order to maximize the responsivity. The tuning can be done by carefully adjusting the thicknesses of the cavity layers. The conditions for a maximum responsivity are for normal incidence (with l, m and n being integer numbers) [243]:

(i) passivating oxide layer: $d_{\mathrm{PO}} = l \cdot \lambda/(2\bar{n})$;
(ii) SOI layer: $d_{\mathrm{SOI}} = m \cdot \lambda/(2\bar{n})$;
(iii) buried oxide layer: $d_{\mathrm{BO}} = n \cdot \lambda/(4\bar{n})$.

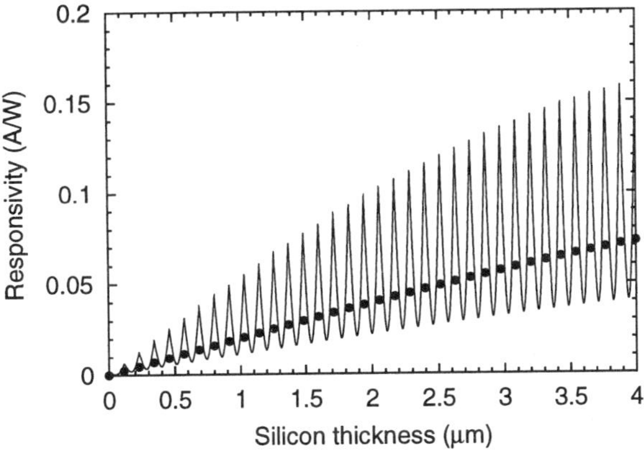

Fig. 4.3. Responsivity of a trench isolated lateral SOI PIN photodetector with a light-sensitive area of 50 % and with a half-wavelength top oxide layer versus SOI layer thickness. The *dotted curve* belongs to a nonresonant photodetector [243]

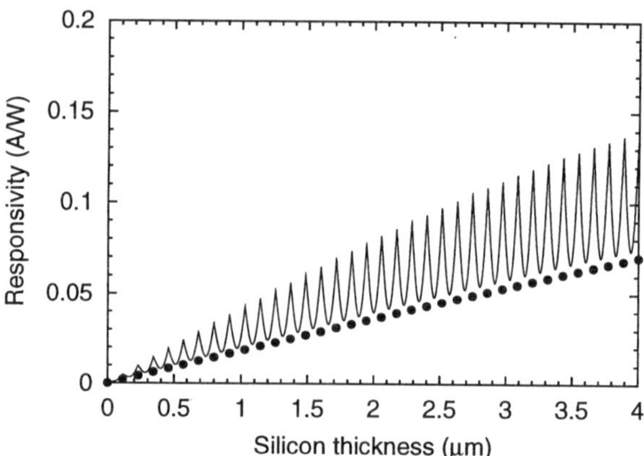

Fig. 4.4. Responsivity of a trench isolated lateral SOI PIN photodetector with a light-sensitive area of 50 % and with a quarter-wavelength top oxide layer versus SOI layer thickness. The *dotted curve* belongs to a nonresonant photodetector [243]

The responsivity of such a resonant structure depends strongly on the thickness of the active layer d_{SOI} (see Fig. 4.3). In the best case, the responsivity of such a resonant structure is more than doubled, but in the worst case, the responsivity of the SOI detector is reduced below that of a nonresonant detector (where no interference effects are taken into account and only reflection at the top oxide to SOI layer interface is considered in addition to the absorption of the SOI layer). The silicon layer incremental thickness between the best case and the worst case was reported to be only 50 nm. The responsivity for $d_{\mathrm{SOI}} = 2.74\,\mu\mathrm{m}$, for instance, is $0.125\,\mathrm{A/W}$ ($\eta = 0.185$), whereas it is only $0.03\,\mathrm{A/W}$ ($\eta = 0.044$) for $d_{\mathrm{SOI}} = 2.79\,\mu\mathrm{m}$ in comparison to $0.055\,\mathrm{A/W}$ ($\eta = 0.081$) for a nonresonant detector with $d_{\mathrm{SOI}} = 2.79\,\mu\mathrm{m}$. Therefore, an extremely good thickness control of less than $\pm\,0.5\,\%$ is required for a typical active layer thickness of a few micrometers. An alternative design was suggested by Ghioni et al. [243], if such a strict control cannot be guaranteed:

(i) passivating oxide layer: $d_{\mathrm{PO}} = (2l+1)\cdot\lambda/(4\bar{n})$;
(ii) SOI layer: $d_{\mathrm{SOI}} = m\cdot\lambda/(2\bar{n})$;
(iii) buried oxide layer: $d_{\mathrm{BO}} = (2n+1)\cdot\lambda/(4\bar{n})$.

Then a maximum responsivity of $0.105\,\mathrm{A/W}$ ($\eta = 0.155$) for $d_{\mathrm{SOI}} = 2.74\,\mu\mathrm{m}$ is obtained and the minimum responsivity of $0.057\,\mathrm{A/W}$ ($\eta = 0.084$) for $d_{\mathrm{SOI}} = 2.79\,\mu\mathrm{m}$ (see Fig. 4.4) is a little larger than the responsivity of $0.055\,\mathrm{A/W}$ for the nonresonant detector. A variation by a factor of approximately two, however, for the responsivity and therefore for the quantum efficiency remains for this design, also, because a thickness tolerance of less than 50 nm for a SOI layer thickness of several micrometers can hardly be obtained by BESOI.

MSM photodetectors were also fabricated in SOI layers to reach very high frequencies. MSM photodetectors on thin-film SOI [187, 188, 191] eliminate the slow contributions from the depth to the photocurrent for red and IR light. Regions in Si with low electric fields (compare with Fig. 3.68) are avoided due to the SOI material supporting a fast transient response. A pulse response time of 3.2 ps FWHM for $\lambda = 780$ nm was measured for a SOI-MSM detector with a SOI layer thickness of 0.1 μm [191]. With the optimistic estimate $f_{3dB} \approx 0.45/\tau_{FWHM}$, the record of 140 GHz for Si detectors was reported [191]. This large bandwidth is at the expense of the responsivity, of course, which was only 12 mA/W for $\lambda = 633$ nm in this case [191]. Continuous-wave operation of lasers for ultrashort pulse generation of down to 0.15 ps is not possible and the responsivity for $\lambda = 780$ nm was not determined in [191]. It is, however, much smaller than that measured for $\lambda = 633$ nm with a CW laser diode. A backside reflector was shown to enhance the responsivity [192].

Another way to improve the field distribution in a MSM detector without decreasing the responsivity was chosen in [244]. Trenches were etched into a 6 μm thick SOI layer (Fig. 4.5).

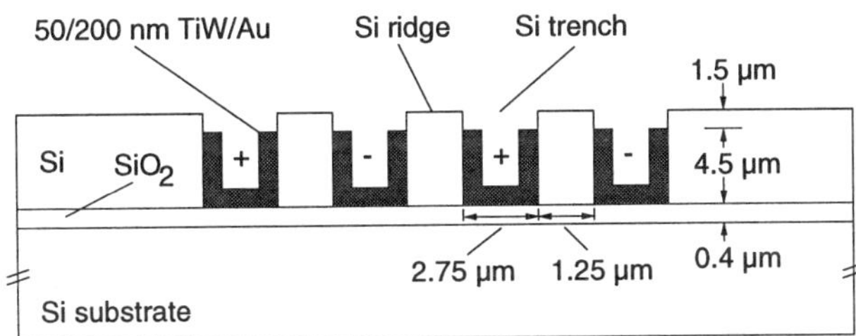

Fig. 4.5. MSM photodiode with a trench structure [244]

The Si ridges between the trenches had a width of 1.25 μm. A homogeneous field distribution is obtained down to a depth of 6 μm. TiW/Au electrodes contacted the sidewalls of these Si ridges. A responsivity of 0.12 A/W ($\eta_e = 0.19$) and a pulse response time of 23 ps FWHM were measured for this advanced MSM detector for $\lambda = 790$ nm and $U_{MSM} = 5$ V [244].

Another approach to an even more efficient SOI photodetector will be described here. Levine et al. [245] suggested increasing the quantum efficiency by the use of a rough surface of the SOI layer (see Fig. 4.6). Thereby, an absorption enhancement can be obtained due to random scattering of the light at the rough SOI layer to top oxide interface. A large part of the incident light can be trapped inside the SOI layer with its high index of refraction by total internal reflection. The effect of an increased light intensity in a medium with complicated nonspherical and nonplanar surface was described earlier in [246]. The absorption enhancement is illustrated in Fig. 4.6 and can be explained thus: The critical angle $\theta_c = \arcsin(\bar{n}_2/\bar{n}_1) = 23.1°$ for total reflection is rather small, because $\bar{n}_2 = \bar{n}_{SiO_2} = 1.45$ and $\bar{n}_1 = \bar{n}_{Si} = 3.7$ differ strongly. Therefore, a large amount of the incident light is reflected at the silicon to buried oxide interface and there is a high probability that another

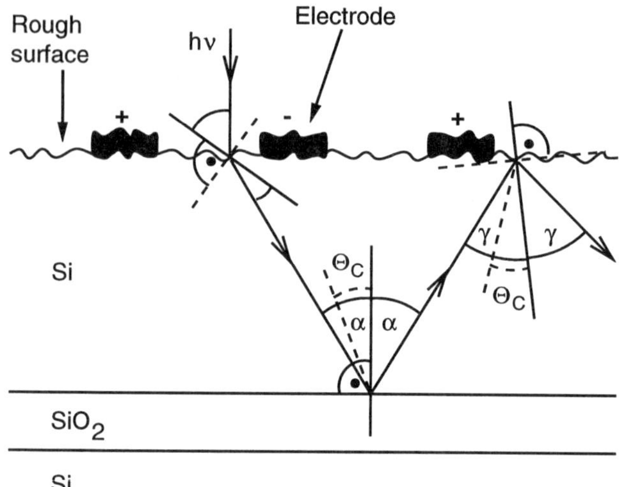

Fig. 4.6. Cross section of an absorption-enhanced MSM photodetector in a SOI layer [245]

reflection is possible at the surface of the SOI layer. Accordingly, the path of the light in the SOI layer is much longer than the thickness of the SOI-layer and the absorption enhancement results.

Levine et al. [245] used a SIMOX (Separation by IMplantation of OXygen, [241]) wafer with a Si layer thickness of 0.4 µm and increased this small SOI layer thickness by epitaxially growing 4 µm silicon on top. The SOI layer obtained was P-type doped to a level of approximately $10^{15}\,\mathrm{cm}^{-3}$. The 5-inch wafers were then roughened by etching off approximately 1 µm using a helicon high density plasma source with the highly Si/SiO_2 selective gas mixture of 69 % HBr, 22 % He, 5.5 % O_2 and 3.5 % SF_6 at a pressure of 5 mTorr. The remaining SOI layer thickness was approximately 3.4 µm. The surface features were determined by a scanning electron microscope (SEM) to have a size of approximately 0.1 µm, which is the scale of roughness needed to optimally randomize the scattered light with a vacuum wavelength of 880 nm ($\lambda_{\mathrm{Si}} = \lambda_0/(2\bar{n}_{\mathrm{Si}}) = 0.12\,\mu\mathrm{m}$, $\bar{n}_{\mathrm{Si}} = 3.7$). Metal-Silicon-Metal (MSM) photodetectors with an area of $100 \times 100\,\mu\mathrm{m}^2$ were then processed with a metal finger width w of 1.6 µm and a finger spacing s of 4.4 µm, i. e. with a light sensitive portion $f = s/(s+w) = 0.73$. The detectors with a roughened SOI layer surface showed a responsivity of 0.19 A/W ($\eta = 0.268$) at a bias of 5 V compared to a responsivity of 0.07 A/W ($\eta = 0.099$) for reference detectors with a smooth SOI layer surface. The reference detector, however, had a SOI layer thickness of 4.4 µm. The corrected responsivity for a SOI photodetector with a thickness of 3.4 µm was given as 0.053 A/W. The quantum efficiency of the SOI MSM photodetector with the roughened surface for $\lambda = 880\,\mathrm{nm}$ was 26.8 % compared to 7.4 % of the directly comparable detector with the smooth surface and the same SOI-layer thickness. The rough surface thus resulted in a factor of 3.6 for the absorption enhancement.

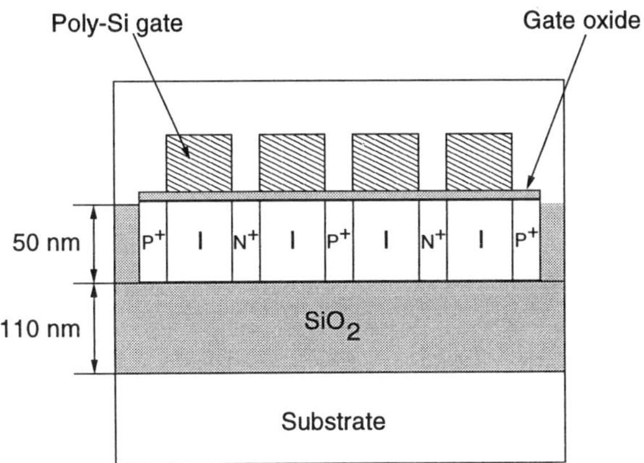

Fig. 4.7. Cross section of an avalanche photodiode in a thin-film SOI layer [247]

The rise and fall times of the photocurrent of the SOI MSM photodetector with the rough silicon surface were approximately 0.2 ns for a bias of 3 V and for a square wave light modulation of 1 GBit/s. The bit-error rate was not reported. One weak point not mentioned in [245], therefore, may be a large dark current due to the rough Si surface, which may result in peaks in the electric field.

It can be concluded that despite several attempts to increase the quantum efficiency, the main disadvantage of SOI photodetectors is the low quantum efficiency or responsivity. Therefore, internal amplification of the photocurrent is highly desirable.

The responsivity of a lateral PIN photodiode in a thin-film SIMOX layer was increased by the exploitation of the avalanche effect for internal amplification of the photocurrent [247]. Figure 4.7 shows the structure of this photodetector for application in local area networks (LANs) such as the Gigabit Ethernet or Fibre Channel with a wavelength in the range of 780–850 nm. The N^+ and P^+ regions are 1 μm wide and the I-regions were 0.6 to 3.0 μm wide. The N^+-polysilicon gates with a thickness of 200 nm make the 50 nm thick silicon layer with an N-type doping concentration of 3×10^{17} cm^{-3} fully depleted at a supply voltage of 2.0 V. The polysilicon gates also can be used to adjust the dark current of the photodetector. On top of the gates, several micrometers of oxide were deposited. The capacitance of a 100-square μm photodiode was only 0.2 pF due to the thin SOI layer and the 110 nm thick buried oxide.

A photodetector with a size of 60×60 μm^2 for the use together with an optical fiber having a 50 μm core diameter was fabricated in a 0.25 μm CMOS technology. Without an antireflection coating, a responsivity of 0.4 A/W at $\lambda = 850$ nm was achieved at a reverse bias of only 2.0 V. A high electric field of 1–2×10^5 V/cm exists in the I-regions at the P^+ side for this voltage bringing about an avalanche effect. The photocurrent is amplified due to the rise of the ionization rate in Si under a high electric field of more than 1×10^5 V/cm [248] achieving a responsivity of the 50 nm thick photodetector similar to that of a thick vertical PIN photodiode. A dark current of $0.2\,\text{pA}/\mu\text{m}^2$ at a polysilicon gate bias of $-0.8\,\text{V}$ was reported for the photodetector. The amplifier used together with the lateral PIN avalanche photodiode in the OEIC for LAN applications will be discussed in Sect. 12.4.5 (Fig. 12.44).

4.2 Phototransistors in SOI

A MOSFET fabricated on SOI can be used as a phototransistor exploiting its operation in the lateral bipolar mode [249] with the gate floating (Fig. 4.8). In this floating gate case, a depletion region is induced under the gate, and this depletion region separates the photogenerated electron–hole pairs effectively. The electrons drift towards the front gate and are then swept to the drain along the silicon/gate oxide interface. The holes drift towards the buried oxide and accumulate in the body, as shown in Fig. 4.8. This results in an increase in the body potential and leads to the bipolar turn-on [250].

The barrier between source and body for the electrons near the silicon/gate oxide interface is given by $\Psi_{\text{BE,f}} = \Phi_{\text{bi}} - V_{\text{BS}} - \Phi_{\text{c}}$ where Φ_{c} is the potential difference between the front and back gate. Φ_{bi} is the built-in potential of the source/body junction. V_{BS} can be regarded here as photo-induced equivalent body-source bias. The barrier of the holes accumulated in the neutral body on the other hand is only $\Psi_{\text{BE,b}} = \Phi_{\text{bi}} - V_{\text{BS}}$. The device, therefore, behaves like a heterojunction bipolar transistor [250,251]. Due to the heterojunction behavior, Ψ_{BE} is not constant along the source/body

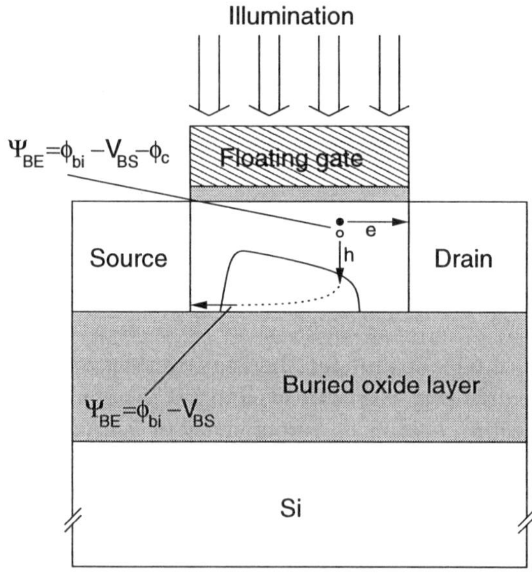

Fig. 4.8. Cross section of a phototransistor in a thin-film SOI layer [249]

junction and it depends on the light intensity [249]. Like heterojunction transistors, the SOI NMOSFET LBJT has a high current gain.

A special test structure with a body contact was used to measure the ratio of drain to body current $I_D/I_B = \beta$, which is the current gain. A value of 100 was obtained for a channel length L of 0.7 µm. For $L = 0.1$µm, β was found to be 4000. The floating gate causes the advantage that the dark current is only composed of the leakage current of the SOI NMOSFET LBJT [249].

The SOI photodetector was fabricated on SIMOX wafers with a silicon film thickness of 200 nm and a buried oxide thickness of 110 nm. The process used was a typical N^+-polysilicon gate non-fully depleted SOI CMOS process [252]. No addition of process steps was necessary for the fabrication of the SOI NMOSFET LBJT. The final silicon film thickness and gate oxide thickness were 165 and 8.4 nm, respectively. The gate length was varied. For a $W/L = 10$ µm/0.4 µm device, the photocurrent saturates at $V_{DS} \approx 0.1$ V, which is a very low voltage required to operate the device as a photodetector. As a result, ultra low power operation in possible. The photocurrent is approximately proportional to the light intensity. The photocurrent depends on the channel length L and increases with decreasing L as the current gain increases with decreasing L. The dark current was about 7 pA for a 10 µm wide device with $L = 0.4$ µm and 60 pA for $L = 0.2$ µm. The spectral response showed a maximum near 500 nm. A responsivity R of 289 A/W was reported for the SOI NMOSFET LBJT with $L = 0.1$ µm. For $L = 0.7$ µm, $R \approx 10$ A/W and for $L = 0.4$ µm, $R \approx 20$ A/W, respectively, were reported.

Let us summarize: as the gate length is reduced, the responsivity of the floating gate SOI NMOSFET LBJT increases. Therefore, this device is scalable and capable of delivering better performance as the minimum feature size of CMOS processes decreases. This is contrary to conventional optical sensors that produce lower signals when the sensor dimensions are reduced.

Such an advantageous behavior of SOI NMOSFET photodetectors also was reported for a negative gate bias, i. e. for the front channel being in accumulation [253]. The measured response time for optical pulses was 19 µs for this SOI NMOSFET photodetector with a channel length of 0.7 µm for $V_{GS} = -2$ V and $V_{DS} = 2$ V. This long response time, however, does not seem very reliable since large parasitic capacitances were reported in the external circuit for the measurement of the transient response [253].

4.3 Optical Modulators in SOI

SOI makes it possible to integrate a Fabry–Perot cavity as already mentioned above. Such a cavity on SOI can be used as an optical modulator for infrared light. Such a modulator was realized by Xiao et al. [254]. The schematic cross section of the SOI modulator is shown in Fig. 4.9.

Infrared laser light with a wavelength of 1.3 µm can be introduced into the modulator cavity via a single-mode fiber. The light with this wavelength

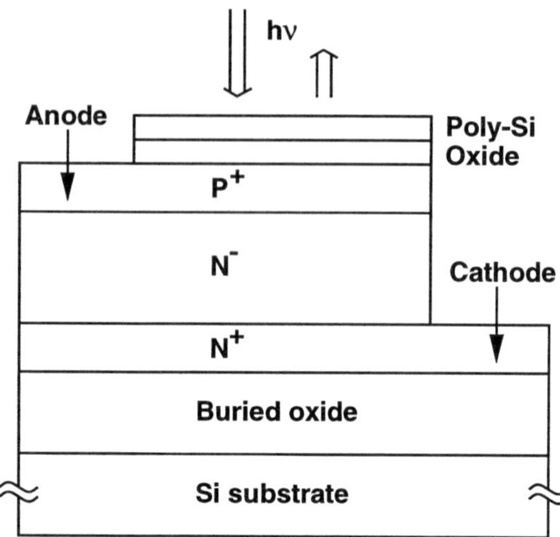

Fig. 4.9. Cross section of a SOI infrared optical modulator [254]

is practically not absorbed in silicon and is partially reflected back and forth inside the cavity. The cavity can be tuned to resonance by optimizing the thickness of the different layers in the modulator structure. The reflectance is minimized at resonance. The modulator electrically consists of a PIN diode. This PIN diode is biased in order not to conduct current in the case of resonance, i. e. in the case of minimum reflectance. Forward biasing the PIN diode results in a large concentration of carriers inside the 'intrinsic' (N^-) region. The refractive index $\bar{n}$ is known to depend on the concentration of free carriers [255]:

$$\Delta\bar{n} = -\frac{\lambda^2 q^2 n}{8\pi^2 \epsilon_0 c^2 \bar{n} m_\mathrm{e}}. \tag{4.1}$$

The change in the refractive index $\Delta\bar{n}$ is proportional to the concentration of electrons n. According to this change, the phase of the laser light in the cavity is varied and the Fabry–Perot cavity is shifted to off-resonance. In such a way, the phase modulation due to carrier density modulation is converted to an intensity modulation of the reflected laser light.

Xiao et al. [254] reported a 40 % modulation depth for the modulator structure shown in Fig. 4.9 for laser light with the wavelength of 1.3 μm. The current density in the PIN diode was switched between 0 and 2.25 A/cm^2 for this modulation depth.

5. Silicon Power Devices

After the discussion of micrometer and sub-micrometer silicon technology with respect to its applicability in the field of optoelectronics, optoelectronic power devices, which can have a diameter of more than 100 mm and which fill complete silicon wafers, are described in this chapter in order to cover the entire field of silicon optoelectronics.

5.1 Light-Activated Thyristors

Thyristors are commonly called semiconductor-controlled rectifiers (SCRs). They are power devices, which exhibit a bistable characteristics in the forward direction and can be switched from a high-impedance state, low-current OFF-state into a low-impedance, high-current ON-state. In order to obtain the OFF-state again, the anode current has to become smaller than the so-called holding current, which happens, for instance, during the zero-crossing of the voltage in AC power lines. The detailed device principles and the first working thyristor devices were reported by Moll et al. [256]. Thyristors are now available with current ratings of more than 5000 A and voltage ratings extending above 10 000 V. The interested reader can find a comprehensive treatment on the operation and fabrication technology of thyristors in [257,258].

The schematic cross section of a light-activated thyristor is shown in Fig. 5.1. The main differences compared to an electrically triggered thyristor are the lack of a gate electrode and a central gap in the cathode metallization in order to enable the penetration of the light. The gaps in the N2 regions are cathode shorts, which are known from conventional thyristors to improve the dV/dt capability. The equivalent circuit on a transistor device level is given in Fig. 5.2, where the symbol of a light-activated thyristor is also included.

Light-triggered thyristors, which are also called Light-Activated Switches (LASs), possess several advantages compared to electrically triggered thyristors. The LASs provide a good isolation technique between the high-voltage power circuits and the low-voltage electronic circuits, which have to provide the firing pulses for the gates of the power switches. The optoelectronic trigger circuits for LASs are much simpler than electrical trigger circuits and consist of fewer and cheaper components. LASs are furthermore immune against

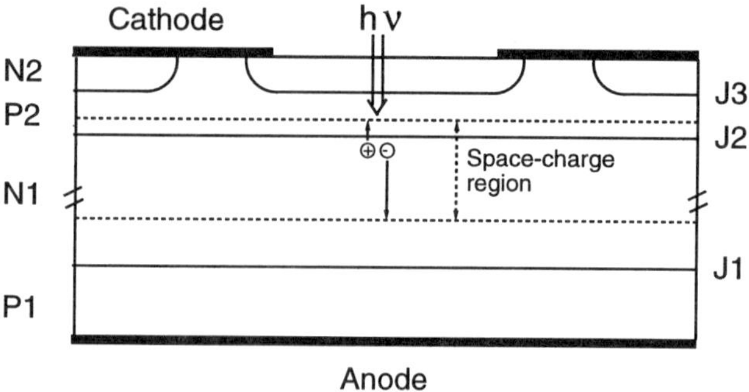

Fig. 5.1. Simplified device structure of a light-activated thyristor [259]

electrical noise. LASs, therefore, make a system like a high-voltage direct current (HVDC) transmission system more reliable.

The optoelectronic triggering of a LAS is due to electron–hole pair generation in the space-charge region of the reverse-biased junction J2 (see Fig. 5.1). These electron–hole pairs will be separated by the electric field within several nanoseconds, which can be considered as instantaneously compared to the turn-on time of typical thyristors which is of the order of microseconds. A thyristor can be switched on in forward direction only, i. e. when the anode is positive with respect to the cathode. Therefore, the photogenerated holes are transported to the P2 region, whereas the photogenerated electrons are transported to the N1 region. Both regions are supplied with equal amounts

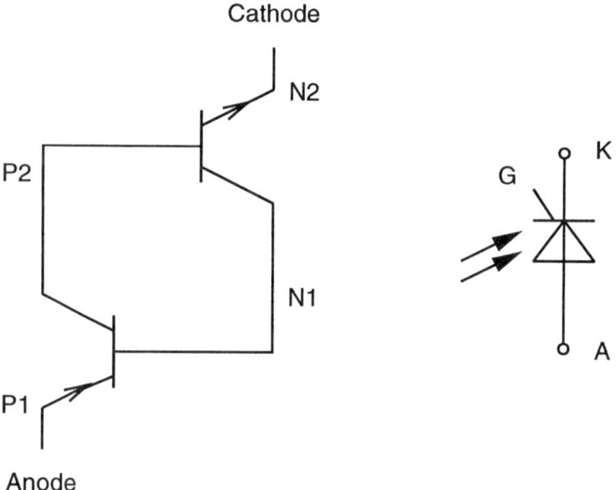

Fig. 5.2. Equivalent circuit of a light-activated thyristor on a transistor level. Also shown is the symbol of the light-activated thyristor, where the gate electrode is drawn in order to keep the symbol different from that of a photodiode

of majority carriers almost without any time delay to the carrier generation, so that at the moment of light incidence the anode current increases abruptly by the amount of the photocurrent. This photocurrent, however, is amplified in the two-transistor P-N-P-N structure (see Fig. 5.2) with regenerative action. The holes being transported to the P2 region make the base of the NPN transistor more positive and its collector current consequently increases, which is the base current of the PNP transistor. At the same time, the electrons being transported to the N1 region make the base of the PNP transistor more negative. In turn, its collector current increases, which is the base current of the NPN transistor. Therefore, the anode current increases after the delay caused by the transit time of the injected minority carriers through the base regions.

Analogously to the threshold gate current in the case of electrical triggering, the LAS possesses a light threshold level for turn-on in the case of optical triggering. This light threshold level and the minimum photocurrent, which belongs to this value, correspond to a stationary current in the thyristor, for which the sum of the common-base current amplification factors α for the two transistors shown in Fig. 5.2 is equal to one. Therefore, the regenerative action of the feedback current will start a short time after the anode current has exceeded the value of this stationary current and lead to a rapid increase of the anode current. The thyristor switches to its ON-state. It should be mentioned explicitly that the gate electrode of a LAS is not used similarly to the base of a bipolar phototransistor in the simplest case of its usage.

The following circumstances had to be considered for the development of light-activated thyristors: (i) The optical output power of cheap semiconductor lasers or light emitting diodes (LEDs) for firing was limited to several milliwatts. (ii) The core diameter of optical fibers usually was only approximately 100 μm. (iii) Light with a wavelength $850\,\text{nm} < \lambda < 1.0\,\mu\text{m}$ has to be used for the optical triggering in order to generate sufficient electron–hole pairs at the junction J2.

The first two aspects actually support each other, because a very small light power of less than 1 mW is sufficient to turn a LAS on, when the area of light incidence, which determines the initial turn-on area, is less than $10^{-2}\,\text{mm}^2$. Therefore, the optical power density in the initial turn-on area can be very high. Once the light power turns on the initial area, the regenerative action of the device will enlarge the turned-on area, and finally the full device will be on. Already when the anode current prevails over the photocurrent, the light power can be turned off without switching the anode current off. The LAS will switch fully to the ON-state. In recent years high power laser diodes have been developed and plastic optical fibers with larger core diameter are now available. The first two aspects, therefore, are no more important for the development of LASs.

The third aspect has a consequence on the construction of some LAS devices: For thyristors capable of blocking several thousands of volts, the PN

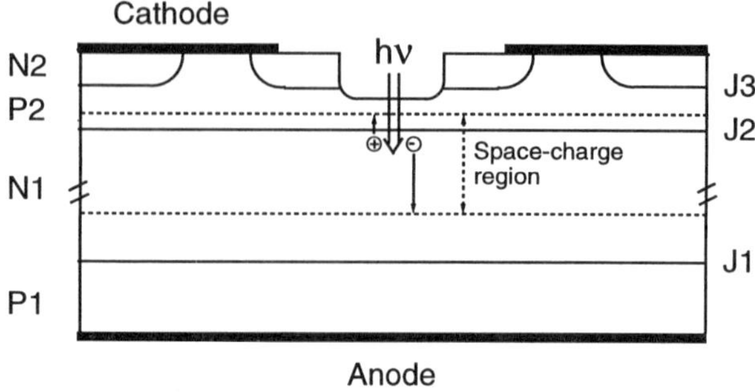

Fig. 5.3. Simplified cross section of a light-activated thyristor for very high-voltage applications

junction J2 can be at a depth of approximately 100 μm [260]. The penetration depth of light with a wavelength in the above mentioned range is too small for such a junction depth and a central groove is etched into the silicon and passivated in an appropriate manner [260, 261]. The cross section of such a high-voltage light-activated thyristor is shown in Fig. 5.3.

The light-activated thyristor reported in [260] having a diameter of 115 mm and being capable of switching 6 kV and 2000 A, for instance, needs a light-triggering power of only 10 mW. This device has a dV/dt capability of 3000 V/μs, a dI/dt capability of more than 300 A/μs and a turn-off time of less than 400 μs.

5.2 Light-Activated Triac

Triacs are bi-directional thyristors [262, 263]. They are power devices which exhibit bistable characteristics in forward and reverse directions. They are therefore useful in AC applications in order to use both the positive and negative half-waves.

The triac structure is more complicated than the structure of a conventional thyristor. There are two more regions (N3 and N4) in addition to the P1-N1-P2-N2 structure of a thyristor (see Fig. 5.4).

The triac actually can be considered to consist of two thyristors in anti-parallel. To understand the triggering for the forward direction, i. e. in the

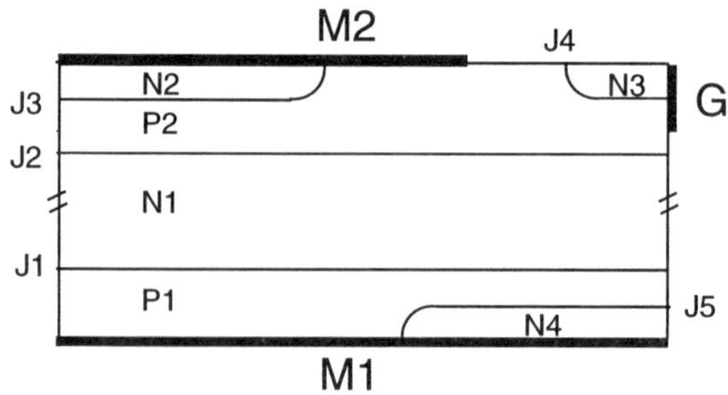

Fig. 5.4. Simplified device structure of a triac

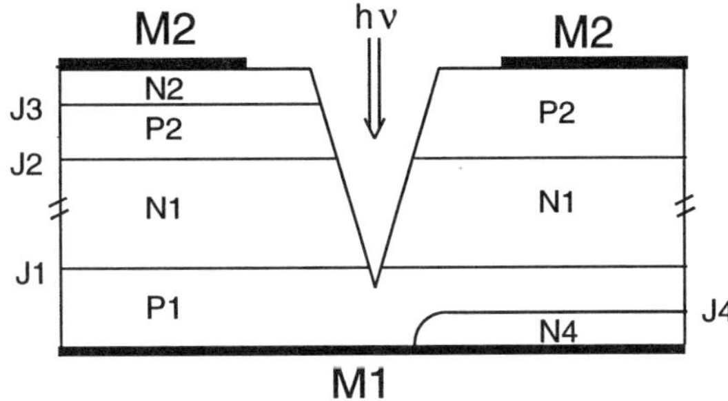

Fig. 5.5. Simplified device structure of a light-activated triac [264]

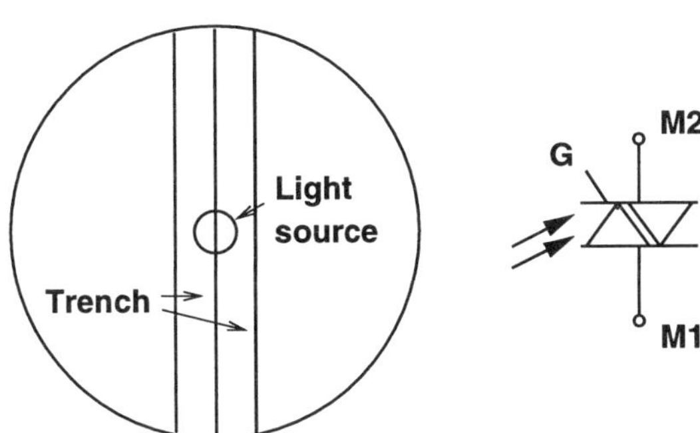

Fig. 5.6. Simplified top-view [264] and symbol of the light-activated triac. The gate electrode is drawn in order to keep the symbol different from that of a diac

case that the electrode M1 is positive with respect to M2, is simple: the gate current (small positive voltage at G) is supplied through the P2 region and the main current is carried through the left side of the device, the P1-N1-P2-N2 section. The behavior is identical to that of a conventional thyristor. To understand the triggering for the reverse direction, i. e. in the case that the electrode M1 is negative with respect to M2 is also easy: the gate current (small negative voltage at G) is supplied through the N3 region and the main current is carried through the right side of the device, the N4-P1-N1-P2 section. For the reverse direction the thyristor being complimentary to the one shown in Fig. 5.1 is active.

When light is used for the triggering, the N3 region and the gate electrode are not necessary. The resulting cross section of a light-activated triac is shown in Fig. 5.5. The top view of the light-activated triac can be seen in Fig. 5.6. For triggering in the forward direction, the light has to reach the N1-P2 junction, as explained above for the light-activated thyristor. The P1-N1-P2-N2 structure is then turned on like the light-activated thyristor described above. For triggering in the reverse direction, the light should reach the P1-N1 junction. Then the N4-P1-N1-P2 structure, which corresponds to the conventional thyristor shown in Fig. 5.1, is triggered analogously to the light-activated thyristor described above.

A 500-A, 1200-V, 40-mm diameter light-triggered triac (LTTriac) based on the principle displayed in Fig. 5.5 has been realized [264]. The minimum light power of a light-emitting diode placed in the V-shaped groove for triggering was 15 mW. An ON-state voltage of less than 1.6 V, a commutating $\mathrm{d}V/\mathrm{d}t$ capability of 100 V/µs and a commutating $\mathrm{d}I/\mathrm{d}t$ capability of 50 A/µs was obtained. These high commutating capabilities were reported to be due to the groove (see Fig. 5.6). The groove suppresses a mutual interaction between the two thyristor sections. The integration of two thyristors on a single chip results in only half of the structure being used at any time. Therefore, there is no advantage in the utilization of the die area compared to two light-activated thyristors. The main advantages of the LTTriac are its good matching of the output characteristics as well as the elimination of one package and of additional external connections.

6. SiGe Photodetectors

SiGe alloys allow the integration of infrared detectors on Si. The addition of Ge to Si also increases the absorption coefficient in the spectral range from 400–1000 nm, allows a reduction of the detector thickness, and, therefore, enables faster detectors than with pure Si. To exploit the advantages of SiGe alloys for Si-based photodetectors, however, the problems associated with the lattice constant mismatch and energy band discontinuities have to be understood. This chapter gives an overview of these aspects and describes several examples of SiGe photodetectors.

6.1 Heteroepitaxial Growth

The growth and physical properties of $Si_{1-x}Ge_x$ heteroepitaxial layers on Si have been investigated for more than 20 years [265–269] finally producing a technology which is entering high-volume and large-scale manufacturing of heterojunction bipolar transistors (HBTs) [270] and SiGe-HBT-BiCMOS circuits [271, 272]. Optoelectronic applications of $Si_{1-x}Ge_x$/Si devices also have been developed. Medium-wavelength (1.3–1.55 μm) photodetectors for optical communication [273–275], 2–12 μm infrared photodetectors for two-dimensional focal plane arrays for thermal imaging and night vision [276–278], optical waveguides [279], and infrared light emitters for chip-to-chip optical interconnects [280, 281] have been suggested.

On the way to introducing optoelectronic devices into VLSI and ULSI to obtain a silicon-based *superchip* [282, 283], however, the following significant limitations of the SiGe/Si material system have been discovered [284]:

(i) Due to the large lattice constant of Ge, which is about 4.2 % larger than that of Si, a critical thickness of $Si_{1-x}Ge_x$ layers on Si for thermal stability against misfit dislocation generation due to lattice mismatch exists [285]. This critical thickness is only about 15 nm for a Ge fraction x of 0.2 with a resulting bandgap change of approximately 150 meV [286]. Due to this low critical thickness, the quantum efficiency of normal incidence photodetectors relying on photogeneration across the $Si_{1-x}Ge_x$ bandgap is severely limited.

(ii) When a $Si_{1-x}Ge_x$ layer with a thickness greatly exceeding the critical thickness is grown directly onto Si, the layer relaxes but forms a dislocation density at the surface of order $10^{12}\,\text{cm}^{-2}$ [287]. This dislocation density reduces the carrier mobility and increases leakage currents significantly and is far too large to allow economic yields of devices in production [272]. Although relaxed $Si_{1-x}Ge_x$ buffer layers with the bulk lattice constant of the $Si_{1-x}Ge_x$ alloy on top of the Si substrate can be introduced and pseudomorphic heterostructures, which have a lateral lattice constant larger than that of Si, can then be grown on top of the buffer [288, 289], these relaxed buffers still reduce the threading dislocation density only to below $10^4\,\text{cm}^{-2}$ for $Si_{0.8}Ge_{0.2}$ [287]. This density is already comparable to many III/V systems [272]. More recently, a number of annealing steps during growth have reduced the dislocation density to $10^2\,\text{cm}^{-2}$ for $Si_{0.8}Ge_{0.2}$ and $x < 0.2$ [290]. Dislocations, however, still pose a problem for higher Ge fractions and future large-scale optoelectronic applications, therefore, remain uncertain.

(iii) The dopant diffusion during thermal processing for device fabrication often degrades the nanometer precision in the placement of dopant atoms required for heterojunction devices, which can be achieved during the initial growth of heterostructures at temperatures below 700 °C [291, 292].

(iv) Under all combinations of strain, $Si_{1-x}Ge_x$ is an indirect-bandgap material, resulting in inefficient emission of light and inefficient detection due to a small absorption coefficient compared to pure Ge, GaAs, or InGaAs, at least for $x < 0.75$. Despite attempts to create a direct bandgap material by the *zone-folding* principle in short-period Si/Ge superlattices [293], no breakthrough in such an approach has been achieved.

SiGe light emitters will be discussed in Sect. 9.2. Let us continue with light absorption of SiGe alloys (point (iv)) in the next section and with an example of an infrared Ge detector on Si exploiting the capability of absorption in Ge at 1.3 μm in Sect. 6.3. Then SiGeC for the solution of problems (iii) and (ii) will be discussed in Sect. 6.4. Multi-quantum well structures for the solution of problem (i) will be described in Sects. 6.5 and 6.6.

6.2 Absorption Coefficient of SiGe Alloys

The exchange of Si atoms by Ge atoms increases the absorption coefficient. Furthermore the bandgap reduces with increasing Ge fraction and wavelengths longer than 1100 nm can be detected. Figure 6.1 shows the absorption coefficient for Ge fractions of 20, 50, and 75 %.

It is advantageous to use SiGe detectors instead of Si detectors, when longer wavelengths, a thinner absorbing layer, a higher quantum efficiency and/or a higher speed of the detector are needed. In the following, several examples of SiGe receivers using these advantages will be described.

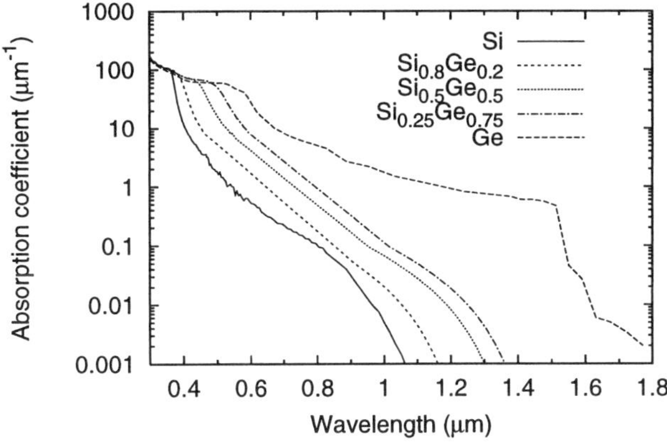

Fig. 6.1. Absorption coefficient for SiGe with different compositions [294]

6.3 SiGe IR Photodetector

The integration of Ge photodetectors on Si substrates is interesting because of the lower bandgap of Ge enabling Si-based detectors for 1.3 µm and to a somewhat less advantageous extent for 1.55 µm [295–298]. Due to the large lattice mismatch between Si and Ge of about 4 %, the most effective way to fabricate high-quality SiGe and Ge layers on Si substrates is to implement graded composition buffer layers [299]. With the increase in Ge composition, however, the surface roughness generally increases. The surface roughness is caused by the influence of strain and misfit dislocations on local growth rate in SiGe mesas. Growth on miscut Si (100) substrates reduces surface roughness and dislocation densities. Thermal mismatch between the Si and Ge expansion coefficients ($\alpha_{Si} = 3.55 \times 10^{-6}\,K^{-1}$ and $\alpha_{Ge} = 7.66 \times 10^{-6}\,K^{-1}$ at 750 °C), however, still leads to undesirable tensile stresses during cooldown from the growth temperature, which can form micro-cracks or residual tensile strain and dislocations.

High-quality Ge layers on optimized relaxed buffers (ORBs) by introducing an intermediate chemical mechanical polishing (CMP) step at $Si_{0.5}Ge_{0.5}$ in the graded structure, therefore, have been developed [300]. The CMP step liberates dislocations and creates the necessity to nucleate new dislocations. An optimized relaxation of the graded buffer results in such a way, where existing threading dislocations are more effectively used to relieve stress. Samples grown on ORB with the CMP step at 50 % Ge and final pure Ge, showed a decrease in the threading dislocation density by a factor of 5–8 to $\approx 2 \times 10^6\,cm^{-2}$. This reduction in threading dislocation density led to a record low dark current density of $0.15\,mA/cm^2$ in the Ge mesa photodiodes [275] shown in Fig. 6.2.

The PN photodiodes were fabricated in Ge layers grown on an ORB. The Ge layers were in situ doped with PH_3 and B_2H_6 to obtain N- and P-type layers, respectively. The persistent PH_3 in the UHV/CVD reactor resulted in

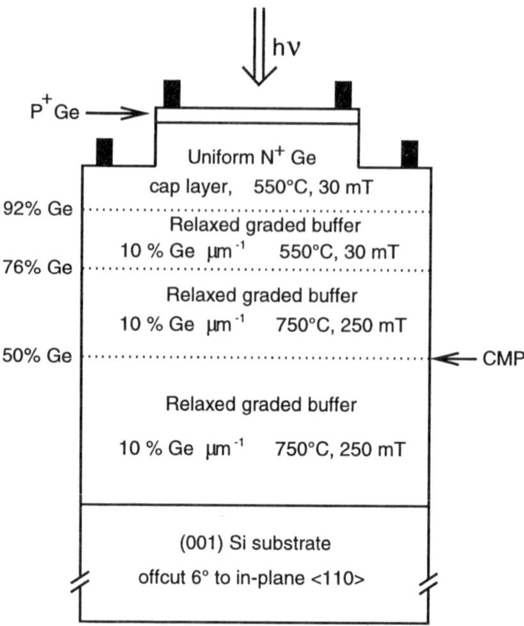

Fig. 6.2. Ge photodiode on SiGe/Si [275]

a graded PN junction. The peak P and N doping levels were $1.18 \times 10^{18}\,\mathrm{cm}^{-3}$ and $1.3 \times 10^{18}\,\mathrm{cm}^{-3}$, respectively. The contacts to the N-type Ge layer were made by etching and patterning different-sized square mesas with sides ranging from 95 to 250 μm. The contact to the P-type Ge was structured on top of the mesas. Ti/Pt contacts were used for the N- and P-type Ge. The series resistances of the Ge diodes were 55 Ω for the 95 μm diodes and 26 Ω for the 250 μm diodes. An ideality factor for $U_{\mathrm{forward}} < 0.3\,\mathrm{V}$ of 1.1 was found. For a reverse bias of $-1\,\mathrm{V}$, the reverse current density ranged from 0.15 to $0.22\,\mathrm{mA/cm^2}$ being almost two orders of magnitude lower than in [301] for Ge diodes integrated on Si substrates. At a reverse bias of $-3\,\mathrm{V}$, the 95 μm diodes had a dark current of 0.45 μA. For the 250 μm diodes a dark current of 0.7 μA was measured. A responsivity of 0.133 A/W ($\eta_e = 12.6\,\%$) was measured at the Ge photodiodes integrated on ORB SiGe/Si structures. These values are reasonable for Ge photodiodes without ARC and with a narrow depletion region of 0.24 μm because the absorption length $1/\alpha$ for $\lambda = 1.3\,\mu\mathrm{m}$ is about 1.4 μm. A better device design with ARC and PIN photodiode structure can improve the quantum efficiency considerably. The bandwidth of 2.3 GHz has been estimated for the PN-Ge diode with the 0.24 μm depletion region [275].

One aspect should be mentioned to show the difficulty of Ge photodiode integration on Si. When we add the layer thicknesses of the buffer shown in Fig. 6.2, we obtain a height of 9.2 μm plus 1.5 μm for the top N^+ and P^+ layers. The metal interconnects from the Ge photodiode to circuits in the Si substrate, therefore, will cause severe problems.

6.4 SiGeC

The basic idea to overcome the limitations of SiGe alloys listed above in Sect. 6.1 is to add carbon (C) to the $Si_{1-x}Ge_x$ structures. The lattice constant of $Si_{1-y}C_y$ is smaller than that of Si offering the possibility to compensate for the larger lattice constant of $Si_{1-x}Ge_x$ and to reduce the mismatch to Si by the growth of $Si_{1-x-y}Ge_xC_y$ layers. Substitutional carbon levels up to 5 % have been achieved by advanced growth techniques at temperatures of 400–650 °C [302,303], despite the very low solid solubility of less than 10^{-4} at all temperatures. The ability to adjust the strain in pseudomorphic layers because of the small size of the carbon atom, which compensates for the strain produced by 8–10 Ge atoms, has been verified [284]. $Si_{1-y}C_y$ layers on (100) Si have been shown to be under tensile strain, and the compressive strain for low C fractions in $Si_{1-x-y}Ge_xC_y$ on (100) Si has been reduced compared to $Si_{1-x}Ge_x$ [303]. Zero strain or even tensile strain has been observed for higher carbon fractions. Due to the reduction of strain as carbon is added in compressively strained $Si_{1-x-y}Ge_xC_y$ on (100) Si, the increase of the critical thickness as C is added (Fig. 6.3) has been demonstrated [304]. Let us conclude here that dislocation densities can be reduced due to the incorporation of small C fractions in $Si_{1-x-y}Ge_xC_y$ layers.

The bandgap of $Si_{1-y}C_y$ or $Si_{1-x-y}Ge_xC_y$ does not increase as fast with y as one might expect from the large bandgap of SiC or diamond [305,306] due to large lattice distortions near C sites. Although the bandgap increases as C is added and the strain is reduced, the bandgap does not increase as fast as it would if the strain were reduced solely by reducing the Ge fraction [284]. A $Si_{1-x-y}Ge_xC_y$ layer, therefore, has a lower strain and a larger critical thickness than a $Si_{1-x}Ge_x$ layer with the same bandgap. Fig. 6.3 shows this effect

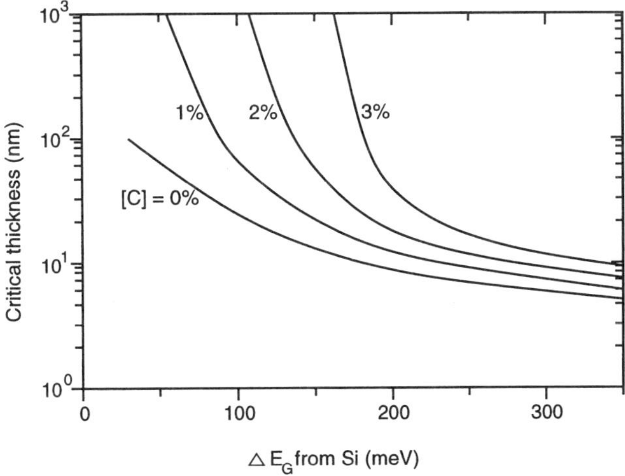

Fig. 6.3. Critical layer thickness of $Si_{1-x-y}Ge_xC_y$ for several compositions [304]

for the case that similar elastic constants for $Si_{1-x-y}Ge_xC_y$ and $Si_{1-x}Ge_x$ can be expected. For a layer with a desired bandgap of 100 meV less than that of Si, the critical thickness without C, i. e. of $Si_{0.86}Ge_{0.14}$, would be about 25 nm, for instance. With 1 % carbon, i. e. $Si_{0.82}Ge_{0.17}C_{0.01}$, the critical thickness can be increased to about 70 nm.

For the application of $Si_{1-x-y}Ge_xC_y$ layers in devices, not only the bandgap itself but also the alignment of conduction and valence bands across the heterojunction interface is important. There is little conduction-band offset in $Si_{1-x}Ge_x$/Si interfaces if relaxed $Si_{1-x}Ge_x$ buffers are not used. The absence of a conduction-band offset [307] has been found for compressively strained $Si_{1-x-y}Ge_xC_y$ on (100) Si although there is no common agreement on this topic. Tensile-strained $Si_{1-y}C_y$ grown commensurate on a (100) Si substrate, however, possesses a smaller bandgap than Si and the conduction band offset is increased allowing the fabrication of modulation-doped devices, which exploit two-dimensional electron gases with the electrons confined to the $Si_{1-y}C_y$ layer. $Si_{1-y}C_y$ infrared photodetectors, therefore, also seem feasible, although the effect of the C on the mobility of the $Si_{1-y}C_y$ layers is not yet known [284].

Many heterojunction and superlattice devices require the dopants to be placed with nanometer precision during the growth with respect to the heterojunctions, but the dopant atoms can diffuse during subsequent thermal processing for device fabrication. Especially during the integration of devices onto silicon integrated circuits, which is the final goal of $Si_{1-x}Ge_x$, $Si_{1-y}C_y$, and $Si_{1-x-y}Ge_xC_y$ research, some thermal processing is inevitable. Here the diffusion of the P-type dopant boron is the most critical problem. Small amounts of substitutional C fortunately can strongly reduce the extent of boron diffusion, especially when the boron diffusion results from Si self-interstitial injection during oxidation or implant annealing [308, 309]. Substitutional carbon is known to act as a sink for silicon self-interstitials, which are required for boron diffusion. The retarding effect of C on boron diffusion can be exploited already with carbon concentrations of 10^{19} cm^{-3}. In a HBT with a boron doped $Si_{0.795}Ge_{0.2}C_{0.005}$ base, for instance, the out-diffusion occurring in a $Si_{0.8}Ge_{0.2}$ base has been suppressed and the HBT characteristics has been considerably improved [309]. The reduction of boron diffusion in $Si_{1-x-y}Ge_xC_y$/Si superlattices also may allow the development of new optoelectronic devices.

6.5 SiGe Waveguide Photodetector

A SiGe planar photodetector for Si-based OEICs was fabricated in a SOI-layer (see Fig. 6.4) in order to increase the quantum efficiency and to enable a high frequency operation at $\lambda = 980$ nm [310].

The absorption coefficient $\alpha = 0.0065\,\mu m^{-1}$ of silicon is very small at $\lambda = 980$ nm, which leads to a 1/e-penetration depth of 154 µm. A SiGe al-

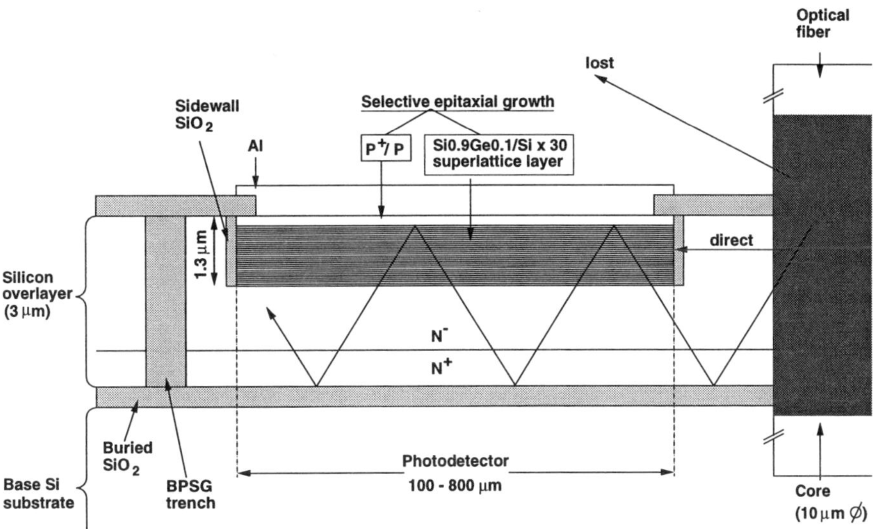

Fig. 6.4. SiGe/Si waveguide photodetector on SOI [310]

loy is known to have a larger absorption coefficient and a smaller penetration depth. For the alloy $Si_{0.9}Ge_{0.1}$, the absorption coefficient is approximately $0.016\,\mu m^{-1}$ [294] and the 1/e-penetration depth is $62.5\,\mu m$, which is still too large for a vertical light incidence in a vertical photodiode, which shall operate in the GHz range. Therefore, a vertical structure with a *horizontal* light incidence from a single-mode fiber is chosen (see Fig. 6.5) for the high-frequency capability. In order to obtain a large quantum efficiency, total reflection in a SOI waveguide is exploited. The external quantum efficiency was improved by a factor of 6 to a value of 29 % by the SOI material compared to bulk Si due to multiple reflection in a 200 μm long waveguide detector. There is, however, a complication with strain, when a "thick" SiGe layer for an efficient absorption layer is needed. The strain is caused by the large difference in the size of Si and Ge atoms, which results in a large difference in the lattice constants. In a SiGe layer thicker than the so-called critical layer thickness, the strain is released by the formation of misfit dislocations. These extended defects cause large reverse currents of PN junctions. A SiGe/Si superlattice structure with many SiGe layers each thinner than the critical layer thickness for the absorption layer, therefore, was employed to prevent a problem with strain due to the difference in the lattice constants of Si and SiGe.

The fabrication of the SiGe/Si planar photodiode in a SOI-layer (Fig. 6.4) was rather complex as a consequence of the above mentioned aspects:

(i) A bonded (100)-oriented SOI substrate with a 1.0 μm thick Si layer doped with arsenic at a concentration of $1 \times 10^{19}\,cm^{-3}$ and with a buried oxide thickness of 0.5 μm serves as the starting material for the fabrication of the SiGe/Si waveguide detector.

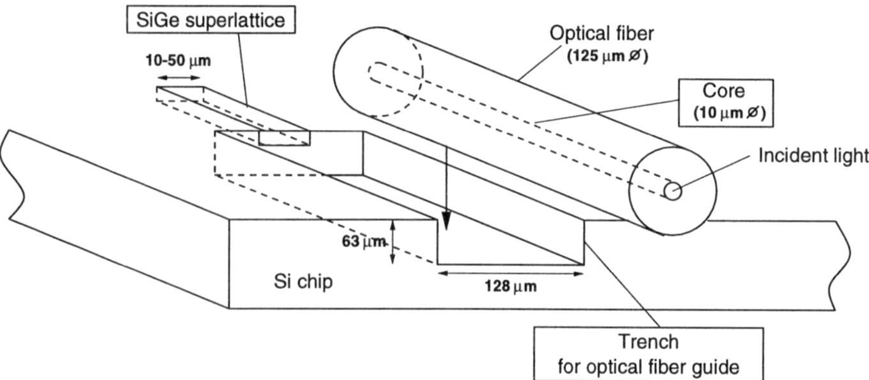

Fig. 6.5. SiGe/Si waveguide photodetector with horizontal light incidence [310]

(ii) A 2.0 μm thick N-Si layer with a doping concentration of $1 \times 10^{17}\,\mathrm{cm}^{-3}$ is grown epitaxially on the SOI layer.

(iii) The boro-phospho-silicate glass (BPSG)-filled trench isolation down to the buried oxide is made in order to isolate the N^+-Si cathode of the photodetector from the other area. In order to reduce the parasitic capacitance of the interconnects, a 1.0 μm thick field oxide layer is made instead of the usually only approximately 0.5 μm thick field oxide. A recessed LOCOS process was used in order to keep the surface of the wafer planar.

(iv) A 1.3 μm deep groove was made in the Si overlayer by dry etching, where the SiGe/Si superlattice could be formed later.

(v) A 0.2 μm thick CVD-SiO_2 side-wall layer in the groove is formed in order to isolate the SiGe/Si superlattice from the N-Si layer.

(vi) Then an intentionally undoped 3 nm $Si_{0.9}Ge_{0.1}$/32 nm Si 30 periods superlattice absorption layer with a total thickness of 1050 nm with an actual boron concentration of approximately $1 \times 10^{15}\,\mathrm{cm}^{-3}$, a 0.1 μm thick P-Si buffer layer with a doping concentration of $1 \times 10^{18}\,\mathrm{cm}^{-3}$, and a 0.2 μm thick P^+ anode contact layer with a doping concentration of $1 \times 10^{20}\,\mathrm{cm}^{-3}$ were grown selectively and successively on the N layer in the groove at a temperature of 680 °C. A special cold wall UHV/CVD system was used for this purpose [311].

(vii) Ti-silicide/TiN/Al was used for the metallization.

(viii) A trench with a depth of 63 μm and with a width of 128 μm for the optical fiber was formed by reactive ion etching (RIE) with $Cl_2/SF_6 = 0.09$ at a temperature of −50 °C, resulting in a high etching rate of 2 μm/min and in almost perpendicular trench side walls.

A dark current density of only 0.5 pA/μm^2 was reported for the SiGe/Si waveguide detector. The capacitance was 0.11 pF and 0.088 pF at a detector bias of 1 V and 10 V, respectively. A high-frequency photoresponse of 10.5 GHz at 5 V for $\lambda = 980$ nm was found for the above described SiGe/Si planar photodiode on SOI with an area of $10 \times 100\,\mu m^2$ [310]. For a detector length of 200 µm a quantum efficiency of 29 % was measured. The combination of a very large bandwidth and of a rather high quantum efficiency was achieved for this advanced and highly sophisticated photodetector. The technology needed for its fabrication, however, far exceeds the possibilities of standard CMOS and BiCMOS technologies.

6.6 Long-Wavelength Infrared SiGe Photoconductor

Another example of a SiGe photodetector built on SOI will be treated next. It is a vertical cavity longwave infrared (LWIR) SiGe/Si photodetector for thermal imaging in the absorption window in air from 8 to 12 µm [312]. It consists of a superlattice of SiGe absorption layers and Si barrier layers. The SiGe absorption layers possess a lower bandgap than the Si layers. There are offsets in the valence and conduction bands as a consequence. The detection mechanism of a LWIR SiGe/Si photodetector is infrared absorption in the degenerately doped P^+-SiGe layers followed by internal photoemission of photoexcited holes over the heterojunction barrier. The cutoff wavelength λ_c of a LWIR SiGe/Si detector is determined by the heterojunction barrier $q\Phi_B$ and is given by

$$\lambda_c = \frac{1.24}{q\Phi_B}\,, \tag{6.1}$$

where λ_c is in µm and Φ_B is in eV.

The heterojunction barrier $q\Phi_B$ is determined by the valence band offset $\triangle E_V$ between the SiGe alloy layer and the Si barrier layer, and by the Fermi level in the degenerately doped SiGe layer (Fig. 6.6).

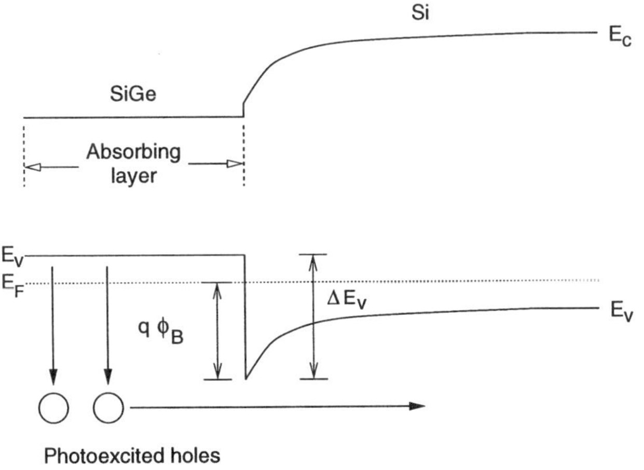

Fig. 6.6. SiGe/Si valence and conduction band discontinuities [277]

The heterojunction barrier $q\Phi_B$ is given by

$$q\Phi_B = \triangle E_V - (E_V - E_F) \; . \tag{6.2}$$

The energy band alignment of the SiGe/Si heterostructure has been studied extensively, and the bandgap difference splits approximately 0.9/0.1 between the valence and conduction band offsets [268]. The bandgap of strained SiGe alloys, and correspondingly the SiGe/Si valence band offset $\triangle E_V$, can be tailored by varying the Ge composition. Figure 6.7 shows the SiGe/Si valence band offset $\triangle E_V$ and corresponding minimum cutoff wavelength λ_c assuming $E_V - E_F = 0$. The two curves in this figure were calculated from the experimental data reported in [268] for Ge compositions of 0.1 to 0.4. The cutoff wavelength of a heterojunction LWIR detector can be tailored over a wide IR range, for instance from 5 to 22 μm with a Ge composition from 0.4 to 0.1. The tailorable cutoff wavelength can be used to optimize the trade-off between the LWIR response and the cooling requirements of the detector.

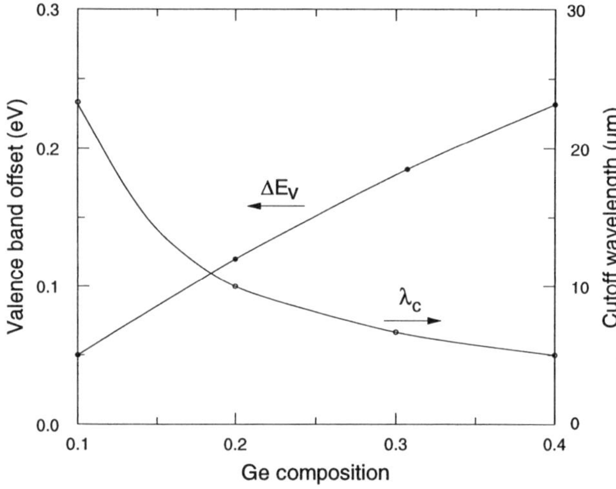

Fig. 6.7. SiGe/Si valence band offsets and corresponding cutoff wavelength as functions of the Ge fraction [277]

In order to increase the quantum efficiency a multi-quantum-well structure with many SiGe absorption layers and Si barrier layers is used (Fig. 6.8). The narrow-bandgap SiGe layers function as quantum wells in confining carriers. The hole density in undoped SiGe layers, therefore, is increased when the Si barrier layers are P-type doped similar to the so-called super-injection effect in III/V heterojunctions [1]. The P-type doping of the SiGe layers during epitaxy, however, is also possible and useful. Effective absorption of long-wavelength infrared radiation occurs in the SiGe layers. The photon energy is transferred to the holes and holes are lifted over the energy band discontinuity increasing the conductivity in the LWIR photoconductive detector according to:

$$\sigma = q\mu_{\mathrm{n}}n + q\mu_{\mathrm{p}}p \,. \tag{6.3}$$

The density of hot holes p is increased in the LWIR photoconductor by the absorption of infrared light. The first (electron) term may be neglected when the enhancement of the conductivity is due to IR light absorption at the holes.

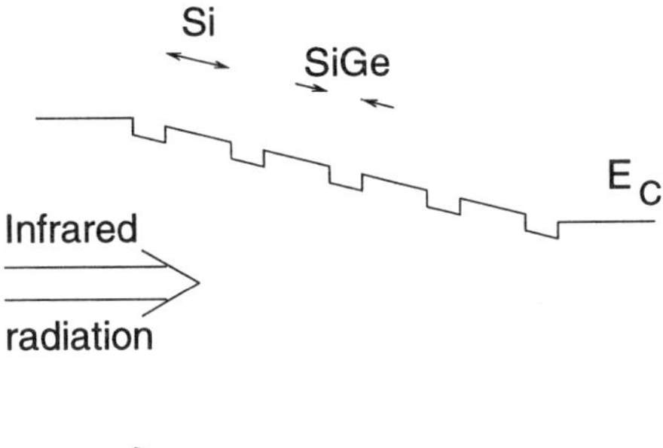

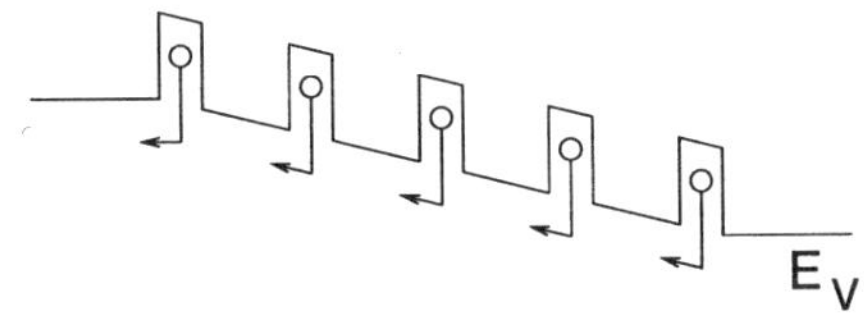

Fig. 6.8. Principle of a SiGe/Si multi-quantum-well LWIR photoconductive detector

The heterojunction LWIR detector offers a higher quantum efficiency compared with silicide Schottky-barrier detectors. One reason is the narrow band of occupied hole states in the P^+-SiGe layer of the heterojunction LWIR detector due to its semiconductor band structure [277]. In Schottky detectors, photons can excite carriers from states far below the Fermi energy. These carriers do not gain sufficient energy to overcome the barrier to the silicon. Only a small fraction of the photoexcited carriers near threshold — namely those originating from the states near the Fermi energy — can exceed the Schottky barrier energy. The Fowler dependence results, where the quantum efficiency rises only slowly with photon energy above the potential barrier (see Sect. 3.5.7). The narrow band of absorbing states in the P^+-SiGe layer of the heterojunction LWIR detector, in contrast, leads to a sharper turn-on, which in turn results in higher responsivities close to the cutoff wavelength. This property avoids the weakness of Schottky detectors, in which the Fowler dependence provides reasonable quantum efficiencies only at photon energies well above the barrier height. As a consequence, a Schottky detector must be designed with a much lower barrier than the desired energy response, leading to considerably lower operating temperatures to reduce the dark current to acceptable values.

The SOI technology is used here in order to fabricate a buried silicide mirror with a high reflectivity and to isolate this silicide from the substrate (Fig. 6.9).

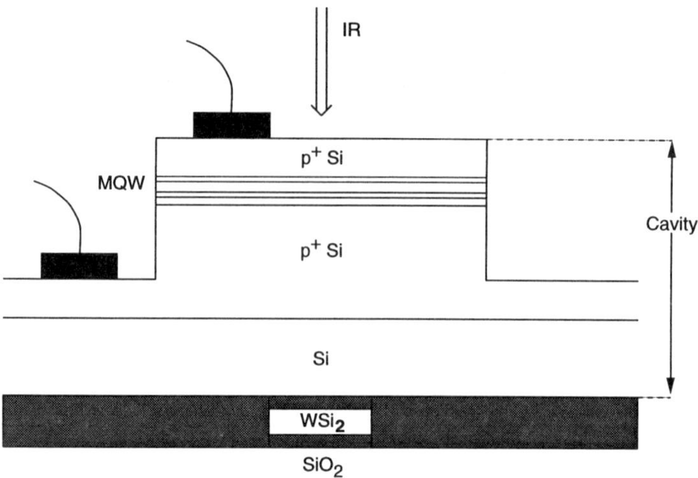

Fig. 6.9. SiGe/Si multi-quantum-well LWIR photoconductive detector on SOI [312]

The 0.3 μm thick tungsten silicide ($WSi_{2.7}$) layer was deposited on a Si wafer by an industry-standard chemical vapor deposition (CVD). An oxide layer with a thickness of 1 μm was subsequently deposited on the silicide. This wafer (device wafer) was then bonded with this silicide/oxide surface to a handle wafer. Then the device wafer was thinned in order to obtain a silicon-silicide-on-insulator (S^2OI) substrate. A 16-period P-$Si_{0.86}Ge_{0.14}$/Si resonant cavity quantum-well infrared photodetector (QWIP) was grown epitaxially on this S^2OI substrate (Fig. 6.9) by low pressure CVD at 650 °C using a $H_2/SiH_4/GeH_4$ mixture at a pressure of 20 Pa. The 16 SiGe quantum wells were each 5 nm thick and doped P-type with an area density of $1.5 \times 10^{12}\,cm^{-2}$ using B_2H_6. The nominally undoped Si barrier layers were 55 nm thick. The P^+-Si contact layer had a doping concentration of $3 \times 10^{18}\,cm^{-3}$. An unpassivated mesa-isolated photoconductive detector with a diameter of 1 mm was fabricated by wet etching. Ohmic contacts were formed by deposition, patterning and alloying of Al.

A nonresonant reference device, a 30-period P-SiGe/Si QWIP, was fabricated on a bulk P^+-Si wafer. In this reference detector, the incident radiation passes through the QWIP region only once. The multiple-quantum-well in this nonresonant detector was designed in order to obtain the maximum responsivity near $\lambda = 9$ μm. In order to achieve resonantly enhanced absorption for incident 9 μm radiation, the cavity of the detector in Fig. 6.9 was designed to be 5.9 μm thick. In such a way a standing optical wave for incident radiation with $\lambda = 9$ μm is obtained.

The measured photoresponse spectrum of the nonresonant detector had its maximum between 9 and 12 μm with a half-width-at-half-maximum (HWHM) of approximately 3 μm due to the SiGe quantum wells. The photoresponse spectrum of the resonant detector had its maximum at $\lambda = 9.2\,\mu m$ with a much smaller HWHM of approximately 0.8 μm. The peak responsivity of the resonant detector was 20 mA/W at a bias of 0.2 V compared to a value of 2.5 mA/W at a bias of 0.5 V for the nonresonant detector. The resonant cavity enhanced the peak responsivity by a factor of 8 reducing, however, the spectral width of the responsivity. For thermal imaging, the most interesting application of longwave IR detectors, the responsivity to a black-body radiation R_{bb} is important. A R_{bb} value of 9.1 mA/W for the resonant detector compared to a value of 1.6 mA/W for the nonresonant detector was measured. The R_{bb} value of the resonant detector was 5 times larger, accordingly. The detector reached dark current densities of below 10^{-6} A/cm^2 when it was cooled to 30 K.

The authors stated that the described device is comparable to III-V QWIPs in its performance. In their opinion, the new device offers a route to large-area monolithic focal plane arrays (FPAs) for the 8 to 12 μm band, where the QWIPs are integrated by epitaxial growth on the Si readout circuit avoiding the hybridization of N-GaAs/AlGaAs FPAs to Si circuits.

6.7 SiGe/Si PIN Hetero-Bipolar-Transistor Integration

SiGe photodiodes were described above and in [310,312], however, the amplifiers in these references employed pure Si transistor devices. SiGe/Si technology has developed rapidly in recent years and it has been demonstrated that it outperforms Si technology in terms of the speed of transistors [313]. Therefore, it is advantageous to combine SiGe/Si photodetectors with SiGe/Si transistors [314].

Figure 6.10 shows the cross section of the first monolithically integrated SiGe/Si PIN photodiode and heterojunction bipolar transistor (HBT) front-end photoreceiver. The PIN-HBT structure was grown by one-step molecular beam epitaxy (MBE). Table 6.1 contains the layer compositions and doping concentrations of the SiGe-OEIC. The emitter and collector layers consist of

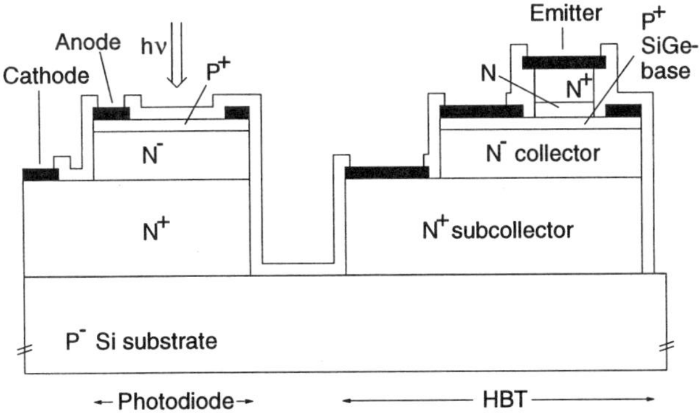

Fig. 6.10. Schematic cross section of a SiGe/Si PIN HBT photoreceiver [314]

Table 6.1. Layer compositions and doping concentrations of a SiGe-Si PIN-HBT OEIC

Layer	Material	Type	Doping (cm^{-3})	Thickness (nm)
Emitter Contact	Si	N^+	1×10^{19}	200
Emitter	Si	N	2×10^{18}	100
Spacer	$Si_{0.9}Ge_{0.1}$	I		1
Base	$Si_{1-x}Ge_x$ (x:0.1→0.4)	P^+	5×10^{19}	30
Spacer	$Si_{0.6}Ge_{0.4}$	I		10
Collector	Si	N^-	1×10^{16}	250
Subcollector	Si	N^+	1×10^{19}	1500
Substrate	Si	P^-	2×10^{12}	540 μm

Sb-doped Si. The base layer of the double heterojunction NPN HBT structure is a $Si_{1-x}Ge_x$ alloy with a smaller bandgap than that of Si. The Ge mole fraction in the base layer is graded from $x = 0.1$ at the emitter side to $x = 0.4$ at the collector side to accelerate the electrons traveling through the base towards the collector by a quasi-electric field. Spacer layers on both sides of the base minimize the effect of outdiffusion during epitaxy and processing.

The PIN photodiode is formed by the Si N^+-subcollector, by the Si N^- collector, and the SiGe P^+ base layers of the HBT (see Fig. 6.10). The I-absorption layer is formed by the Si N^- collector due to the one-step MBE growth, which provides advantages over regrowth such as better planarity, simpler processing, and higher yield [314]. The MBE growth temperatures for the collector and emitter layers were 415 °C. The base was grown at 550 °C. The growth rate was only 0.2 nm/s at these low temperatures.

The devices were isolated by mesa formation. The mesa size of the photodiode was $12 \times 13\,\mu m^2$. After MBE, the emitter contact was defined by evaporation, followed by emitter mesa formation with SF_6- and O_2-based dry and KOH-based wet etching. Minimal undercut and base over-etch were obtained by the two-step etch procedure, resulting in a low base access resistance. Then the base-collector mesa was formed by dry etching. The collector and PIN cathode contacts were formed by evaporation on the exposed highly doped subcollector layer. Another etch step of the subcollector layer was applied for the separation of the devices. After this mesa isolation a PECVD SiO_2 layer was deposited and the pad contacts were opened.

The dark current of the photodiode was about 0.1 μA at $U_{PIN} = 4$ V and 1 μA at $U_{PIN} = 9$ V. For the PIN photodiode with an I-layer thickness of 0.25 μm, a bandwidth of 450 MHz for $\lambda = 850$ nm and $U_{PIN} = 9$ V was measured. The PECVD SiO_2 layer with a thickness of 1.1 μm served as an antireflection coating leading to a measured responsivity of 0.3 A/W and an external quantum efficiency of 43 % for $U_{PIN} = 5$ V. Here, the SiGe anode did not increase the quantum efficiency by a significant amount compared to a Si anode. For a large enhancement of the quantum efficiency a high Ge fraction in the intrinsic zone of the PIN photodiode would have been necessary, which of course would introduce a high density of dislocations.

The bandwidth of 450 MHz was ascribed to the slow diffusion of carriers generated in the substrate and in the subcollector [314], because of the small I-layer thickness. We should not, however, accept this explanation, because the responsivity value of 0.3 A/W is too high for this explanation. It can be understood only when the cathode current, which is the sum of the subcollector/P^--substrate diode photocurrent and of the P^+-SiGe-anode/N-collector/N^+-subcollector diode photocurrent, was measured. The bandwidth then was not limited by carrier diffusion from the P^- substrate but by drift in the space-charge region of the subcollector/P^--substrate diode. The space-charge region extended far into the low doped P^--substrate with $N_A = 2 \times 10^{12}\,\mathrm{cm}^{-3}$ and the measured bandwidth of 450 MHz for $U_{PIN} = 9$ V seems possible. The reported increase in the responsivity and in the bandwidth with increased reverse bias also supports the new explanation with drift instead of carrier diffusion.

7. Detectors Based on a Resonant Cavity

In the SOI chapter, rather inefficient resonant cavity detectors with quantum efficiencies up to 27 % were described. Here, a different approach, which also is appropriate for Si detectors with a bandwidth of several GHz, for the improvement of the quantum efficiency will be introduced. The efficiency improvement of the detectors can be achieved by an enhancement of the reflection coefficients on the top and bottom of the Si device layer by implementation of so-called Bragg reflectors. In addition to the efficiency enhancement for selected wavelengths, the efficiency for other wavelengths is reduced to near zero by the implementation of the Bragg reflectors. An example of a resonant cavity enhanced Si detector with a quantum efficiency of 65 % at 700 nm, with an optical bandwidth of 20 nm, and with a bandwidth of up to 5 GHz will be discussed.

7.1 Principle of Resonant Cavity Enhanced Detectors

The performance of the optoelectronic device can be enhanced by placing the active device structure inside a Fabry–Perot resonant micro-cavity whose mirrors are formed by Bragg reflectors. Such a resonant cavity enhanced (RCE) device benefits from the wavelength selectivity and the large increase of the resonant optical field introduced by the cavity. The increased optical field allows RCE photodetector structures to be thin and therefore faster, while simultaneously increasing the quantum efficiency at the resonant wavelengths. Off-resonance wavelengths, however, are rejected by the cavity.

We will derive an expression for the quantum efficiency of a RCE device. Figure 7.1 shows the schematic cross section of a RCE device. The top and bottom mirrors are made of $\lambda/4$ stacks of semiconductor or insulator materials, which provide a large refractive index contrast. In simple designs, the top mirror can be the native semiconductor to air interface, which due to the large refractive index difference at the boundary, provides a reflectivity $\bar{R}_1$ of approximately 30 %. The active layer, where absorption takes place, is placed between the two mirror structures and is characterized by its thickness d and absorption coefficient α. The field reflection coefficients of the top and bottom mirrors are $r_1 e^{-i\Psi_1}$ and $r_2 e^{-i\Psi_2}$, respectively, with $\bar{R}_1 = r_1^2$ and $\bar{R}_2 = r_2^2$. Ψ_1 and Ψ_2 denote phase shifts due to light penetration into the

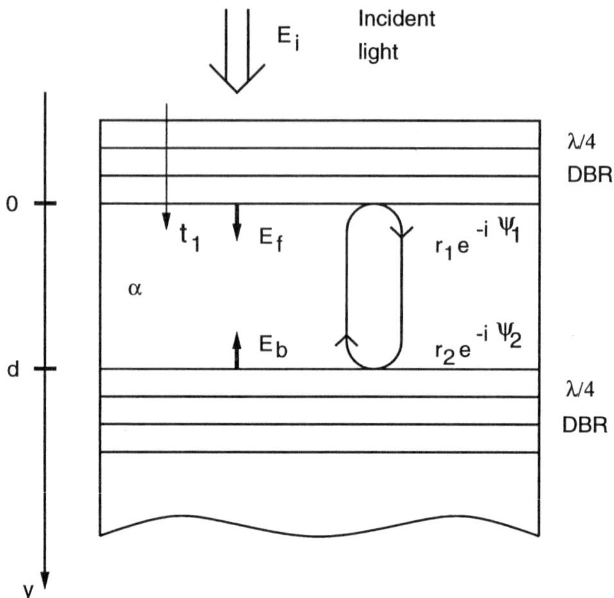

Fig. 7.1. Schematic cross section of a resonant-cavity enhanced photodetector with top and bottom distributed Bragg reflector (DBR) [315]

mirrors. The portion $t_1 E_{\mathrm{i}}$ of the incident lightwave electric field component E_{i} is transmitted by the top mirror. In the cavity, the downward traveling wave E_{f} at $y = 0$ (Fig. 7.1) can be obtained through a self-consistent consideration [315]. E_{f} is the sum of the transmitted field and the feedback after a round trip in the cavity:

$$E_{\mathrm{f}} = t_1 E_{\mathrm{i}} + r_1 r_2 \mathrm{e}^{-\alpha d} \mathrm{e}^{-\mathrm{i}(2\beta d + \Psi_1 + \Psi_2)} E_{\mathrm{f}}, \tag{7.1}$$

with $\beta = 2\bar{n}\pi/\lambda_0$, where $\bar{n}$ is the refractive index of the absorption region and λ_0 is the vacuum wavelength. It should be mentioned that $\exp^{-\alpha d}$ and not $\exp^{-2\alpha d}$ has to be used for the way forth and back, because the electric field and not the optical power or intensity, which are proportional to $|E|^2$, is calculated here. In the phase term, however, $2\beta d$ has to be used.

Equation (7.1) can be solved for E_{f}:

$$E_{\mathrm{f}} = \frac{t_1}{1 - r_1 r_2 \mathrm{e}^{-\alpha d} \mathrm{e}^{-\mathrm{i}(2\beta d + \Psi_1 + \Psi_2)}} E_{\mathrm{i}}. \tag{7.2}$$

The upward traveling wave at $y = d$ can be expressed as the portion of E_{f} arriving at the bottom mirror and being reflected with the factor r_2:

$$E_{\mathrm{b}} = r_2 \mathrm{e}^{-\alpha d/2} \mathrm{e}^{-\mathrm{i}(\beta d + \Psi_2)} E_{\mathrm{f}}. \tag{7.3}$$

The optical power inside the resonant cavity is given by:

$$P_s = \frac{\bar{n}}{2\eta_0}|E_s|^2, \tag{7.4}$$

for $s = f$ or b. The light power P_l being absorbed in the active region can be obtained from the incident power P_i in the form:

$$\begin{aligned} P_l &= (P_f + P_b)(1 - e^{-\alpha d}) \\ &= \frac{(1 - r_1^2)(1 + r_2^2 e^{-\alpha d})(1 - e^{-\alpha d})}{1 - 2r_1 r_2 e^{-\alpha d}\cos(2\beta d + \Psi_1 + \Psi_2) + (r_1 r_2)^2 e^{-2\alpha d}} P_i . \end{aligned} \tag{7.5}$$

With the assumption that all the photogenerated carriers contribute to the detector photocurrent, the quantum efficiency η is the ratio of the absorbed power to the incident optical power:

$$\begin{aligned} \eta &= \frac{P_l}{P_i} \\ &= \frac{1 + R_2 e^{-\alpha d}}{1 - 2r_1 r_2 e^{-\alpha d}\cos(2\beta d + \Psi_1 + \Psi_2) + R_1 R_2 e^{-2\alpha d}} \\ &\quad \times (1 - R_1)(1 - e^{-\alpha d}). \end{aligned} \tag{7.6}$$

Since β depends on the wavelength, η is a periodic function of the inverse wavelength. The quantum efficiency η is enhanced periodically at the resonant wavelengths, which are determined by:

$$2\beta d + \Psi_1 + \Psi_2 = 2m\pi, \tag{7.7}$$

with $m = 1, 2, 3....$ The spacing of these maxima, i.e. of the resonant wavelengths or of the cavity modes, is defined as the free spectral range.

On the right hand side of (7.6), the first term represents the cavity enhancement effect. This term becomes unity for $R_2 = 0$ giving η for a conventional detector. Conventional photodetectors provide roughly constant η across a broad wavelength range while RCE photodetectors can be designed to have significantly improved η at specific wavelengths.

For the sake of completeness, a distributed Bragg reflector (DBR) will be briefly described. The quarter-wave superlattice reflector contains alternate layers of high and low refractive index and each layer thickness is equal to $\lambda/4$. The light reflected at each interface, therefore, interacts constructively at the surface. The reflection coefficient of an entire DBR stack depends on the difference in the refractive indices and on the number of periods in the stack. The reflection coefficient can be made quite large. Values of almost 1 can be achieved. The design wavelength λ_d and the superlattice parameters are related by

$$\lambda_d = 2(\bar{n}_H d_H + \bar{n}_L d_L) \tag{7.8}$$

where d represents layer thickness and the subscripts H and L denote layers with high and low refractive indices, respectively [316]. The spectral bandwidth $\triangle\lambda$ of the superlattice stack is given by

$$\triangle\lambda = \frac{4}{\pi}\frac{\bar{n}_{\mathrm{H}} - \bar{n}_{\mathrm{L}}}{\bar{n}_{\mathrm{H}} + \bar{n}_{\mathrm{L}}}\lambda_{\mathrm{d}}. \tag{7.9}$$

The reflectance R depends on the refractive indices

$$R = \left(\frac{1 - \bar{n}^*}{1 + \bar{n}^*}\right)^2, \tag{7.10}$$

where

$$\bar{n}^* = \left(\frac{\bar{n}_{\mathrm{L}}}{\bar{n}_{\mathrm{H}}}\right)^{2p+1}. \tag{7.11}$$

Here p is the number of periods. Equation (7.11) is valid for $(p+1)$ layers of the first refractive index and p layers of the second refractive index. Each layer has the thickness $\lambda_{\mathrm{d}}/4$. These equations are valid for normal incidence.

It should be mentioned that for thin active layers, the standing-wave effect (SWE) can further improve the quantum efficiency [315]. In the following example, however, SWE is not important, because the active layer is relatively thick.

7.2 Example of a Resonant Cavity Enhanced Silicon Detector

A resonant cavity is the basis of the Si detector reported by Neudeck et al. [317] and shown in Fig. 7.2.

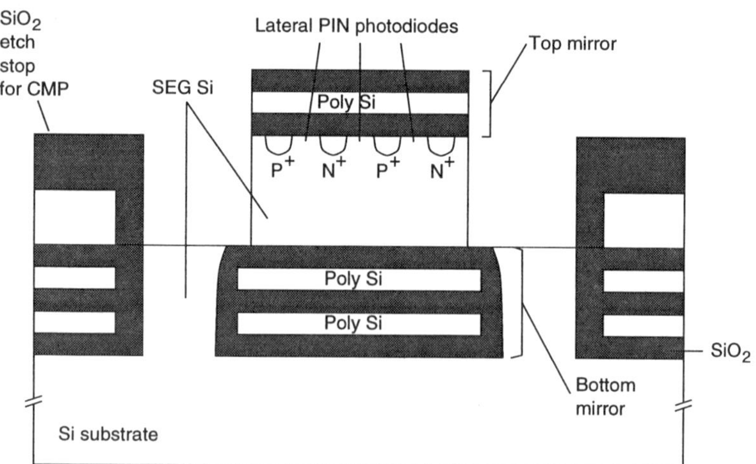

Fig. 7.2. Schematic cross section of a SEG resonant-cavity photodiode with a Bragg reflector [317]

The bottom mirror uses the principle of Bragg reflectors known from vertical-cavity-surface-emitting-lasers (VCSELs), for instance. The bottom mirror was formed by low pressure chemical vapor deposition (LPCVD) of three pairs of quarter-wavelength SiO_2/polysilicon layers on a P-type (100)-oriented substrate with a resistivity of 3–7 Ωcm. The combination of SiO_2 and polysilicon is well suited for the distributed mirror due to the large difference in the refractive indices. With this bottom mirror, a reflectivity of more than 99 % was attained in the wavelength range from 600 to 900 nm. After the formation of the bottom mirror, two $20 \times 160\,\mu m^2$ trenches with a separation of 40 μm were then etched into the SiO_2 and polysilicon layers by reactive ion etching (RIE). The exposed Si substrate in the trenches was then used as a seed window for the selective epitaxial growth (SEG). Oxide spacers were formed prior to the SEG by LPCVD oxide deposition, patterning, and RIE in order to prevent the polysilicon from nucleating during the SEG process. The top polysilicon layer served as an etch-stop layer for the spacer RIE etch. The top polysilicon was removed after spacer formation between the two trenches, where the resonant cavity will be completed. The SEG was done in a standard commercial cold-wall reduced pressure chemical vapor deposition (RPCVD) system. The SEG was combined with an epitaxial lateral overgrowth (ELO) technique [318, 319] in order to grow single crystalline Si over the bottom mirror. The SEG was performed at 970 °C using H_2 carrier gas with SiH_2Cl_2 and HCl for the control of the selectivity and growth rate. For this growth technique, the epitaxial growth starts from the Si seed areas and first grows vertically in the trenches and then laterally over the bottom mirror. The process is continued until the growth fronts meet. Chemical mechanical polishing (CMP), finally, was used to smooth the surface and to establish the desired Si thickness of about 1 μm over the bottom mirror. The 1 μm thick oxide on the left and right mirror stacks (see Fig. 7.2) is used for stopping the CMP. The SEG Si then was patterned and etched to form a Si island being separate from the Si seeds and the substrate.

Lateral interdigitated PIN photodiodes then were fabricated on these SEG islands on the bottom mirrors. As and BF_2 implants and anneals were used to form the N^+ and P^+ fingers. One and a half pairs of quarter-wavelength SiO_2/polysilicon layers as the top mirror of the resonant cavity then were deposited and patterned. Metallization finally completed the detectors.

The detector devices with an area of $20 \times 20\,\mu m^2$, with a finger width of 1 μm and a finger spacing of 3 μm showed dark currents of 3 nA at a bias of 5 V and of 0.8 μA at 40 V. The quantum efficiency η of the detectors strongly depends on the wavelength. Six peaks of η between 570 and 820 nm were found. Between these peaks, η was almost equal to zero. The maximum quantum efficiency was 65 % at 700 nm. The full-width at half-maximum (FWHM), however, was only about 20 nm. This small FWHM value leads to the conclusion that this detector will suffer from a Si thickness variation of the resonant cavity, i. e. the quantum efficiency will decrease strongly, when the Si thickness is not controlled accurately. The photodiode exhibited a bandwidth of more than 5 GHz at a bias of 48 V. A remark, however, should be made on this large bias value. The dark current of the photodiode already exceeded a value of 1 μA for a reverse bias of 48 V. Such a high value, therefore, seems unrealistic for applications, especially because the operating temperature can be higher than room temperature causing an even larger dark current. A measured bandwidth of 1.5 GHz at a bias of 10 V with a reverse current of less than 10 nA, therefore, seems more realistic. Nevertheless, this detector possesses the largest quantum efficiency compared to others with a comparable speed [244, 320]. Moreover, the authors stated that the device process is completely compatible with Si CMOS or bipolar process [317]. We, however, should be aware of the high additional process complexity necessary for the integration of the RCE detector in a bipolar, CMOS, or BiCMOS process placing it far from a near-standard-process OEIC.

8. III-V Semiconductor Materials on Silicon

GaAs- and InP-based lasers on Si substrates are highly desirable for the realization of OEICs and applications in long-range optical data transmission, in optical interconnects, and in optical computing. GaAs, InP, and related compound materials on Si are also interesting for the realization of very fast photoreceiver circuits. In particular, effective light emitters and very fast photoreceivers consisting of III-V materials can be combined with VLSI or ULSI silicon circuits in hybrid OEICs. This combination seems advantageous due to the following aspects: III-V materials and technologies are not as highly developed as silicon material and silicon technologies. Only 10^4–10^5 transistors can be integrated on one III-V chip compared to 10^8 and even more transistors on Si chips. III-V light emitters are much more effective and faster than Si-based light emitters. III-V photodetectors allow much higher data rates than Si photodetectors.

The possibilities of growing III-V layers on Si, flip-chip mounting of III-V chips on Si chips, and III-V wafer bonding to Si will be described in this chapter.

8.1 Heteroepitaxial Growth

Much research in integrating optoelectronic emitter and receiver functions on Si has been performed involving the heteroepitaxial growth of III-V semiconductor materials on Si. The most important parameters for heteroepitaxy are the lattice constants of substrate and deposited material. Table 8.1 lists the lattice constants of some important semiconductor materials. The lattice constant of Si is much smaller than that of Ge and of the most important III-V materials and compounds.

The mismatch of Si and III-V lattice constants is up to 4 % for Ge and GaAs. The direct deposition of Ge and GaAs on Si, therefore, results in epitaxial layers with a very high density (10^8–$10^9\,\mathrm{cm}^{-2}$) of threading dislocations. InGaP on Si causes similar densities of dislocations on Si [321]. Such high densities of dislocations degrade the properties and long-term stability of column IV and mismatched III-V devices on Si. For instance, dislocations cause high leakage currents of PN junctions and high threshold currents of laser diodes. Several ternary compounds with fixed bandgap

Table 8.1. Bandgap and lattice constants of important semiconductor materials at 25 °C

Material	E_g (eV)	a (nm)
Si	1.12	0.54307
Ge	0.66	0.56461
GaAs	1.424	0.56533
InAs	0.360	0.6058
InP	1.351	0.5869
GaP	2.261	0.5451

values can be grown lattice matched on common substrates. For instance $In_{0.53}Ga_{0.47}As$ can be grown lattice matched on InP. InGaP can be grown lattice matched on Ge, for example. Quaternary alloys provide more freedom for bandgap engineering. For many very interesting device applications of heteroepitaxial growth and material combinations, lattice matched ternary and quaternary alloys, however, are not available or should be avoided. Special growth techniques, like the implementation of strained-layer superlattices, low-temperature buffer layers, and the use of thermal cycling [322], therefore, have been investigated. However, none of these techniques has been capable of lowering the density of threading dislocations in GaAs on Si below $10^8\,cm^{-2}$ [323]. Low-mismatched interfaces via composition grading, therefore, remain for an improved III-V heteroepitaxial growth on Si. Low-mismatched interfaces involve glissile 60 ° dislocations. If such interfaces are driven to complete relaxation, the misfit is released and all residual dislocations are glissile. The glissile dislocations have the potential to be removed through subsequent processing.

Low-mismatched SiGe/III-V interfaces, therefore, can be exploited, because compositionally graded Ge_xSi_{1-x} layers can be grown on Si in order to obtain larger lattice constants on Si. It is possible to achieve completely relaxed Ge_xSi_{1-x} layers with $0.1 < x < 1$ whereby the threading dislocation densities could be kept low in the range of 10^5–$5 \times 10^6\,cm^{-2}$ [321]. These relaxed SiGe layers can be used as templates for lattice-matched III-V growth. Ge fractions of 0.75 to 1.0 are of interest for GaAs and InGaP growth on SiGe/Si.

In [321], for instance, a red-emitting InGaP LED was fabricated on GaAs/Ge/SiGe/Si. The LED operated at room temperature. The full width at half maximum (FWHM) of the optical emission spectrum was 52 meV with the maximum at a wavelength of 655 nm. The current–voltage characteristics of the InGaP LED proved a reverse current of less than 1 μA at 6 V reverse bias. As in all GaAs growth on the structures graded to pure Ge, misfit dislocations at the GaAs/Ge interface were observed due to the small mismatch in lattice constants. The threading dislocation density was $5 \times 10^6\,\mathrm{cm}^{-2}$.

Recently, the quality of the relaxed buffers has been further improved. The threading dislocation densities have been reduced to about $2 \times 10^6\,\mathrm{cm}^{-2}$ involving chemical-mechanical polishing (CMP) in pure Ge on Si [300] supporting the integration of nearly all electronic and optoelectronic devices on Si. For lasers integrated on Si, however, a further lowering of the dislocation density is necessary. Another method for controlling dislocations promises success for such a purpose. Very low dislocation densities were achieved by patterning [324]. SiGe was grown on Si mesas. The advantage of patterning is simply that the dislocation has to travel a shorter distance to exit the film [323]. The dislocation density decreases with decreasing mesa area. For a 210 nm thick $Si_{0.85}Ge_{0.15}$ layer on Si mesas with a size of $25 \times 25\,\mu\mathrm{m}^2$ no threading dislocations have been observed.

These advanced growth techniques, however, are rather complex and have not been applied so far to most of the reported III-V devices grown on Si. Several examples of III-V devices grown on Si will now be described.

GaAs LEDs as light emitters and GaAs photoconductors (PCs) as photodetectors were implemented in an optically coupled three-dimensional (3D) memory system for ultra fast parallel processing in computation [325]. Figure 8.1 shows the cross section of the 3D memory chip.

This 3D integration technique is especially interesting due to the small sizes of the LEDs and photoconductors of $5 \times 5\,\mu\mathrm{m}^2$ and $10 \times 10\,\mu\mathrm{m}^2$, respectively [326], which are much smaller than bondpads. The photoconductors had a thickness of $2\,\mu\mathrm{m}$ and the vertical spacing between LEDs and

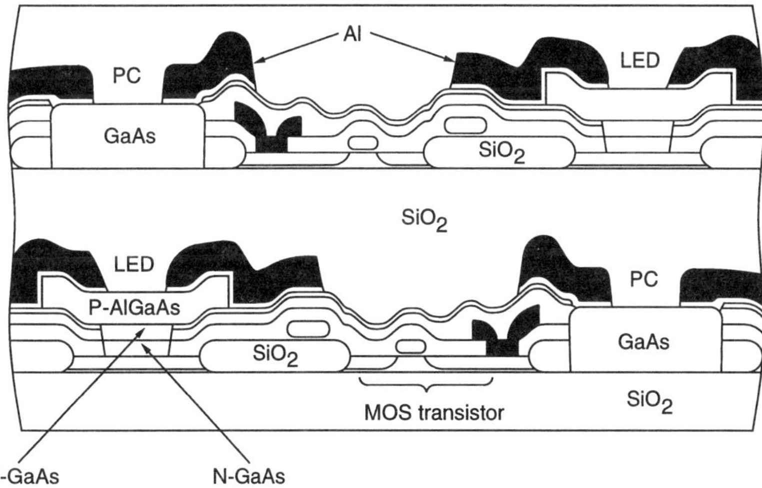

Fig. 8.1. Schematic cross section of a three-dimensional LED-photoconductor-CMOS integration [326]

photoconductors was 5 μm. After completing the CMOS circuits in a 2 μm static random access memory (SRAM) process, the LEDs emitting at 870 nm and the photoconductors were fabricated in a GaAs-on-Si technology using heteroepitaxial growth of GaAs on Si [327, 328]. Silicon wafers with LEDs and photoconductors then were thinned and glued to each other [329]. The LED light emitted downwards goes through the thin Si substrate where it is absorbed only partially because of the Si thickness of 0.5 μm. Most of the emitted light, therefore, reaches the photoconductor in the layer beneath.

Two photoconductors were used in a flip-flop as sensors and load resistors at the same time. Two LEDs, controlled by the LED control line, also were present with two driver NMOS transistors in a memory cell [326]. A block of 512 data bits was transferred in parallel through four memory layers within 16 ns, being equivalent to a data rate of 128 Gb/s.

Edge-emitting lateral-cavity lasers grown on Si have been investigated in [330–333]. Room temperature continuous-wave (CW) operation of GaAs-based lasers on Si has been reported. However, reliable GaAs-based lasers on Si have been hindered by a high density of dislocations and strain due to the differences in the lattice constants and thermal expansion coefficients of GaAs and Si [330–333].

Vertical cavity surface-emitting lasers (VCSELs) on Si seem to be advantageous compared to edge-emitting lasers, because they allow wafer-scale testing, high-density two-dimensional array fabrication, ultrafast parallel optical information processing, and monolithic integration with other optical or electronic devices [334–336]. Most important, however, is that the strain and dislocations can be reduced in the VCSELs on Si with small active volumes (compare with patterning mentioned above), which results in more reliable lasers on Si. In [335], an AlGaAs-GaAs VCSEL on Si emitting at about 875 nm with a pulsed threshold current (I_{th}) of 125 mA, corresponding to a threshold current density (j_{th}) of 60 kA/cm^2 at 300 K was demonstrated. Here, however, an improved approach reported in [337] will be described. The schematic cross section of the VCSEL grown on Si is depicted in Fig. 8.2.

An N^+-Si substrate oriented 2° off <100> towards <110> was used. The AlGaAs/GaAs VCSEL structure resulting in an emission wavelength of 840 nm was grown using metalorganic chemical vapor deposition (MOCVD) at atmospheric pressure at 750 °C by the conventional two-step growth technique [337]. The structure consists of a 0.85 μm thick N^+ GaAs buffer layer, 20 pairs of a quarter-wavelength N^+-AlAs/N^+-GaAs (71 nm/59 nm) multilayer distributed Bragg reflector (DBR), a 0.46 μm thick lower N-$Al_{0.7}Ga_{0.3}As$ cladding layer, a 70 nm thick lower N-$Al_{0.3}Ga_{0.7}As$ confining layer, a 9 nm thick GaAs single quantum well (SQW) active layer, a 70 nm thick upper P-$Al_{0.3}Ga_{0.7}As$ confining layer, a 0.34 μm thick upper P-$Al_{0.7}Ga_{0.3}As$ cladding layer, and a 80 nm thick P^+-GaAs contact layer.

Thermal cycle annealing was performed five times by varying the substrate temperature between 350 and 850 °C in order to reduce the threading

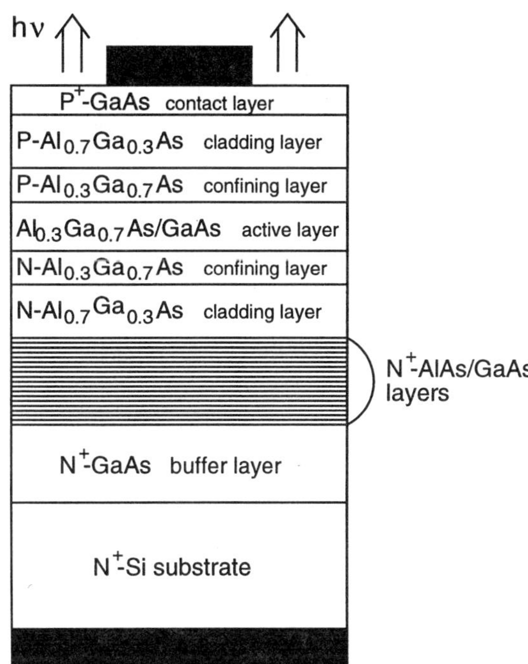

Fig. 8.2. Schematic cross section of an AlGaAs/GaAs VCSEL grown on a Si substrate by MOCVD [337]

dislocation density [322]. Au-Sb/Au was used for the contact on the N^+-Si substrate. Nonalloyed Au-Zn/Au of $40 \times 40\,\mu m^2$ for the top mirror and the electrical contact was formed by photoresist patterning and lift-off. It should be mentioned that these contact materials are not compatible with Si technology, where the recombination centers Au and Zn have to be strictly avoided. One issue of III/V-laser integration on Si will be the use of compatible metallization.

The device was tested with pulses of 100 ns at 1 MHz and 300 K. The light output characteristics was investigated by measuring the light emitted around the edges of the Au-Zn/Au metallization.

Transmission electron microscopy (TEM) was applied to determine the dislocation density. The dark spot density (DSD) was measured by the electron-beam-induced current (EBIC) method. Some dislocations originating at the GaAs/Si interface were confined into the thermal-cycle annealed N^+-GaAs layer. However, many of these dislocations propagated into the active layer. A DSD of $2.5 \times 10^7\,cm^{-2}$ was determined.

The current–voltage characteristics of the VCSEL grown on Si revealed a turn-on voltage of 1.3 V and a series resistance of 26 Ω. A threshold current I_{th} of 79 mA corresponding to a threshold current density j_{th} of $4.9\,kA/cm^2$ was determined for the lasing wavelength of 840.3 nm with a FWHM of 0.28 nm for pulsed operation at 300 K [337]. The improvement compared to [335] was probably due to the lower dislocation density and the different laser structure.

The implementation of a $Al_{0.55}Ga_{0.45}P$ on the N-type Si substrate and of a $Al_{0.5}Ga_{0.5}As$ layer below the N^+ buffer layer shown in Fig. 8.2 with a smooth heterointerface improved the performance of the laser [322]. At an emission wavelength changed to 860 nm, an average threshold current density of 1.83 kA/cm^2 and an internal quantum efficiency of 83 % was reported [322].

A monolithically integrated AlGaAs/InGaAs laser diode and a GaAs MESFET for fast modulation of the laser diode have been grown on a Si substrate [331]. In addition, an N-Si layer on top of the P-type Si substrate was exploited as a monitor photodiode for the laser diode (Fig. 8.3).

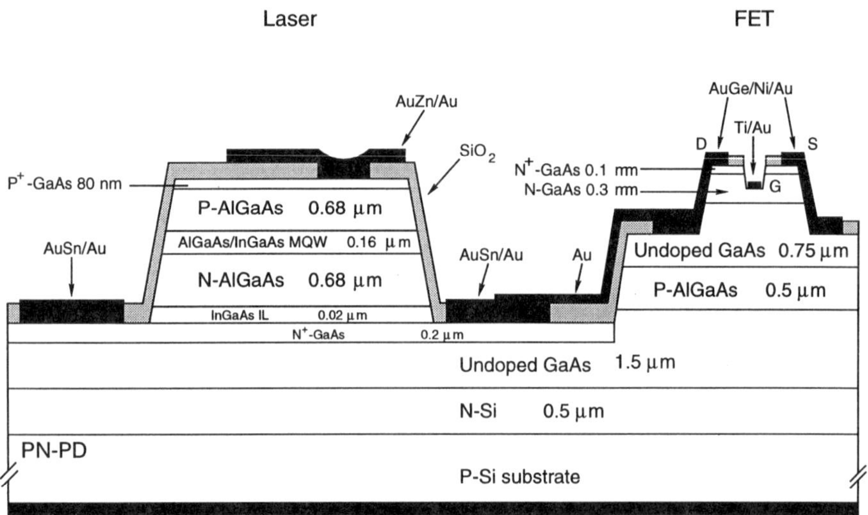

Fig. 8.3. Schematic cross section of a laser and a MESFET monolithically integrated on a Si substrate [331]

The AlGaAs/$In_{0.02}Ga_{0.98}As$ multi-quantum-well (MQW) laser diode was grown selectively on the Si substrate with an $In_{0.08}Ga_{0.92}As$ intermediate layer (IL) for strain relief [331] after the MESFET has been fabricated [332]. On the IL, the N-type $Al_{0.7}Ga_{0.3}As$ cladding layer, a 61 nm thick N-type $Al_{0.3}Ga_{0.7}As$ confinement layer, the MQW with three 9 nm thick $In_{0.02}Ga_{0.98}As$ active layers and two $Al_{0.3}Ga_{0.7}As$ barrier layers with a thickness of 5.5 nm, a 61 nm thick P-type $Al_{0.3}Ga_{0.7}As$ confinement layer, the P-type $Al_{0.7}Ga_{0.3}As$ cladding layer, and a 80 nm thick P^+ GaAs contact layer were grown by MOCVD. The emission wavelength was not reported, however, it should be about at 850 nm according to the composition of the active layers. In this wavelength region, Si is appropriate as a detector material and this opportunity has been exploited for the monolithic integration of a monitor photodiode. A threshold current of 34 mA and a threshold current density of 1.7 kA/cm^2 as well as an internal quantum efficiency of 64 % were reported for the laser diode, however, a rapid degradation was also admitted [331]. The MESFET with a gate length of 2.5 µm and a gate width of 400 µm had a threshold voltage of −2.2 V and a transconductance of 90 mS/mm. A laser diode current to monitor diode photocurrent efficiency of 1.9 % was reported above laser threshold.

We can summarize that heteroepitaxial growth on Si still has to be improved in order to enable long-term stable and reliable III-V laser diodes with low threshold currents grown on Si. According to the growth problems, another method of III-V device integration on Si has become important. This flip-chip hybridization method will be described next.

8.2 Flip-chip Hybridization Technique

Sometimes it is highly desirable to combine a III/V detector with a silicon amplifier or with a chip in a different III/V compound material. Heteroepitaxy, if possible at all in principle, is not possible in every fabrication line. An alternative is to connect the III/V chip via bond-wires with the Si chip. Usually the bondpads, which are used on both chips for providing the area where the bond-wires are bonded to, have a size of approximately $100 \times 100\,\mu m^2$. These large bondpads introduce a large parasitic capacitance and the bond-wire represents an inductance. For high-speed optical receivers and data rates of more than several Gb/s, the bond-wires and bondpads cannot be implemented. The so-called flip-chip interconnection technique [338] can be used instead. This flip-chip technique used gold bumps with a diameter of approximately 30 µm on bondpads of approximately the same size and thereby reduces stray capacitance and inductance, when the chips are carefully aligned face-down (Fig. 8.4).

The pads are located directly over the junction area of the photodiode in order to realize a minimum parasitic capacitance. In this flip-chip technique,

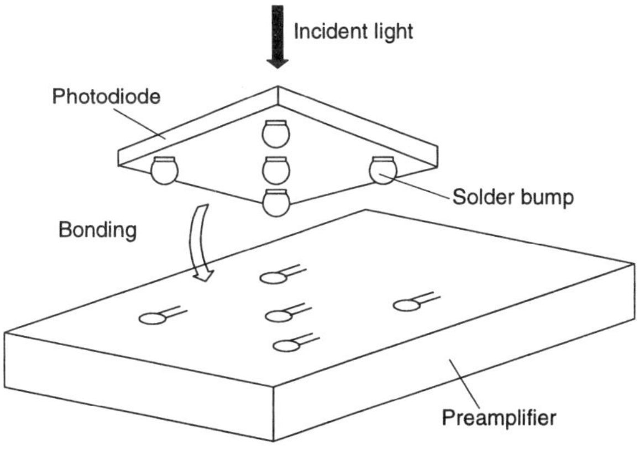

Fig. 8.4. Schematic view of a photodetector chip and of a preamplifier chip which are flip-chip bonded [339]

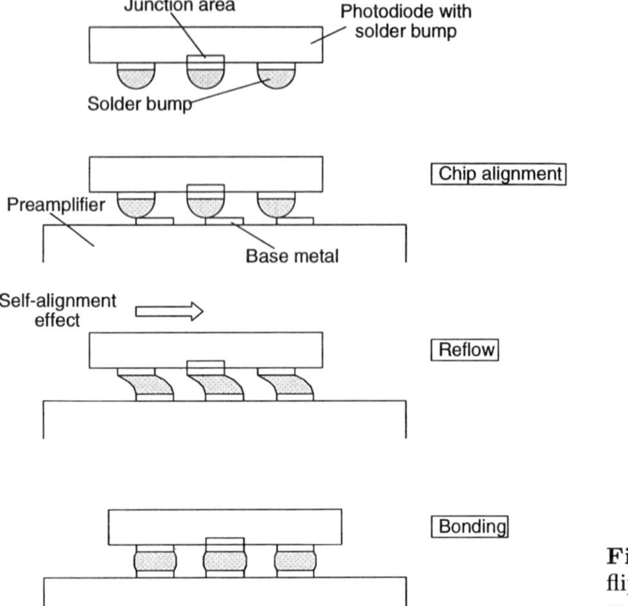

Fig. 8.5. Self-alignment flip-chip bonding by the molten solder [339]

the bumps had to cover the pads perfectly and there were strict requirements for the alignment to an accuracy of only a few micrometers.

An improved flip-chip interconnection technique, which permits loose chip alignment because of the self-alignment effect of the molten solder [340], was developed [339]. The solder, moreover, enables bonding at a lower temperature ($\approx 300\,°C$) than the gold bump bonding technique and, therefore, reduces damage to the devices, like an increased dark current of InGaAs photodiodes for instance. In [339], solder bumps with a diameter of 26 μm were used. The solder bump for contacting the photodiode is fabricated over the junction area of the photodiode (Fig. 8.5).

The solder bump fabrication is shown in detail in Fig. 8.6. The self-alignment bonding by the molten solder is depicted in Fig. 8.5. The surface tension of the molten solder moves the chip and accomplishes precise chip positioning. An alignment accuracy of approximately 10 μm is sufficient due to the self-alignment effect during the reflow process. A positioning error of less than ± 1 μm, which is almost equal to the tolerance of the bump diameter, was reported.

The junction area diameter of the InGaAs PIN photodiode chip was 20 μm, the base metal diameter was 20 μm, and the solder bump diameter was 26 μm [339]. Almost the same junction capacitance of 60 fF at a photodiode bias of 5 V was measured for the InGaAs PIN photodiode with and without a solder bump fixed on the junction area. It was concluded that no parasitic capacitance, caused by the solder bump, affected the photodi-

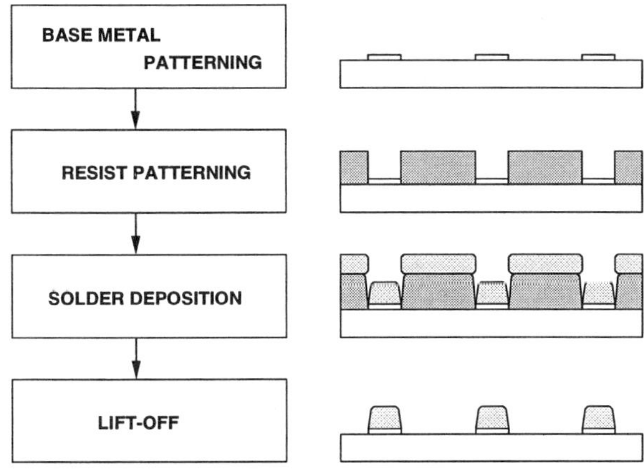

Fig. 8.6. Solder bump fabrication for flip-chip bonding [339]

ode junction capacitance. A frequency response of more than 22 GHz at $\lambda = 1.55\,\mu m$ was demonstrated for the InGaAs PIN photodiode flip-chip bonded to a high-speed GaAs metal-semiconductor-field-effect-transistor (MESFET) preamplifier.

Figures 8.7–8.10 show further applications of the flip-chip technique. In Fig. 8.7 a four-channel AlGaAs laser array with an emission wavelength of 830 nm is flip-chip bonded to a silicon motherboard [341]. The silicon motherboard carries the transmission lines for the laser electrodes. Alignment trenches are etched into the laser chip for self-aligned mounting with the help of alignment mesas on the Si motherboard. V-grooves are etched into the (100)-oriented Si motherboard for the fibers to achieve maximum precision. The Si is etched anisotropically with KOH at 80 °C to a depth of 20 µm for the connecting plane and 80 µm for the fiber grooves. In such a way the light from the edge emitting lasers can be coupled into the optical multimode fibers with a cladding diameter of 125 µm and with a core diameter of 50 µm. (111) crystal planes are obtained for the side walls of the alignment mesas and of the V-grooves for the optical fibers. The etched wafer is oxidized thermally to a thickness of 0.5 µm prior to metallization to avoid direct metal to silicon contacts. At the ends of the transmission lines opposite to the laser electrodes the indium bumps with a thickness of 8 µm are electroplated. For the mounting of the laser chip, the Si motherboard is placed on a heating chuck. The laser chip carried by vacuum tweezers, then is positioned upside-down over the alignment mesas and set down in order to exploit the self-positioning effect of alignment mesas and trenches. The chuck is heated to 180 °C for 30 s for forming the solder connections.

Figure 8.8 shows the hybrid integration of a laser array on a silicon driver and photodetector array chip [279]. The $m \times m$ array of GaAs vertical-cavity surface-emitting lasers (VCSELs) were bump-solder-bonded to a silicon chip to form a hybrid Si OEIC smart-pixel chip for an application in optical computing.

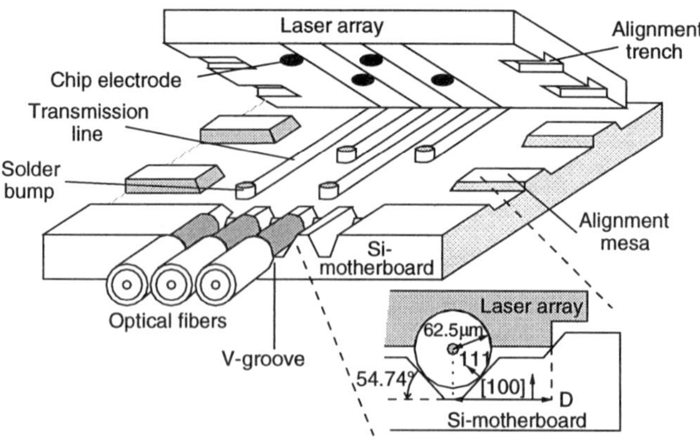

Fig. 8.7. Laser diode array bonded to a Si motherboard and self-alignment scheme for fiber coupling [341]

Figure 8.9 shows the flip-chip integration of a GaAs-AlGaAs multi-quantum-well (MQW) modulator , which also could be used as a PIN photodiode, to a CMOS chip. The GaAs-AlGaAs MQW device had a responsivity of 0.45–0.50 A/W for $\lambda = 850\,\mathrm{nm}$ when used as a detector. The MQW devices had a size of $20 \times 45\,\mu\mathrm{m}^2$ and a dark current of about 10 nA at a reverse bias of 10 V. The MQW devices could also be used as a reflection-mode modulator. The biasing of the MQW device determines whether incident light is reflected or absorbed. In [342], the MQW device was used in the absorption mode as a PIN photodiode with a reverse bias of 10 V. The advantage of the

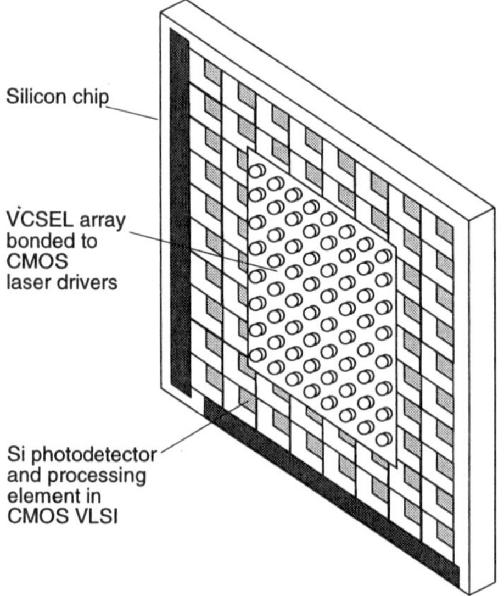

Fig. 8.8. VCSEL array bonded to CMOS laser driver and photodetector chip [279]

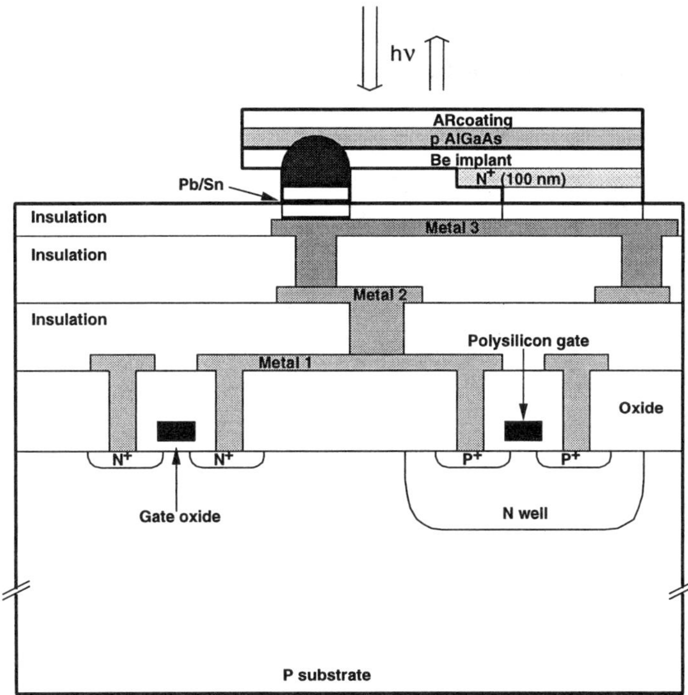

Fig. 8.9. Cross section of a hybrid integrated CMOS AlGaAs multi-quantum-well modulator [342]

GaAs-AlGaAs device was its large absorption coefficient, enabling a thinner absorption layer and a larger bandwidth compared to a silicon photodiode.

The GaAs-AlGaAs device was flip-chip bonded to bondpads in the third metal layer of a 0.8 μm CMOS chip using Pb/Sn bumps. An opening in the uppermost passivation layer allowed for deposition of the Ti-Ni-Au wetting metal alloys and solder. After the alignment and bonding of the silicon and GaAs chips, epoxy was flowed in-between the chips to act as an etch-protectant, and the substrate of the GaAs chip was removed to expose the MQW device [343]. This technique is called substrate removal technique.

The GaAs chip was located directly over active CMOS circuits. Together with a 3-stage CMOS inverter amplifier in a 0.8 μm technology, a bit rate of 375 Mb/s was obtained [342]. An improved 4-stage inverter amplifier enabled a bit rate of 625 Mb/s [344, 345]. The speed of these hybrid OEICs was limited by the CMOS amplifiers. The intrinsic i. e. drift-limited speed of GaAs photodiodes would be sufficient for multi-gigabit operation.

The flip-chip bonding technique has been improved and used for the integration of vertical-cavity surface-emitting laser (VCSEL) arrays with gigabit-per-second CMOS drivers [346]. The VCSELs have been flip-chip bonded di-

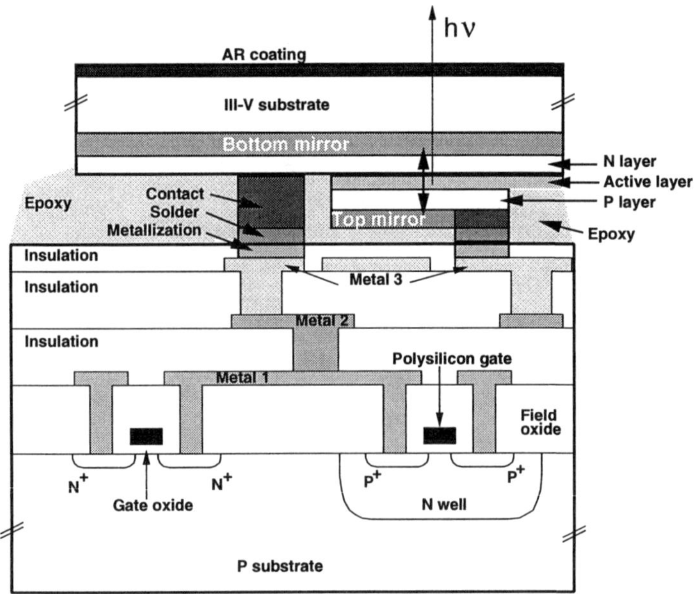

Fig. 8.10. Cross section of a hybrid integrated CMOS VCSEL chip [346]

rectly to the third metal level on a CMOS chip in 0.5 μm CMOS technology (see Fig. 8.10).

It has been reported that this VCSEL bonding does not interfere with the operation and the layout of the underlying CMOS circuits. This is due to the fact that the flip-chip bonding is accomplished using $10 \times 10\,\mu m^2$ bumps with an ultra low pad and bump capacitance of less than 10 fF.

As for the GaAs MQW modulator array to CMOS flip-chip bonding [343, 347] described above, the exposed pads for flip-chip bonding on the CMOS chip have been first metallized with Ti-Ni-Au metals to improve adhesion to the top-level Al on the chip. PbSn solder was then deposited on the CMOS chip. Subsequently, the bottom-emitting VCSEL array on the GaAs chip was flip-chip bonded to the CMOS circuits using two coplanar Au contacts per laser. Epoxy was wicked in between the chips to act as an adhesive, finally.

For the experiments described in [346], 2×10 VCSEL arrays operating at a wavelength of 970 nm were chosen. This enabled the exploitation of transmission through the III-V substrate which could be left intact after the flip-chip integration. The VCSEL lasers with an aperture of 10 μm had Au contacts of $10 \times 10\,\mu m^2$ and the center-to-center spacing of the lasers was 125 μm. The VCSELs [348] had laser threshold voltages of 1.4–1.5 V and threshold currents between 0.9 and 1.0 mA at room temperature, which are good values for application in optical interconnect technology. Within the drive capability of the CMOS drivers, continuous-wave output powers of 4.3 mW were achieved, confirming that high-quality electrical and thermal contact was obtained. At 80 °C, the maximum optical output power reduced to 1 mW due to the temperature dependence of the VCSEL emission efficiency and the increasing threshold current.

A simple two-transistor CMOS VCSEL driver (Fig. 12.8) based on a current-shunting principle, which provides a low-area, tunable power circuit with a measured small-signal bandwidth of 2 GHz in 0.5 μm CMOS technology, has been implemented. For a bit-error rate of less than 10^{-11} at 1.25 Gb/s, the total power consumption of one driver and laser was about 17.5 mW at an optical power of −6.9 dBm. This low-power, high-speed operation demonstrates the utility and potential of the flip-chip bonding technique for optical interconnect technology.

8.3 Wafer Bonding

Wafer bonding is a possibility to avoid the problems associated with lattice mismatch and the resulting misfit dislocations of thick heteroepitaxially grown layers. Wafer bonding enables the fusion of layers with strongly differing lattice constants. Almost all investigated combinations of two semiconductor wafers [349] can be directly bonded or can be bonded to oxidized silicon provided the wafers are properly polished and prepared. Direct bonding of III-V and Si wafers, therefore, allows the fabrication of devices, which maximize the advantages of each material [350–355].

The major problem in bonding of dissimilar materials is the thermal stress originating from a mismatch in thermal expansion coefficients (see Table 8.2). When the two bonded wafers of different materials with large diameters are thick and a high-temperature anneal is used to strengthen the bond, the pair therefore will usually bend severely. If the thermal stresses become sufficiently high, they may be released by debonding or generation of misfit dislocations, for instance. If the temperature and the thickness of one film are below critical values misfit dislocations, however, can be avoided. The general practice for bonding dissimilar materials, therefore, is to anneal the bonded pair at a temperature that is as low as possible to achieve a bonding strength high enough to allow subsequent thinning. The thickness of the resultant film should be below a critical value at the maximum device processing temperature to avoid dislocation generation [356].

The bonding of GaAs to Si and of InP to Si is especially interesting for the fabrication of optoelectronic and microwave devices. The GaAs and InP films on the Si substrate, furthermore, can be used as buffer layers for the growth of other compound semiconductors.

In contrast to the case of epitaxial growth of GaAs on Si, the direct wafer-bonding approach can avoid the generation of threading dislocations in the GaAs layer caused by the 4.1 % difference in the lattice constants of the two materials when high annealing temperatures and long annealing times are not applied [357]. Due to the three times higher thermal conductivity of the Si

Table 8.2. Some properties of important semiconductor materials at 25 °C

Property	Si	GaAs	InP	SiO_2
Crystal structure	Diamond	Zincblende	Zincblende	Amorphous
Lattice constant (nm)	0.54307	0.56532	0.58687	N/A
Bandgap (eV)	1.17	1.424	1.344	≈ 8.0
Thermal conductivity ($W\,cm^{-1}\,K^{-1}$)	1.5	0.455	0.7	0.015
Thermal expansion coefficient (K^{-1})	2.6×10^{-6}	6.8×10^{-6}	4.8×10^{-6}	0.56×10^{-6}

substrate, a GaAs device fabricated in a GaAs film bonded to the Si substrate can dissipate much more power than a GaAs device fabricated in a GaAs substrate. The 160 % mismatch in the thermal expansion coefficients of GaAs and Si, however, presents a challenge for GaAs/Si direct bonding because of thermal stresses induced in the bonded pair when annealed to strengthen the bond. Bonding of GaAs to Si has already been investigated [349, 358].

The as-received 638 μm thick GaAs and 525 μm thick Si wafers were bonded at room temperature in a microcleanroom [358]. It has been found that GaAs/Si wafer pairs may debond during annealing at temperatures above 160 °C owing to the large difference in their thermal expansion coefficients.

At higher bonding temperatures, a hydrogen atmosphere and a weight on the bonded pair may help. During annealing at 600 °C in hydrogen with a weight on top of the bonded pair, the inferior oxide on the GaAs is removed by the hydrogen and covalent bonds form at the bonding interface similar to those in heteroepitaxial growth [350]. Due to the high surface diffusion coefficient of Ga atoms, the microgaps originating from the surface roughness at the interface vanish [359].

It has been recognized that deposition of a thin SiO_2 layer on the GaAs surface makes the bonding of GaAs wafers to Si much easier [356]. This aspect has been exploited in the following experiment: The transfer of a thin, single-crystalline (100) GaAs layer from hydrogen-implanted wafers with a diameter of 3 inch onto a Si substrate with the same diameter was demonstrated by wafer bonding and layer splitting [360]. A process recently developed at LETI in Grenoble, France, for the fabrication of SOI [361] was used for the SiO_2-SiO_2 GaAs/Si bonding. This so-called smart-cut process basically consists of four steps for the GaAs film transfer to Si:

(i) Hydrogen is implanted into a GaAs wafer capped by a PECVD SiO_2 layer with an energy of 100 keV and a dose in the range of 5×10^{16} to $1 \times 10^{17}\,cm^{-2}$.

(ii) This GaAs wafer is bonded at room temperature to a Si wafer with a PECVD SiO_2 layer. Before bonding, both wafers are polished to provide a low surface roughness and then cleaned.

(iii) The two bonded wafers are annealed at 400–700 °C. During this heat treatment the implanted GaAs wafer splits into two parts giving rise to the transfer of a thin GaAs film and the remainder of the GaAs wafer, which can be recycled. For the proton implantation energy of 100 keV, the transferred GaAs film has a thickness of 0.8 µm.

(iv) A final polishing is carried out to eliminate the region at the GaAs film surface damaged by implantation.

This technique allows the fabrication of wafer-size bonded GaAs/Si substrates. However, the advantage of the high thermal conductivity of the Si substrate cannot be exploited due to the SiO_2 layer at the bond interface. Vertical current flow also is not possible through the SiO_2 bond interface.

Like GaAs, the compound InP is a semiconductor with a direct bandgap. It is, therefore, interesting to bond InP to Si substrates to integrate InP photonic devices with Si electronics on the same chip. The fabrication of an InP-based laser on a Si substrate will be described in Sect. 8.3.2.

8.3.1 Wafer Bonded Photodetectors

For instance, a high performance InGaAs-PIN-photodetector on a Si substrate was demonstrated for optical telecommunication at $\lambda = 1.55\,\mu m$ by Bell Labs of Lucent Technologies [362]. InGaAs on Si is especially interesting due to the absorption edge at $\lambda \approx 1.7\,\mu m$, its large absorption coefficient at $\lambda = 1.55\,\mu m$ (see Fig. 1.2), and the highest mobility of all the more mature III-V materials and compounds. Si, of course, allows fast ULSI circuits at a low price. Therefore, InGaAs on Si is the most interesting combination for wafer bonding.

A (100)-oriented N^+ Si wafer with a nominally undoped 0.5 µm thick epitaxial Si layer was used for bonding to an $In_{0.53}Ga_{0.47}As$ layer with a thickness of 1.0 µm [363]. This $In_{0.53}Ga_{0.47}As$ layer was grown on top of an InP wafer with two layers. The first of these two layers, a 0.3 µm thick $In_{0.53}Ga_{0.47}As$ layer, was grown lattice matched on the N^+ InP substrate by low-pressure metalorganic chemical vapor deposition (MOCVD) as an etch stop layer [363]. The second of these two layers was a 0.3 µm thick InP layer, which will form the surface layer of the bonded $In_{0.53}Ga_{0.47}As$-Si photodetector. The already mentioned 1.0 µm thick $In_{0.53}Ga_{0.47}As$ absorption layer was grown on this InP layer. All these III-V layers were nominally undoped.

In [363], the optimum fusion conditions were determined to be a bonding temperature of 650 °C in a H_2 ambient. After bonding, the InP substrate and then the $In_{0.53}Ga_{0.47}As$ etch stop layer were removed. Device fabrication began with the deposition of a 0.15 µm thick SiO_2 layer into which a 30 µm circular window was etched down to the InP, which was the top semiconductor layer at this stage of fabrication. The acceptor Zn was diffused

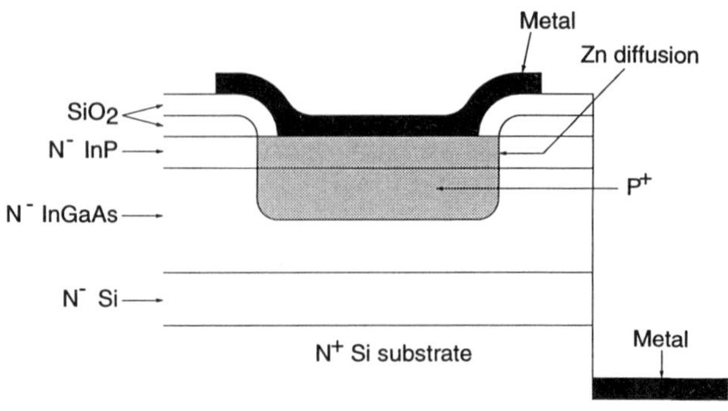

Fig. 8.11. Cross section of a back-illuminated InGaAs-Si detector

through the InP layer to a depth of 0.8 μm in this window area. 0.5 μm of the $In_{0.53}Ga_{0.47}As$ layer was doped P^+ by the Zn accordingly leaving a 0.5 μm thick N^--$In_{0.53}Ga_{0.47}As$ I-layer. After a second SiO_2 deposition, a 20 μm diameter window was opened to the InP where the Au:Be contact was deposited. The contact to the N^+ Si substrate was made by etching a 70 μm diameter mesa and depositing Al (see Fig. 8.11). The InP layer was implemented in order to reduce the dark current I_d. Values for I_d of $10^{-10} A$ to 10^{-9} A for $U_{PIN} = 0$ V to $U_{PIN} = 10$ V were measured. The back-illuminated $In_{0.53}Ga_{0.47}As$-Si PIN detector showed a 100 % quantum efficiency with an experimental uncertainty of 5 %. A −3dB bandwidth of (13.4 ± 0.8) GHz was measured for the detector for $U_{PIN} \geq 1$ V. The bandwidth was limited by the RC time constant with a capacitance of 162 fF and a resistance of 66 Ω according to $f_{RC} = 1/(2\pi RC) = 13.3$ GHz. The value of 66 Ω was the sum of the impedance of the RF line of 50 Ω and 16 Ω contact resistance of the detector. The transit time limited frequency was calculated to be 42 GHz.

A tunneling interface was present between the two bonded semiconductor materials, which, however, did not degrade the internal quantum efficiency and −3dB bandwidth for $U_{PIN} \geq 1$ V [363]. A nearly ideal Si-$In_{0.53}Ga_{0.47}As$ heterointerface, which showed essentially lossless carrier transport with no evidence of charge trapping, recombination centers, or substantial bandgap discontinuity, was obtained for the bonding at 650 °C in H_2.

The bonding temperature of 650 °C suggests that devices fabricated in the Si wafer prior to bonding should not be influenced by the bonding. For the fabrication of $In_{0.53}Ga_{0.47}As$-Si OEICs, however, much development for process integration has to be done.

8.3.2 Wafer Bonded Light Emitters

The integration of III-V optoelectronic functions would facilitate the fabrication of large-scale optoelectronic integrated circuits. The main approach to fabricate III-V optical devices on Si has been heteroepitaxial growth. This approach, however, suffers from a large lattice mismatch, which causes a high density of threading dislocations. Although a long-wavelength laser has been demonstrated [364], the dislocation density in InP/Si systems remains in a rather high region of about $10^7\,\mathrm{cm}^{-2}$ hindering high-performance laser operation. The newer approach of wafer bonding can circumvent the problem of threading dislocations caused by lattice mismatch between Si and the relevant III-V compounds when heteroepitaxial growth is performed.

For wafer bonding, threading dislocations are still an issue, due to different thermal expansion coefficients of Si and III-V compounds (see Table 8.2) and due to the resulting thermal stress, when high annealing temperatures and long annealing times are used. However, wafer bonding can be performed at lower temperatures than heteroepitaxial growth resulting in less thermal stress. Additionally, buffer layers can be adopted to relax the thermal stress, which may cause the threading dislocations at the bonding interface, and to avoid their propagation into the device layer at the annealing temperature. In [351], for instance, an etch-pit density of only $10^4\,\mathrm{cm}^{-2}$ on an InGaAs/InP MQW structure bonded on Si was found, indicating a dislocation density of the same order of magnitude.

A 1.55 μm laser using a substrate fabricated by direct bonding of an InP wafer to a Si wafer [351] has been demonstrated [365]. The procedure of direct wafer bonding is thoroughly described in [356] and is shown in Fig. 8.12 for the InGaAs/InGaAsP MQW laser diode described in [365]. On an N-type (100)-oriented InP substrate, a Be-doped P^+-InGaAs contact layer with a thickness of 0.3 μm, which also acted as an etch-stop layer during the InP substrate removal after bonding, was first grown by molecular beam epitaxy

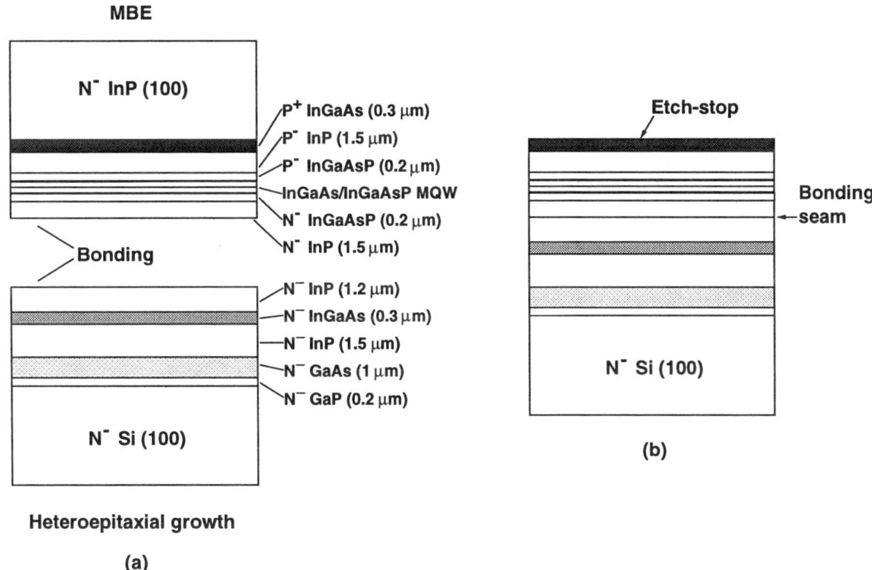

Fig. 8.12. InGaAs/Si wafer bonding scheme [356]

(MBE) followed by a Be-doped P-type InP cladding layer with a thickness of 1.5 μm. An intentionally undoped InGaAs/InGaAsP MQW structure with four 5 nm thick InGaAs and 10 nm thick InGaAsP layers sandwiched between two InGaAsP optical guide layers with a thickness of 0.2 μm was grown. The composition of the active InGaAs quantum layers was chosen in order to result in an emission wavelength of 1.55 μm. The InGaAsP guide layers had a bandgap corresponding to a wavelength of 1.15 μm. The P-type InGaAsP guide layer was doped with Be and the N-type InGaAsP guide layer was doped with Si. Finally a Si-doped N-type InP cladding layer with a thickness of 1.5 μm was grown.

The N-type Si wafer was (100)-oriented, 2° off <110> on which buffer layers consisting of an N-type GaP layer with a thickness of 0.2 μm, a 1 μm thick GaAs layer, a 1.5 μm thick N-type InP layer, a 0.3 μm thick N-type InGaAs contact layer, and a 1.2 μm thick N-type InP layer were grown heteroepitaxially. The buffer layers prevent the possibility of threading dislocations, originating at a Si–InP bond interface during annealing at a high temperature, from propagating into the device layers. The InP surfaces of these two wafers were mirror-polished to reduce the surface roughness formed during epitaxial growth, cleaned in an H_2SO_4:H_2O_2:H_2O solution with a ratio of 3:1:1, and rinsed in deionized water. Then the two wafers were brought into contact at room temperature with a weight of about 300 g/cm^2. After this room-temperature bonding, the wafer pair was loaded immediately into a furnace and annealed at 700 °C for 1 h in a hydrogen ambient. Finally the InP substrate was removed by HCl selective etching, which stopped at the top of the P^+ InGaAs etch-stop layer.

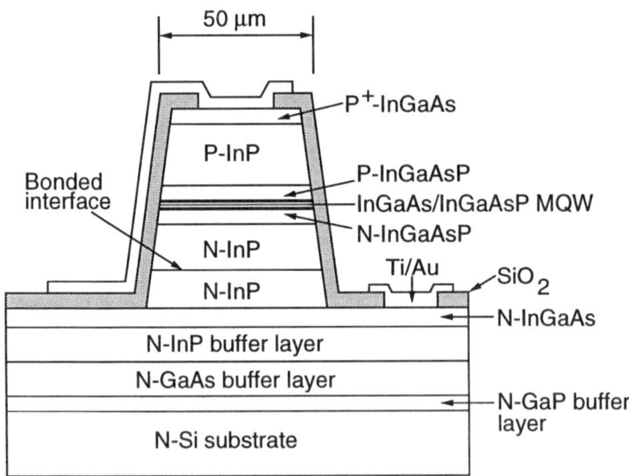

Fig. 8.13. Cross section of a InGaAs MQW laser diode fabricated on a wafer bonded InGaAs-Si substrate [365]

On the bonded InP/Si wafer, the stripe InGaAs/InGaAsP MQW laser shown in Fig. 8.13 was fabricated by wet mesa etching, opening of contact windows, and formation of Ti/Au contacts. The Si substrate was lapped down to a thickness of 60 μm and finally cleaved into laser array bars with a typical resonator length of 330 μm. The mesa with a width of 50 μm, therefore, had a length of 330 μm with the cleaved surfaces as the two mirror facets defining the resonator length of the edge-emitting laser. For comparison, lasers with the same structure were fabricated on an InP substrate. Although the current passes through the N-InP/N-InP bonded interface, the lasers show excellent PN-junction characterisitics and the same series resistance as that of lasers on an InP substrate was obtained for the lasers on bonded InP/Si wafers indicating that no current barrier exists at the bonded interface.

The threshold current density of the lasers was about $1.7\,\mathrm{kA/cm^2}$ for the Si and the InP substrate, corresponding to a threshold current of about 280 mA for the $50 \times 330\,\mu\mathrm{m}^2$ lasers. The lasers fabricated on the bonded InP/Si wafer were tested under pulsed conditions (300 ns at 1 kHz) at room temperature and an optical output power of 2 mW was observed for a current of about 340 mA.

9. Silicon Light Emitters

The emission efficiency of silicon for light is very low, because of the indirect bandgap. The momentum of a phonon is required in order to allow the transition of an electron from the minimum of the conduction band to the maximum of the valence band in Si (see Fig. 9.1). This results in a two-particle process with a low probability.

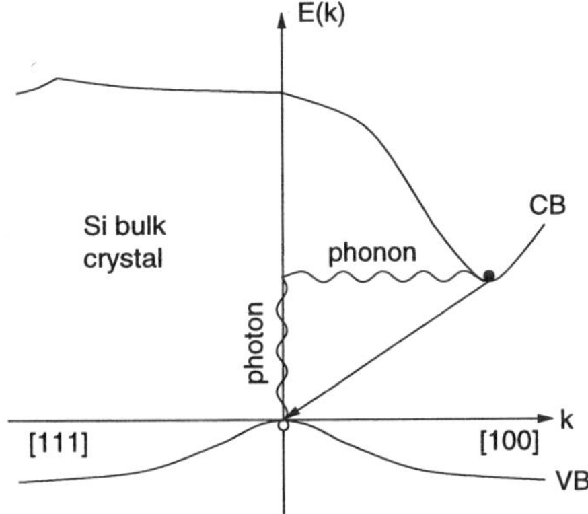

Fig. 9.1. Indirect band–band transition in bulk Si requiring a phonon for the conservation of momentum

However, much effort was made to characterize [366] and to understand [367,368] the light-emission phenomena of silicon devices. Probably even more attempts were made to enhance the emission efficiency of silicon for light. The attempts to utilize quantum effects, silicon-germanium superlattices, erbium-doped silicon, semiconducting silicides, organic polymers on silicon, and liquid crystals on Si chips will be described in the following sections.

9.1 Quantum-Size Silicon

Many investigations of photoluminescence and electroluminescence from nanoporous silicon having their origin in a quantum effect have been performed. After the main technological interest in porous silicon, which was its application in fully isolated porous silicon (FIPOS) [369], it has been observed that the optical properties of porous silicon differ drastically from those of bulk crystalline silicon [370, 371]. From the increased bandgap compared to bulk crystalline silicon, it has been concluded that the formation of porous silicon is associated with a quantum wire effect [370]. Quantum levels exist in silicon wires with diameters of the order of nanometers and radiative electron transitions occur between quantum levels in the conduction band and quantum levels in the valence band (see Fig. 9.2). In such a way the photon energy can be larger than the bandgap energy $E_g = 1.1\,\text{eV}$. Simultanously, strong room-temperature photoluminescence in the visible range at about 1.5 eV was detected from porous silicon [371]. Electroluminescence in the red and orange spectral range stronger than that from bulk crystalline silicon devices was also reported for nanoporous silicon LEDs [372, 373]. However, voltages larger than 5 V had to be applied to these porous Si LEDs [373,374].

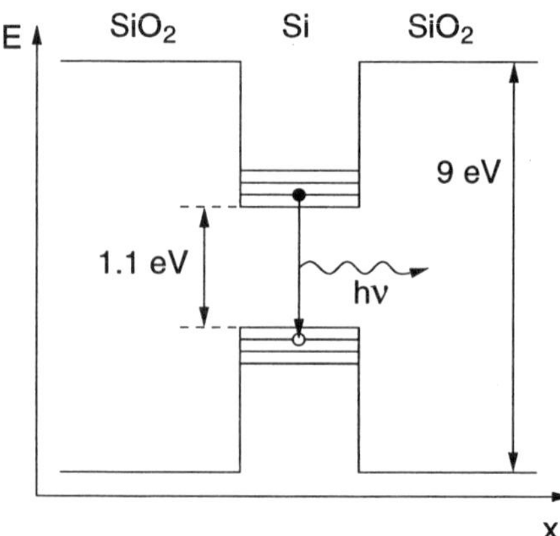

Fig. 9.2. Radiative electron transition in nanocrystalline silicon due to a quantum wire effect

The explanation of the increased radiative recombination rate in nanoporous silicon is based on Heisenberg's uncertainty relation

$$\triangle x \triangle p \geq \frac{\hbar}{2} \,. \tag{9.1}$$

Applied to electrons and holes in the nanoporous silicon with a diameter of the quantum wires in the range from 10–1 nm, the electron and hole momentum, i. e. their wave number obtains a certain probability distribution (see Fig. 9.3). This probability distribution of the momentum results in a certain probability for 'direct band–band' transitions and in the increased radiative recombination rate in nanoporous silicon (see Fig. 9.3).

The distance of the quantum levels from the conduction and valence band edges as well as between the levels $\triangle E$ increases for thinner quantum wires. $\triangle E$ is proportional to d^{-2}, where d is the wire diameter. Accordingly, the photon energy increases for thinner quantum wires.

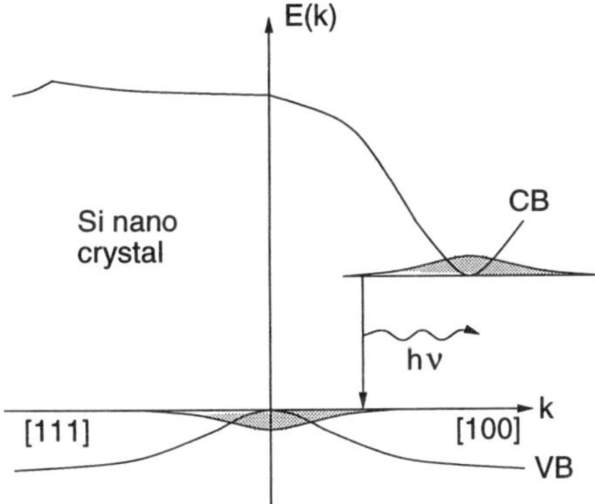

Fig. 9.3. Direct band–band transition in nanocrystalline silicon due to Heisenberg's uncertainty relation in a quantum wire

There is, however, the disadvantage of a high resistance due to the thin quantum wires. Furthermore, nanoporous silicon is difficult to passivate and long term stability is poor due to the simple anodization process usually used for fabricating the porous Si in an HF/ethanol solution. For instance, for the highest continuous-wave (cw) quantum efficiency of 0.18 %, the efficiency fell by a factor of four over a period of 5 h [375]. In addition, the mobility of porous silicon is only about $10^{-4}\,\mathrm{cm^2\,V^{-1}\,s^{-1}}$ [376]. Therefore, a rather slow transient response of porous silicon LEDs can be expected. Indeed, the decay time of the orange-red luminescence component (1.5–1.9 eV) has been found to be of the order of 10 μs at room temperature [377]. The rise time of the light emission from a porous silicon-aluminum Schottky diode has been determined to be 100 ns [378]. The blue-green band (2.3–2.6 eV) is much faster with response times in the 10-ns range [377]. These long decay times make porous Si poorly suited for optical fiber transmission and optical interconnect applications, where rise and fall times below 1 ns are required in the ULSI era. The main field of applications, therefore, may be in displays and in certain optical microsystems, which do not require high speed and high quantum efficiency.

Another example of Si light emitters with a better fabrication and passivation approach has been reported in [379]. Si nanopillars (see Fig. 9.4) were fabricated in this investigation using electron beam lithography and a highly anisotropic silicon etching process based on a mixture of SF_6 and CHF_3 gases. This process resulted in very smooth pillar side walls and bottom surfaces. Immediately after the reactive ion etching process, pillars with a thickness below 0.1 μm were obtained with an aspect ratio of about 40:1. High temperature thermal oxidation and oxide removal were then applied for thinning of the pillars. Finally, a thin oxide was grown. The final thickness of the pillars in the range of 4–6 nm was determined by TEM investigations. Planarization and isolation was achieved by using PolyMethylMethAcrylate (PMMA) to fill the trenches between the nanopillars. PMMA is a good insulator and it is transparent in the visible spectral range. The PMMA was removed from

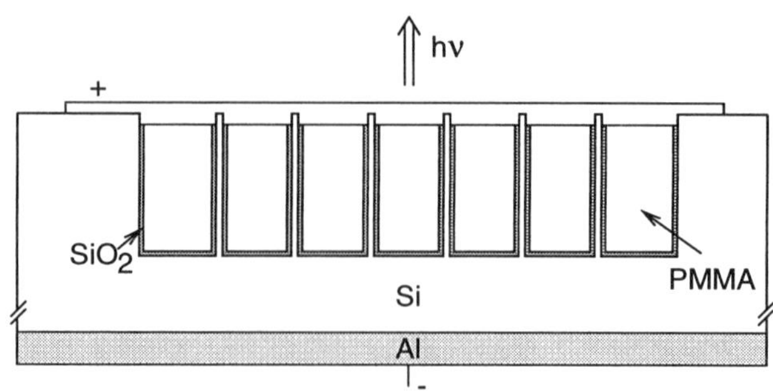

Fig. 9.4. Nanopillar Si LED [379]

the top parts of the nanopillars and a thin semitransparent metal layer was deposited in order to obtain an ohmic contact.

The devices showed rectifying behavior with an onset voltage of about 12 V. The electroluminescence obtained with a voltage of 24 V and a current of 22 mA had its maximum at approximately 650 nm. The device in Fig. 9.4 leaves a large portion of the surface of the silicon unused for light emission due to the small width of the pillars of several nanometers and their distance of more than 100 nm.

Another possibiliy to obtain electroluminescence from Si is to exploit nanocrystallites embedded in a good passivation in order to minimize non-radiative recombination mechanisms. Quantum confinement and bandgap widening have been investigated [380–382] and a law with $\triangle E \propto d^{-m}$ with $1.4 < m < 2$ was suggested for nanocrystallites with a diameter d. Compared to quantum wires, nanocrystallites obtain the same bandgap energy already for larger diameters [380] due to the confinement in three instead of only in two directions. Nanocrystallites, therefore, may allow control of the minimum feature size, and consequently of the bandgap widening and of the emission wavelength, more easily than nanowires.

A controlled surface passivation and size of the nanocrystallites is important as well as their packing density in order to achieve a high quantum efficiency. Low pressure chemical vapor deposition (LPCVD) [383], plasma enhanced chemical vapor deposition (PECVD), and sputtering also used in microelectronics technology have been used to fabricate nanocrystallites. Size control by deposition of Si between silicon dioxide layers [384] has been proposed. Molecular beam epitaxy for the deposition of Si/CaF_2 multilayers also has been used to achieve light emission. In [379], a one-layer deposition of Si by LPCVD on a thin thermally grown SiO_2 film was chosen. The deposition of films with a thickness between 15 and 30 nm above 600 °C led to crystallite sizes between 1.5 and 15 nm. Emission peaked between 600 and 700 nm. A blue shift could be achieved by further oxidation at temperatures between 800 and 900 °C.

Efficient IR spectra also were obtained peaking at 1140 nm whereby spatial confinement of electron–hole pairs in Si nanocrystallites with sizes close to the Bohr exciton diameter was found to be present [384]. The structure of the luminescent device is shown in Fig. 9.5. Carriers are injected from the semitransparent Au contact and from the Si substrate by tunneling through the SiO_2 layer with thicknesses in the range 5–15 nm into the nanocrystalline silicon (nc-Si) layer. A bias voltage of 9 V and a current density of 2.9 A/cm^2 were needed to observe efficient electroluminescence peaking at about 620 nm.

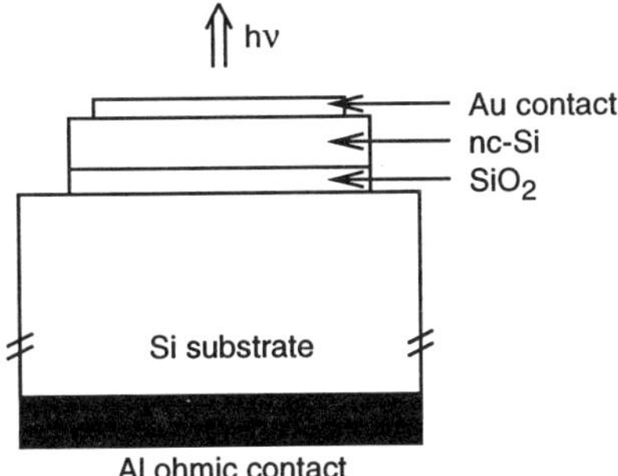

Fig. 9.5. Nanocrystalline Si LED [379]

A silicon device showing visible light emission based on a quantum effect and aiming towards a display application has been integrated into a microelectronic circuit [385]. The authors of this article had recently shown that the thermal and chemical stability of porous silicon can be enhanced by partial oxidation while retaining the desired light emission [386,387]. These improvements in material properties were used in order to demonstrate the successful integration of silicon-based visible light emitting devices (LEDs) into a bipolar microelectronic circuit. The circuit with the Si-based LED is shown in Fig. 9.6. First, however, we will restrict ourselves to the LED, which is based on silicon-rich silicon oxide (SRSO) [386, 387]. In contrast to porous silicon, the active SRSO layer satisfies processing requirements such as tolerance to thermal processing up to 1000 °C and chemical resistance.

The fabrication sequence for the SRSO-based LED will be described in the following. A 10 Ωcm P-type Si wafer with a heavily P$^+$-doped surface layer was anodized in a HF/ethanol solution. The P$^+$ region is transformed into a mesoporous layer with a porosity of about 40 %. This P$^+$ region will be called the transition layer (TL). The underlying P-type Si is transformed into a nanoporous Si layer with a porosity of about 75–80 %. This nanoporous Si layer extends to a depth of 0.5–1.0 μm into the Si substrate. Annealing at 800–900 °C in a dilute oxygen atmosphere (10 % O_2 in N_2) then transformed the as-anodized hydrogen passivated porous silicon into SRSO. A 0.3 μm thick polysilicon film was deposited subsequently by low-pressure chemical vapor deposition at 610 °C, which was then N$^+$ doped and annealed at 900–1000 °C to form the cathode of the LED. Aluminum contacts were finally made to the N$^+$ polysilicon film and the bulk substrate. Electrons and holes can be

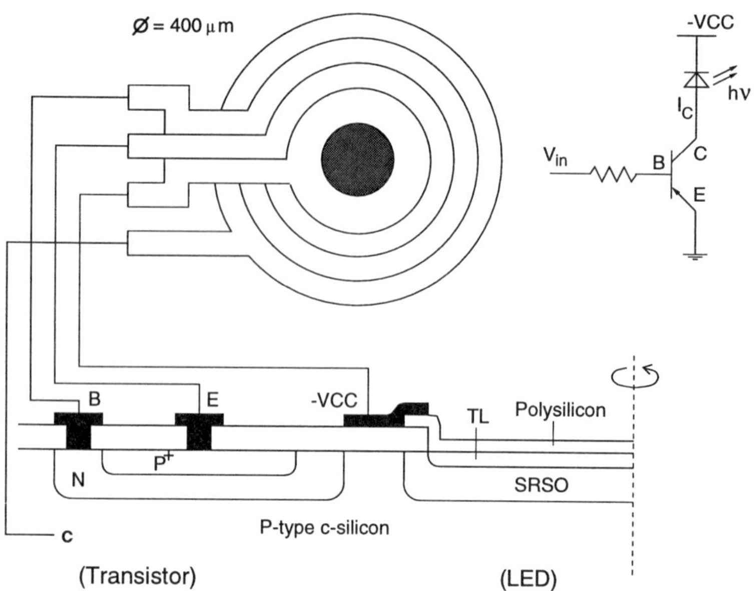

Fig. 9.6. Integrated silicon LED [385]

injected into the semi-insulating SRSO layer in this multilayer structure effectively. A low contact resistance with minimal light absorption is provided by the thin N^+ polysilicon layer. Effective carrier transport from the N^+ polysilicon cathode to the active light-emitting SRSO layer results from the low density of interface states of less than $10^{12}\,\mathrm{cm}^{-2}$ [387] between the polysilicon and the mesoporous transition layer. The surface passivation by the polysilicon layer and the effective carrier injection resulted in an enhanced quantum efficiency of the LED. The SRSO-based LEDs exhibited electroluminescence peaks between 1.7 and 2.0 eV at room temperature with an applied voltage of approximately 2 V and a current density of approximately $10\,\mathrm{mA\,cm}^{-2}$ with a maximum light intensity of approximately $1\,\mathrm{mW\,cm}^{-2}$. The highest external power efficiency was about 0.1 %. The modulation bandwidth exceeded 1 MHz. The SRSO-based LEDs were stable without degradation for several weeks of continuous operation.

After the description of the process sequence and of the properties of a SRSO-based LED, the integration of this LED into a bipolar microelectronic circuit will be discussed (Fig. 9.6): A PNP driver transistor is connected to the LED in a common-emitter configuration. This driver transistor can be used to modulate the light emission. A circular area-efficient design has been chosen. The original bipolar process was modified by including additional lithography steps in order to integrate the SRSO-based LED. The bipolar process used implanted base and emitter regions in an implanted P-well collector on an N-type substrate or a bulk P-type wafer for nonisolated structures. A thermal oxide was grown during the emitter and base drive-in steps for electrical isolation. This oxide was patterned and etched to open windows to bare silicon for fabrication of the LED. A Si_3N_4 layer was deposited by low-pressure chemical vapor deposition and patterned in order to protect the transistor regions during the formation of the anodized regions. The LED process sequence continued with a high-dose BF_2 implant and anneal for the P^+ layer formation, the anodization, and SRSO formation. The N^+ polysilicon layer was then deposited and patterned to form the LED cathode and to serve as a local interconnect (Fig. 9.6). Contact holes were patterned and etched through the Si_3N_4, which remained for electrical isolation, and through the underlying SiO_2 layer to the bipolar driver device. Finally, an aluminum layer was sputtered, patterned, and annealed to form the device contacts and interconnects.

The PNP transistors exhibited a current gain of approximately 100, which was typical for the bipolar process used. The LEDs possessed a rectifying ratio of about 10^5. Bright orange electroluminescence (EL) was observed for the integrated LEDs. The shape of the EL spectrum was independent of the applied current. Neither degradation of the LED nor of the driver transistor characteristics was observed. The drivers were able to amplify low-level base input signals to current pulses of about 100 mA and to modulate the LEDs at frequencies in the kHz range, which did not represent the upper limit of the LED switching speed as mentioned above. The authors of [385] report that further improvement in efficiency and power dissipation is necessary for display applications.

9.2 SiGe and SiGeC

$Si_{1-x}Ge_x$, $Si_{1-y}C_y$, and $Si_{1-x-y}Ge_xC_y$ alloys all have indirect bandgaps [306]. The luminescence without the support of phonons has been exploited nevertheless, because alloy scattering due to the random nature of the alloy helps to conserve the momentum in transitions between the conduction and valence band [388–392]. This process is stronger than phonon-assisted processes, but it is still a weak process [284]. The efficiency of the electroluminescence compared to SiGe quantum wells can be increased in SiGe quantum dots for low injection conditions due to the localization of excitons [393]. For high injection conditions, however, the efficiency decreases due to nonradiative Auger recombination. Another example of a SiGe quantum dot LED with a low quantum efficiency (η =0.14 % at 1.312 μm) is given in [394]. The fabrication of highly efficient optical emitters using $Si_{1-x}Ge_x$, $Si_{1-y}C_y$, and $Si_{1-x-y}Ge_xC_y$ materials, therefore, remains difficult.

There is, however, an aspect which may promise an improvement of the situation of column IV optoelectronic devices. The direct bandgap of Ge is only 0.1 eV above the indirect gap. An alloy of tin (Sn) and Ge probably could reduce this direct gap below the indirect gap resulting in a direct-bandgap semiconductor on a SnGe/Ge/Si structure [395]. The growth of SnGe layers, however, is difficult because of the low solubility of Sn in Ge and because of the surface segregation of Sn during growth of SnGe. Therefore, the required very low growth temperatures for Sn_xGe_{1-x} alloys make the fabrication of Sn_xGe_{1-x} optoelectronic devices on Ge very difficult [284], although a direct bandgap of $Sn_{0.15}Ge_{0.85}$ with an energy of 0.346 eV lower than the indirect bandgap of 0.441 eV has been verified experimentally [396]. Furthermore, this direct bandgap corresponds to a wavelength of 3.5 μm, incompatible with fibers used in optical communication. Photodetectors with such low bandgaps for receiving these low-energy photons usually exhibit high dark currents and high noise at room temperature. The integration of $Sn_{0.15}Ge_{0.85}$ light emitters on Si, therefore, seems rather unlikely.

9.3 Erbium-Doped Silicon

Another approach to a silicon light emitter in the infrared spectral range is the doping of silicon with erbium. Erbium emits light at the same wavelength independently of the host. This emission is due to the excited states of the 4f manifold of trivalent Er. The transition from the lowest excited state of this manifold to the 4f ground state yields light at 1.54 μm [397]. The emission energy is independent of the temperature due to the inner atomic transition. Room temperature electroluminescence has been reported from Er-doped Si, however with a low internal quantum efficiency of approximately 6×10^{-5} [398] although codoping of Er with oxygen has been exploited for an increased efficiency [397]. The quantum efficiency for a reverse bias was found to be larger than for a forward bias. This finding led to the conclusion that the mechanism of exciting the Er is based on hot carrier impact [398].

The first light emitting diodes (LEDs) were made through the implantation of Er at 20 keV during molecular beam epitaxial (MBE) growth [399]. The devices were then implanted with boron to form a junction with the N-type substrate, and mesa structures were etched to define the devices. Electroluminescence of Er was observed at 77 K with forward bias and 1 mA of current. The quantum efficiency of these devices was estimated to be 5×10^{-4} at 77 K.

LEDs operating at room temperature have been made from both high energy (4.5 MeV) and low energy (400 keV) Er implanted Si [400, 401]. Oxygen was coimplanted in these devices to overlap the Er profile. The device processing was kept compatible with VLSI. It was furthermore shown that cross-contamination in device processing due to Er was negligible. The low energy implanted samples produced 2.5 times more light per Er atom than the high energy implanted samples. This difference was probably due to less crystal defects for the low energy implantation. An Er-doped Si LED integrated into a SOI waveguide was reported in [401]. Figure 9.7 shows this edge emitting Si LED.

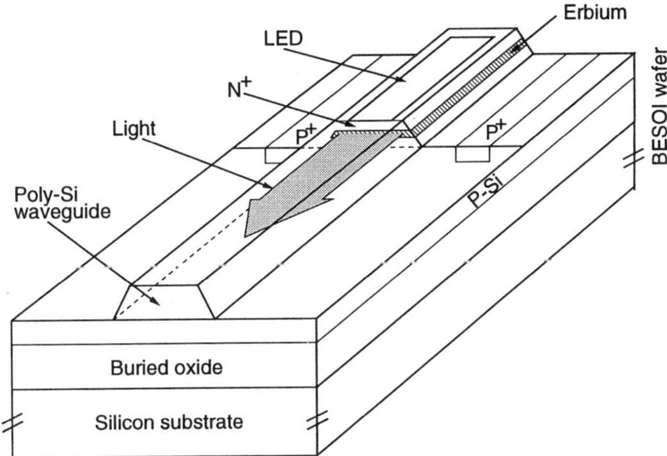

Fig. 9.7. Cross section of an Er-doped Si-LED coupled to a waveguide on SOI [401]

The major obstacle to the use of Er for the fabrication of c-Si LEDs is the strong luminescence quenching towards higher temperature [402]. Compared to a temperature of 50 K, the electroluminescence decreases by more than two orders of magnitude towards room temperature. Another disadvantage of Er is the long lifetime of the excited state of approximately 1 ms due to the inner atomic electron transition. According to this long lifetime, Er-doped Si LEDs cannot be modulated rapidly. An external modulator is needed for high data rates. The inner atomic transition is also the reason for the poor excitation efficiency, because the atomic shells involved in the transition are rather well screened from the surroundings by other shells.

Another way of achieving high concentrations of oxygen is the exploitation of semi-insulating polycrystalline silicon (SIPOS) being used for passivation and enhancement of the breakdown voltage of power semiconductors. SIPOS is a material containing a large concentration of oxygen, but still shows a semiconducting behavior. The microstructure of SIPOS annealed above 800 °C consists of Si nanograins surrounded by oxygen-rich SiO_x shells. The resistivity at room temperature ranges from 10^6–10^9 Ωcm. The combination of its semiconducting properties and its high oxygen content makes SIPOS a very interesting candidate as a host material for Er. Indeed, Er-implanted SIPOS showed intense room temperature photoluminescence (PL) at 1.54 μm [403]. The approximately 1 μm thick SIPOS layers were deposited onto single crystalline (100) Si wafers by low-pressure chemical vapor deposition of SiH_4 and N_2O at a substrate temperature of 620 °C. Varying the flow ratio SiH_4/N_2O resulted in layers with different oxygen contents between 4 and 27 at %. After deposition the wafers were annealed at 920 °C for 30 min in an oxygen ambient. Then, the SIPOS films were implanted with 500 keV Er ions with a dose of 2×10^{15} cm^2. Finally, the samples were annealed in vacuum for 30 min at

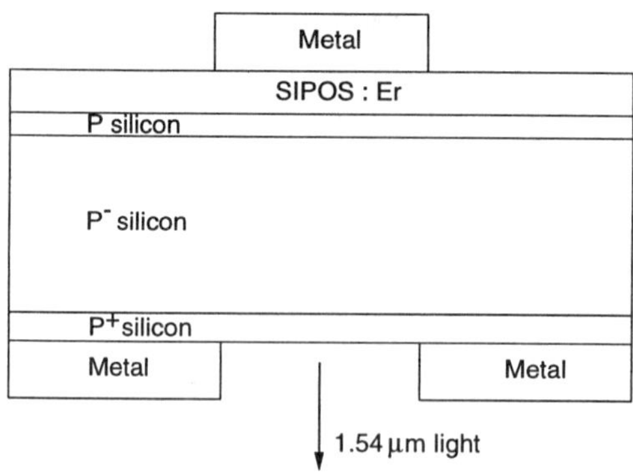

Fig. 9.8. Cross section of an Er-doped Si-LED [404]

temperatures in the range from 300 to 1100 °C. The PL intensity reached its maximum for annealing temperatures between 500 and 700 °C [403].

In order to exploit the potential of Er-implanted SIPOS as a material for electrically activated light emission, a structure shown in Fig. 9.8 has been suggested. Boron-doped Si wafers with a resistivity of 10–20 Ωcm, (100) oriented, and polished on both sides, were implanted with 80 keV at the top with a fluence of $5 \times 10^{13}\,\mathrm{cm}^{-2}$ and at the bottom with $2 \times 10^{15}\,\mathrm{cm}^{-2}$. Rapid thermal annealing (RTA) at 1000 °C for 5 min in N_2 was applied for dopant activation. After removal of the native oxide by a dip in dilute HF, SIPOS was deposited by chemical vapor deposition of SiH_4 and N_2O at 620 °C resulting in a 30 nm thick layer with 27 at % oxygen. Then erbium was implanted with an energy of 35 keV at an angle of 60 ° from the surface normal in order to confine all the erbium in the SIPOS layer. A dose of $5 \times 10^{14}\,\mathrm{cm}^{-2}$ resulting in an Er peak concentration of 1.6 at % and a rapid thermal annealing for 5 min in N_2 at 400 °C showed the most intensive electroluminescence.

A metal electrode with an area of $0.16\,\mathrm{cm}^2$ is placed on top of the SIPOS layer and the reverse side of the wafer is metallized and structured in order to obtain a hole in the back aluminum below the top electrode [404]. The 1.54 μm light, therefore, is emitted from the reverse side of the substrate. It has been estimated that only about 2 % of the generated light could exit through the hole in the back electrode. Electroluminescence has been verified with this structure at room temperature [404]. Due to the high resistivity of the SIPOS layer, the current is confined in the cylindrical volume under the metal contact on SIPOS and is then spread in the silicon substrate.

The intensity of the emitted light was about ten times higher under reverse bias than in forward direction (consider the device as a Schottky diode with the metal as cathode and the P-type substrate as anode). The device had to be operated in full breakdown, i. e. at large reverse currents, in order to achieve clear electroluminescence. Rather high voltages had to be applied due to a threshold voltage of 14.5 V for light emission. A threshold current of 75 mA was reported. Currents of up to 260 mA were used. An important result of this work clearly is that a very small quenching effect of the electroluminescence was observed up to 400 K [404]. It was concluded that the electroluminescence has to be attributed to the impact excitation of the Er atoms by hot carriers injected into the Si:O, Er layer.

9.4 Semiconducting Silicides

Nine semiconducting silicides are known: $CrSi_2$, $MnSi_2$, β-$FeSi_2$, Ru_2Si_3, $ReSi_{1.75}$, OsSi, Os_2Si, Os_2Si_3, and Ir_3Si_5 [405]. Band structure calculations verified the semiconducting nature [406]. β-$FeSi_2$ is the most promising candidate of these silicides for an efficient light emitter.

β-$FeSi_2$ has an orthorhombic crystallographic structure. Both Fe and Si occupy two crystallographically inequivalent sites with slightly differing distances to nearest neighbors. The lowest energy gap at 0.78 eV was found to be rather structure sensitive [407]. A slight decrease in the nearest-neighbor distance induces a change of the gap nature from indirect to direct. A second direct gap has an energy of 0.82 eV. These theoretical results are in good agreement with experimental results.

Optical investigations were performed in order to determine the real (ϵ_1) and imaginary (ϵ_2) parts of the dielectric function $\bar{\epsilon} = \bar{\epsilon}_1 + i\bar{\epsilon}_2 = (\bar{n}_{sc} + i\bar{\kappa})^2$ [406]. The absorption coefficient can be calculated from $\bar{\kappa}$ (compare with (1.15)). The β-phase of $FeSi_2$ is the only semiconducting silicide which showed light emission. Recently, a light-emitting diode emitting at 1.5 μm has been demonstrated [408]. β-$FeSi_2$ has been incorporated into a conventional silicon bipolar junction. The samples have been fabricated using ion-beam synthesis with Fe^+ ion implantation into Si wafers followed by high-temperature annealing. It should be mentioned, however, that light emission from epitaxial layers and single crystals of β-$FeSi_2$ could not be observed down to the lowest temperatures. It may be assumed, therefore, that light emission from ion-beam synthesized samples is due to yet unidentified defects in Si or β-$FeSi_2$. Stress and related deformation in the silicide lattice or emission from β-$FeSi_2$ precipitates may be involved in the light emission found.

There may be several reasons why no light emission from single crystals or epitaxial layers could be observed. The quality of single crystals and epitaxial layers is not very good. Intentionally undoped $FeSi_2$ exhibits carrier densities of 10^{18} to $10^{20}\,cm^{-3}$ at room temperature. Experiments with epitaxial layers have shown that the photoresponse of β-$FeSi_2$ is limited to low temperatures only. Electrical transport measurements have proved the presence of deep acceptor states in the β-$FeSi_2$ layers. These deep defects act as recombination centers and cause very short minority carrier lifetimes. This short lifetime may be much shorter than the radiative lifetime and no radiative recombination takes place. Further improvement of the layer growth technique is necessary in order to reduce the high defect density.

At room temperature, interaction of carriers with acoustic and optical phonons seems to be the dominating carrier scattering mechanism. The maximum room temperature mobilities are between 10 and 40 cm^2/Vs for epitaxial films and single crystals. The electron mobility is much smaller than the hole mobility. These circumstances lead to the conclusion that $FeSi_2$-LEDs may be slow.

A heterojunction is present between Si and $FeSi_2$. The conduction band offset at the interface is about 0.3 eV and there is almost no valence band offset [407]. Effective electron injection from Si into $FeSi_2$ should therefore be possible.

9.5 Organic Polymers on Silicon

Organic polymers for applications in light emitting diodes have been investigated at Kodak since 1987 [409]. In Cambridge, England, research on organic LEDs was started several years later [410]. The main reason for the interest in organic polymers as light-emitting diodes in flat-panel displays is their high quantum efficiency in combination with a low operating voltage and a simple device processing. The simplest design of an organic LED consists of three layers: two electrodes for carrier injection and the light-emitting polymer between the electrodes. One electrode, for instance, can be ITO-coated glass for hole injection and the second electrode can be a low work-function metallic electrode for electron injection.

Conjugated polymers possess a semiconductor-like electron configuration. The macromolecules of these polymers contain alternating systems of single and double bonds. The π-electrons, therefore, are delocalized over the whole molecule chain. Due to their interaction, the discrete energy levels of the single electrons change to bands of continuous levels. Between the energy bands, there are gaps as in inorganic semiconductors. The highest occupied molecular orbital (HOMO) corresponds to the valence band, and the lowest unoccupied molecular orbital (LUMO) to the conduction band. The emission of photons is possible due to recombination of oppositely charged carriers injected from the electrodes.

Since 1989, many different materials have been the focus of intensive research. Some examples of organic LEDs (OLEDs) may be mentioned [411–416]. OLEDs can be formed by vacuum deposition of organic molecules with emission in the red, green, and blue spectral range [414]. Their lifetime already exceeds 10 000 h [414]. The emission spectrum of OLEDs also can be tailored by spin-coating a soluble precursor polymer and subsequently converting it into the conjugated state by thermal treatment [417] and attaching different side groups, thus influencing the energy gap. Quantum efficiencies from 0.7 to 1.2 % were reported for OLEDs [415, 416].

Due to the success in OLED research, the compatibility of OLEDs with CMOS technology has been investigated recently [418]. Accordingly, conjugated polymers with a poly(paraphenylene-vinylene) (PPV) backbone are suited for optoelectronic applications on silicon. P^+ silicon is used as the

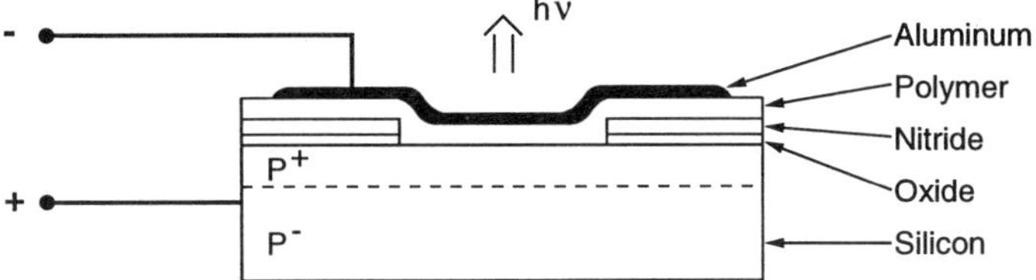

Fig. 9.9. Cross section of an organic LED on Si [418]

anode (see Fig. 9.9), because of the low barrier from the Si valence band to the polymer HOMO. The boron implantation has been performed through a 25 nm pad oxide. Si_3N_4 has then been deposited by LPCVD and structured for local insulation of the LEDs. An annealing step for boron activation followed. The precursor polymer was spun on the wafer immediately after oxide removal with HF. The conjugation was done by a thermal treatment in nitrogen atmosphere at 220 °C for 6 h. The semitransparent aluminum cathode was sputtered through a shadow mask for simplicity in this investigation [418]. Because of the requirements for MOS compatibility, aluminum or titanium is the best choice although calcium with the lowest work function would lead to a more effective electron injection into the polymer layer. The cathode on the PPV can be structured by conventional CMOS process steps ($SiCl_4$ plasma etching) without removing the organic polymer layer. The polymer itself can be etched by reactive ion etching in an O_2 plasma. For completeness, it should be mentioned that the temperatures of standard oxide and nitride deposition steps in CMOS processes may be too high for organic polymers when deposited before these steps. Therefore, OLEDs have to be fabricated after completely finishing the CMOS process.

The PPV-OLED on Si showed an onset of light emission at 5 V with an operation current of 1.5 nA/μm^2 [418]. The peak in the light emission was at about 2.2 eV, i. e. at $\lambda = 550$ nm with a full width at half maximum of 70 nm. Dynamic measurements indicated a low carrier mobility of about 10^{-4} cm^2/Vs. The results reported in [418] are promising for OLED displays integrated on CMOS chips.

9.6 Liquid Crystals

Liquid crystal displays (LCDs) need less electrical power than light emitting diode (LED) displays. LCDs, therefore, are widely used. LCDs are very important particularly in portable applications. Usually LCD modules, which contain chips with input logic and registers and sometimes microcontroller in addition to the actual but separate LCD display device, are used. In order to reduce the weight, volume, and power consumption of portable systems, liquid-crystal-on-silicon (LCOS) technology has been developed [419]. LCOS enables same-die integration of control circuitry and display for substantial overall system power reduction by eliminating parasitic bondpad and package contact capacitances as well as by implementing a power saving control logic configuration. LCOS creates the ability to display high-resolution images.

A 160×120 LCOS microdisplay with a 25 μm pixel size integrated on a 0.8 μm 3-metal silicon backplane has been presented [419]. The display consists of a three-element stack: CMOS silicon backplane chip, liquid crystal material, and transparent conducting cover. Figure 9.10 also shows precision spacers to control the electrode separation, a surface alignment layer, and the cover glass. The display uses a twisted-nematic (TN) liquid crystal (LC) in which the LC molecules twist between the bottom and top electrodes, rotating the polarization of incident light to pass through a laminated polarizer. The molecules orient vertically, when an electric field is applied, destroying the twist, not rotating the polarization of the incident light, resulting in a dark pixel.

Surface topology and optical properties have to be considered in the pixel design in addition to circuit properties. The pixel electrode is formed by a rectangle of metal 3. This electrode also acts as a highly reflective mirror and as a light shield for incident light to reduce leakage-inducing photocurrents in the CMOS substrate. The reflectivity is increased by mirror flatness, which is enhanced by the layout of underlying pixel circuitry. The brightness of the displayed picture at a particular pixel is controlled by applying a

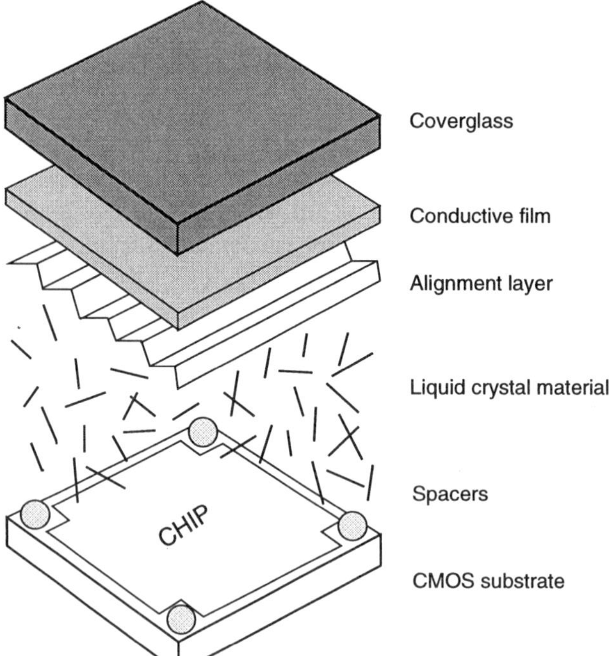

Fig. 9.10. Liquid crystal display on a CMOS chip

voltage between the glass and pixel electrodes resulting in an analog display mode. The higher the voltage, the darker the pixel. This voltage is sampled by a CMOS transmission gate and stored on a capacitor formed by both diffusion/substrate and intermetal dielectric structures. There is an additional nonlinear capacitor formed by the pixel-LC-coverglass structure.

An analog display architecture has been implemented by integrating a 6-bit sampled-ramp digital-to-analog converter (DAC) and a row and column matrix in order to generate and direct the voltage to individual pixels for brightness control. This sampled analog display architecture reduces the power consumption considerably. At 4 MHz, corresponding to a picture frequency of about 200 Hz, and 5 V a power of 7.25 mW was consumed by the LCOS chip while displaying a worst-case black/white checkerboard pattern. The LCOS display with an overall size of $4 \times 3\,\mathrm{mm}^2$ consumed 5.9 mW when displaying a constant uniform image. For a 640×480 LCOS display, which will be driven by a 60 Hz video input corresponding to a 18.4 MHz pixel clock, a power consumption of less than 6 mW at 2 V digital supply has been estimated. Such a low power consumption is realized because the DAC replaces a single pixel-rate, high-voltage, high-capacitance analog signal with several pixel-rate digital signals and associated logic, enabling the use of digital low-power techniques.

10. Integrated Optics

Integrated optics covers a very wide field. Silicon, polysilicon, SiO_2, and Si_3N_4 are among the materials which are important in integrated optics due to the highly developed silicon process technology. Although Si-based elements are only a subdivision of integrated optics, the number of reports describing Si-based elements is very large and only some of them can be discussed here. Among the Si-based elements of integrated optics, examples of waveguides, filters, optical amplifiers, micro-spectrometers, pressure and distance sensors, gratings, grating couplers, and beam splitters fabricated on silicon will be summarized in this chapter. After waveguides, a photonic bandgap filter, and an optical amplifier, the other elements will be described together with their implementation in integrated optical systems. The so-called planar hybrid-integration electronic-module packaging and the Z-axis photonic interconnect approach are included as examples of optical interconnect technology. Furthermore, an optical transceiver based on wavelength division multiplexing will be described.

10.1 Waveguides

10.1.1 Total Reflection

Before some integrated waveguides are described, the basic principle will be briefly explained. Waveguides are based on total reflection of a light ray at the interface of two dielectric layers with different refractive indices, whereby the guiding layer, i. e. the core of the waveguide, has the larger refractive index $\bar{n}_1$ and the bottom and top cladding layers have a lower refractive index $\bar{n}_2$ ($\bar{n}_1 > \bar{n}_2$). Total reflection occurs when the light ray has an incident angle Θ_i between the so-called critical angle Θ_c and 90 ° (see Fig. 10.1).

The critical angle is given by

$$\Theta_c = \arcsin(\bar{n}_2/\bar{n}_1). \tag{10.1}$$

The bottom and top cladding layers have to be sufficiently thick in order to avoid losses for the guided wave, because the electric field penetrates exponentially into the cladding [420]. This penetration, however, only leads

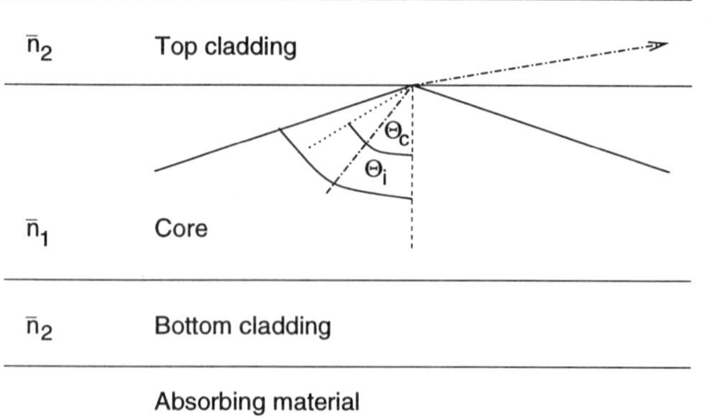

Fig. 10.1. Total reflection in a waveguide

to losses when the cladding is thin and the electric field enters an absorbing medium below the bottom cladding or on top of the upper waveguide cladding layer.

10.1.2 Losses

Waveguides for integrated optical circuits have to fulfill several demands such as (i) low optical attenuation, (ii) high coupling efficiency to optical fibers, and sometimes (iii) monomode propagation. Monomode operation restricts the waveguide cross section to dimensions of the order of a few wavelengths, which can be realized in conventional microelectronic integration. The coupling efficiency from integrated waveguides into optical fibers usually is high. Due to the usually considerably smaller dimensions of the waveguide cross section than the core diameter of optical fibers, the coupling efficiency from optical fibers into integrated waveguides still poses a problem. Waveguide losses are primarily due to material absorption because of incorporation of hydrogen, due to substrate losses by leaky waves into the silicon substrate, and due to scattering losses at rough surfaces. The first and third effects can be reduced by optimizing the deposition processes. In order to reduce substrate losses, the bottom cladding layer has to be made thick enough.

10.1.3 Waveguides for Visible and Infrared Light

The main point of loss reduction is the confinement of the evanescent field distribution of the waveguiding film to areas far away from rough surfaces. Therefore, lateral confinement should not be done by structuring the waveguiding film. Instead, the strip-loaded design is commonly used (Fig. 10.2) according to which a step is etched into the top cladding layer.

Silicon oxynitride (SiO_xN_y also denoted SiON) is a material of high transparency and flexibility because the refractive index is adaptable between 1.46 ($x=2$, $y=0$) and 2.0 (Si_3N_4). The refractive index is correlated linearly to the oxygen content of the material. The refractive index can be controlled by varying the oxygen gas flow between zero and 20 % oxygen relative to the total gas flow in a plasma-enhanced or low-pressure chemical vapor deposition process. Absorption in SiON can be neglected in the wavelength range from about 300 to 1100 nm, in which silicon absorbs. Film thickness, refractive index, and homogeneity can be optimized to variations below 1 % across wafers and from run to run [421].

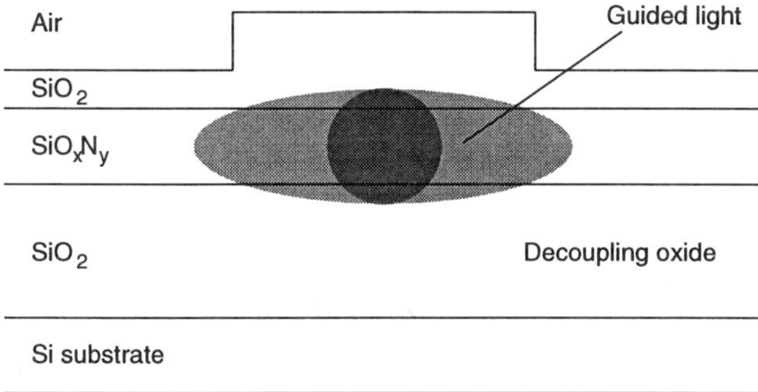

Fig. 10.2. Strip-loaded SiO_xN_y waveguide

The waveguiding film sandwich consists of a thick SiO_2 layer insulating the waveguide region from the silicon substrate, an SiO_xN_y film with higher refractive index and a SiO_2 cover layer (Fig. 10.2). Using the plasma enhanced chemical vapor deposition (PECVD) technique the layer sequence can be deposited within one continuous run, whereas low pressure chemical vapor deposition (LPCVD) starts from the thick thermally grown SiO_2 base layer because of the lower deposition rate [421].

The vertical confinement of the optical wave is based on total reflection due to the higher refractive index in the SiO_xN_y core of the waveguide and the lower refractive index in the SiO_2 cladding layers. The lateral confinement is achieved by a step in the top oxide cladding layer (Fig. 10.2). The strip-loaded waveguide can be designed by the effective index method [422], which yields good agreement with experimental results. An attenuation of 0.2–0.3 dB/cm for $\lambda = 633$ nm can be achieved for strip-loaded waveguides [421].

In [67] a SiON waveguide has been integrated in a 0.8 μm CMOS process. This monomode waveguide is shown in Fig. 10.3. The rib width was 3 μm. The propagation loss was reported to be less than 0.5 dB/cm at 633 nm [68]. Three different possibilities for coupling optical signals into integrated photodiodes (see Fig. 10.4) have been suggested [67]. For all three possibilities the silicon is etched before field oxidation. The first possibility is direct butt coupling applying a mesa structure for the lateral photodetector. The second possibility is leaky-wave coupling by a step-less transition from field oxide to

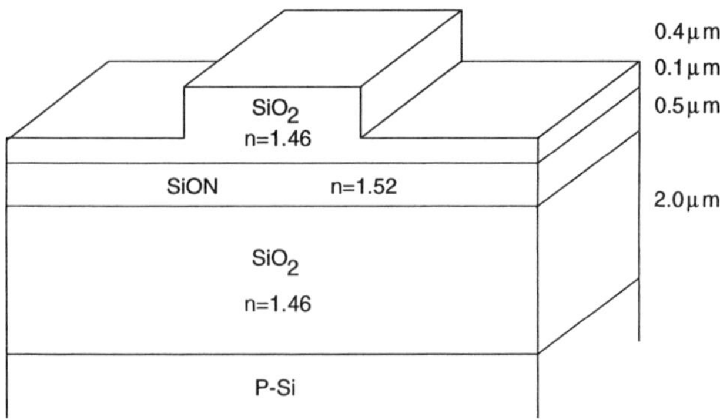

Fig. 10.3. Monomode SiON waveguide [68]

the active light detector area. The third possibility is mirror coupling using total reflection by fabricating a sloped end of the waveguide.

In the case of the first two possibilities, some CMOS process steps have to be changed. In order to ensure optical insulation of the light guiding film, the thickness of the field oxide has to be increased to 2 μm and the phosphorus-silicate-glass (PSG) has to be replaced by SiON [67]. These modifications strongly affect the MOS parameters. In [67], therefore, mirror coupling has been preferred. After the CMOS process was finished and MOS structures were tested, the PSG oxide film was deposited to increase the optical insulation layer thickness. Then the light-guiding SiON film and the oxide cover layer (not shown in Fig. 10.4) were deposited. The cover layer was structured to fabricate ribs by dry etching. Coupling mirrors and bondpad openings were

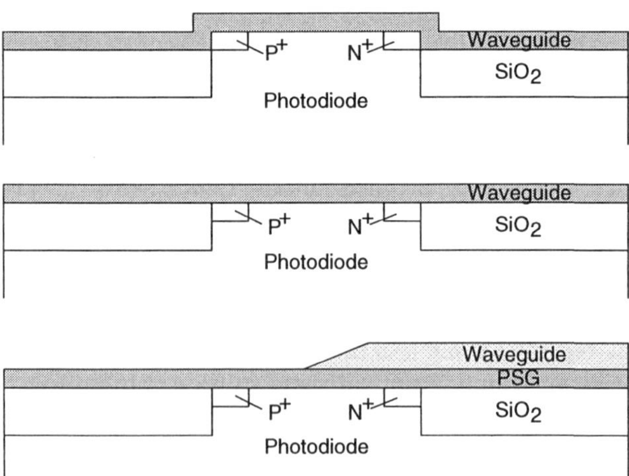

Fig. 10.4. Different possibilities for the coupling of light from integrated waveguides into photodiodes [67]

etched with an O_2/CHF_3 gas mixture for 45° sloped wall formation. Due to the processing of the waveguide after the standard CMOS process, MOS transistor parameters were not affected by the integration of the waveguide.

A SiON strip waveguide with a rib width of 3.5 μm and a height of 0.35 μm has been integrated in an industrial 2 μm CMOS process in [69]. Mirror coupling from the SiON waveguide to an N-well/P-substrate photodiode has been implemented here as it is suitable for the wavelength of 780 nm (see Fig. 10.5). The strip waveguide propagation loss reported for this wavelength was 1 dB/cm. The mirror loss was −3 dB. The photodiode responsivity was 0.5 A/W ($\eta = 70\,\%$).

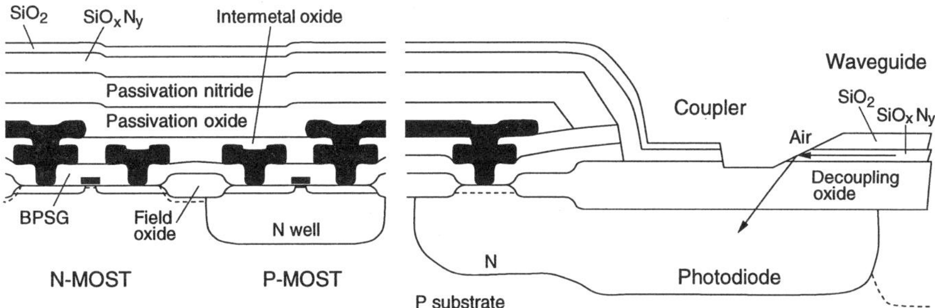

Fig. 10.5. Example of a CMOS-integrated waveguide [69]

Technologies for the optical interconnection in VLSI chips have been developed by using conventional Si LSI technologies [423]. A micrometer-size optical waveguide with a Si_3N_4 core ($\bar{n} = 2.0$) and SiO_2 cladding layers ($\bar{n} = 1.45$) has been fabricated (Fig. 10.6).

The propagation loss of the Si_3N_4 waveguide reduces with increasing wavelength (Fig. 10.7). The propagation loss of the waveguide with a core width of 10 μm was about 3 dB/cm for $\lambda = 940$ nm and for a core thickness of 470 nm (Fig. 10.7). A cross section of the waveguide without a planarization layer is shown in Fig. 10.6.

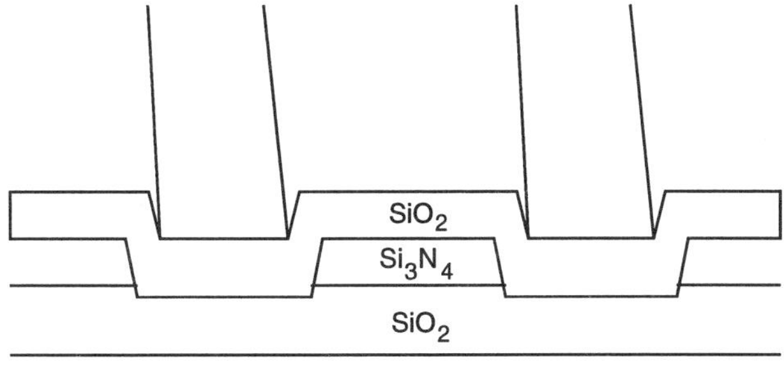

Fig. 10.6. Waveguide with a Si_3N_4 core [423]

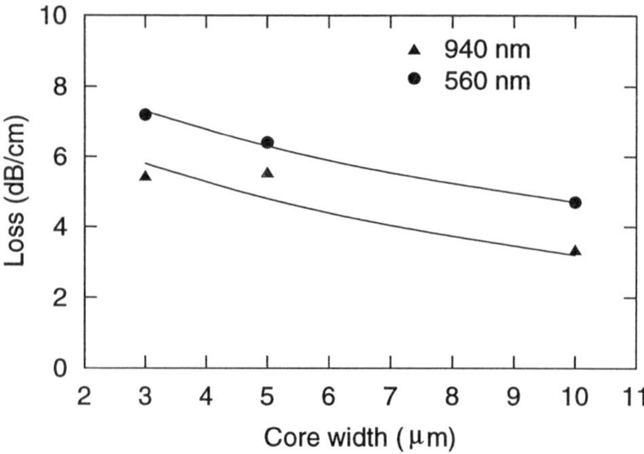

Fig. 10.7. Propagation loss of a Si_3N_4 waveguide with a core thickness of 470 nm [423]

A light direction converter, which changes the light propagation direction from vertical to horizontal for the interlayer connection has also been fabricated (Fig. 10.8). The coupling from the waveguide to a N^+/P-substrate photodiode is also shown in Fig. 10.8.

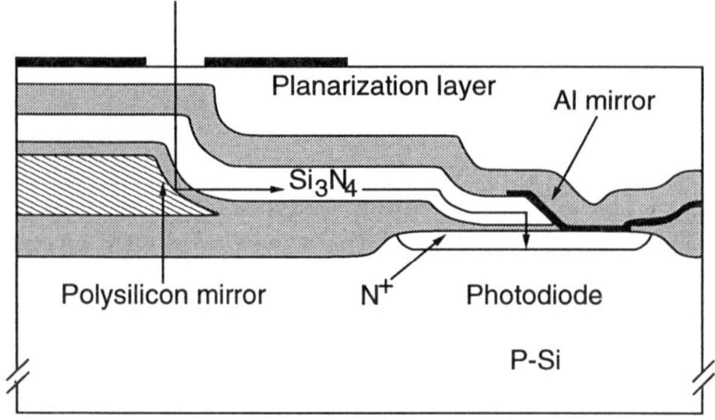

Fig. 10.8. Light direction converter [423]

In order to fabricate the light direction converter, a Si_3N_4 mask is formed on Si and a groove with a depth of 1 μm is prepared by wet chemical etching with HF:HNO_3 (5:95). Then the $SiO_2/Si_3N_4/SiO_2$ waveguide is formed (see Fig. 10.9).

The microlenses in Figs. 10.10 and 10.11 were made from resist and polysilicon, respectively. The resist lens is obtained by patterning of a resist layer by photolithography to produce a rectangular shape and by heat-

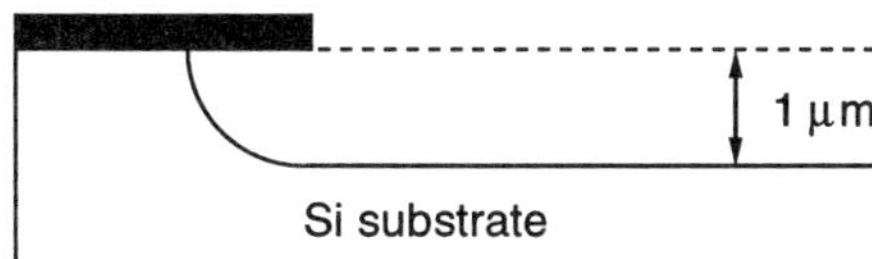

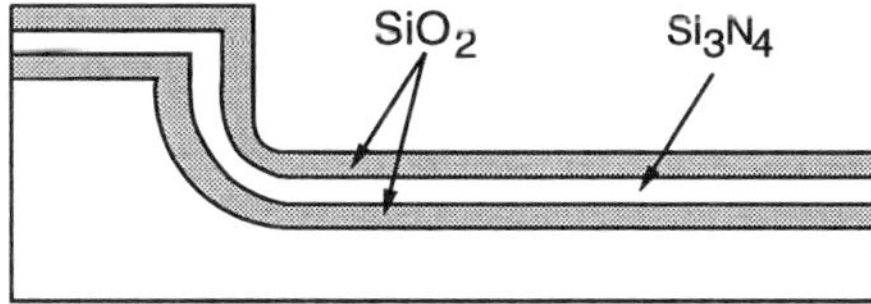

Fig. 10.9. Fabrication process of a light-direction converter exploiting a waveguide with a Si_3N_4 core [423]

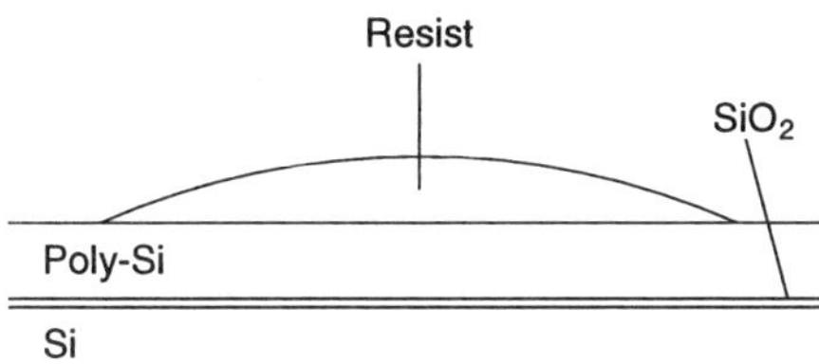

Fig. 10.10. Resist microlens [423]

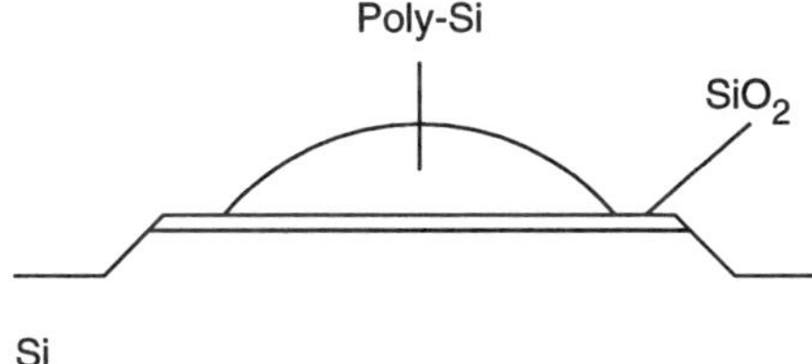

Fig. 10.11. Micrometer-size polysilicon lens [423]

ing to approximately 200 °C to obtain a circular shape. After hardening, the fabrication procedure of the lens is finished.

The fabrication process for the polysilicon lens starts from the resist lens. An etching procedure with a low selectivity is used and the shape of the resist lens is transferred to the polysilicon layer (Fig. 10.11).

10.1.4 Waveguides for Infrared Light

Infrared waveguides for light with wavelengths longer than about 1.2 µm can be realized with Si and SiGe [282]. Let us first consider an SOI infrared (IR) waveguide. Total reflection in Si on SiO_2 occurs because the optical index of refraction of Si is much larger than that of SiO_2 ($\bar{n}_{Si} = 3.5$ and $\bar{n}_{SiO_2} = 1.4$ at $\lambda = 1.3$ µm). Figure 10.12 shows such a waveguide.

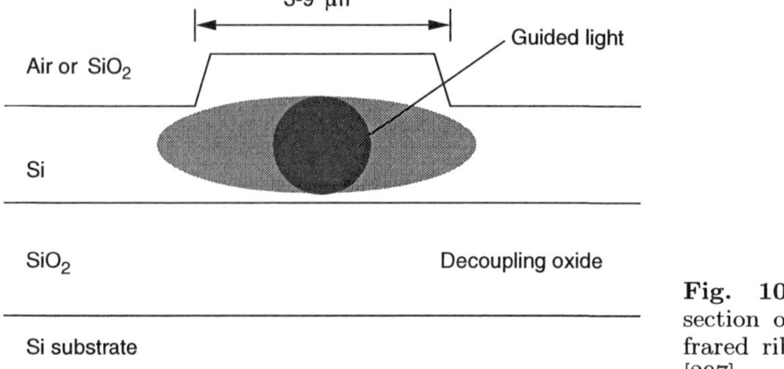

Fig. 10.12. Cross section of a SOI infrared rib waveguide [297]

The rib confines the guided light laterally. In fact, a single-mode waveguide can be obtained with thick SOI layers of 5.2 μm, for instance [297], because the dimensions (rib height: 2.2 μm; rib width: 3–9 μm) can be chosen in such a way that other modes are leaky and, therefore, not guided. Attenuations of about 0.5 dB/cm have been obtained for SOI rib waveguides at $\lambda = 1.3$ μm with an SOI layer thickness of 5–7 μm [297]. For an SOI single-mode waveguide with a top Si layer thickness of 11 μm, loss values around 0.1 dB/cm have been reported [424].

The optical index of refraction of Ge is 4.3, which is much larger than that of Si with 3.5 for $\lambda = 1300$ nm. A very small fraction x of the order of one percent, therefore, is sufficient for $Si_{1-x}Ge_x$ to obtain a waveguide core in a Si substrate. Figure 10.13 shows a SiGe IR waveguide core formed by diffusion of Ge into the Si substrate. Attenuation values of less than 1 dB/cm have been achieved for Ge-indiffused SiGe waveguides with SiGe stripe widths from 4 to 8 μm for $\lambda = 1.3$ μm [297].

Rib waveguides are also possible with SiGe on Si. Attenuation values of about 0.6 dB/cm were achieved at IBM for SiGe on Si rib waveguides for $\lambda = 1.3$ μm and $\lambda = 1.55$ μm [425].

Another type of infrared silicon waveguide was reported in [426]. The cladding layer of this waveguide was formed by a buried $CoSi_2$ layer (see

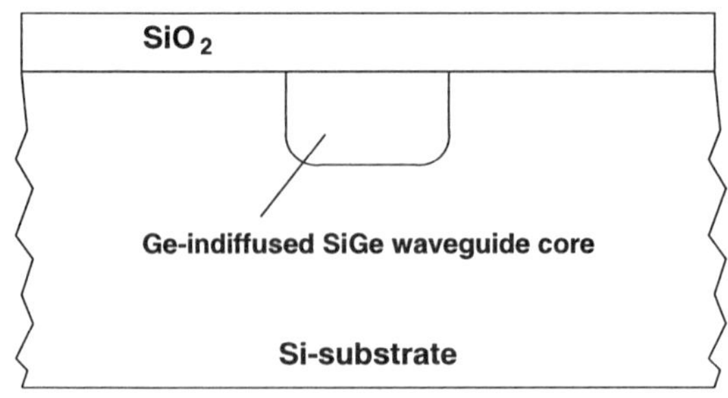

Fig. 10.13. Cross section of a SiGe infrared waveguide [297]

Fig. 10.14). The $CoSi_2$ serves not only as a reflective optical cladding but also as an electrical conductor. Co was implanted into a <100> P-type silicon wafer at an energy of about 180 keV with doses of $(1\text{–}5) \times 10^{17}\,\text{cm}^{-2}$. The implanted wafer was annealed for 2 h at 700 °C to produce a 50 nm thick layer of monocrystalline $CoSi_2$ buried about 0.4 μm below the wafer surface. A sheet resistance of 20 Ω/square was found for the 50 nm thick $CoSi_2$ layer. The cubic fluorite lattice of the buried $CoSi_2$ had a 1.2 % mismatch to the 0.5431 nm diamond lattice of Si. To increase the Si thickness on top of the $CoSi_2$ layer, epitaxy of Si on the 400 nm thick crystalline Si region was performed in a chemical-vapor-deposition (CVD) reactor to produce a Si thickness of 20 μm.

The complex index of refraction $\bar{n}' + i\bar{n}''$ is 0.464+i4.31 at a wavelength of 1.3 μm [426] and differs considerably from the 3.50+i0.0 index of Si. The $Si/CoSi_2$ index step, therefore, provides confinement of light in the Si layer on top of the silicide layer. An attenuation of 2.4 dB/cm was measured for the waveguide structure of Fig. 10.14 with a $CoSi_2$ thickness of 50 nm and a Si thickness of 20 μm.

Stacked Si waveguides shown in Fig. 10.15 were also investigated [426]. The attenuation for the upper and lower waveguide with a Si thickness of

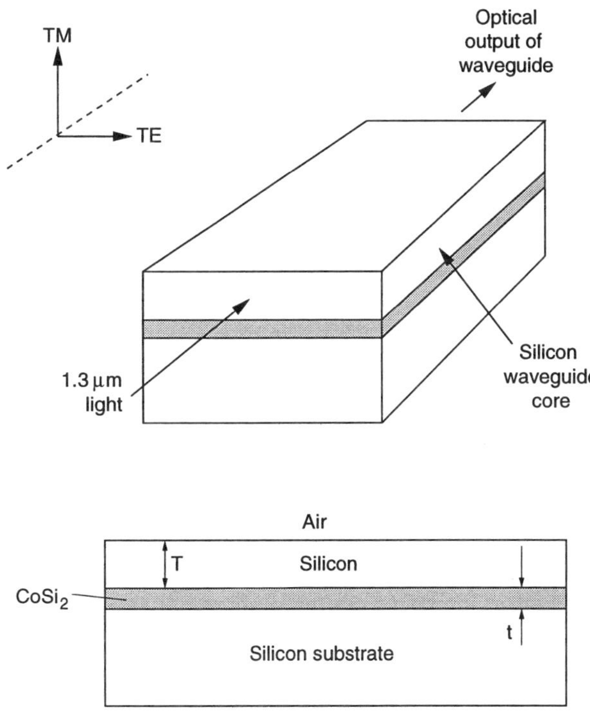

Fig. 10.14. $Si/CoSi_2$ waveguide [426]

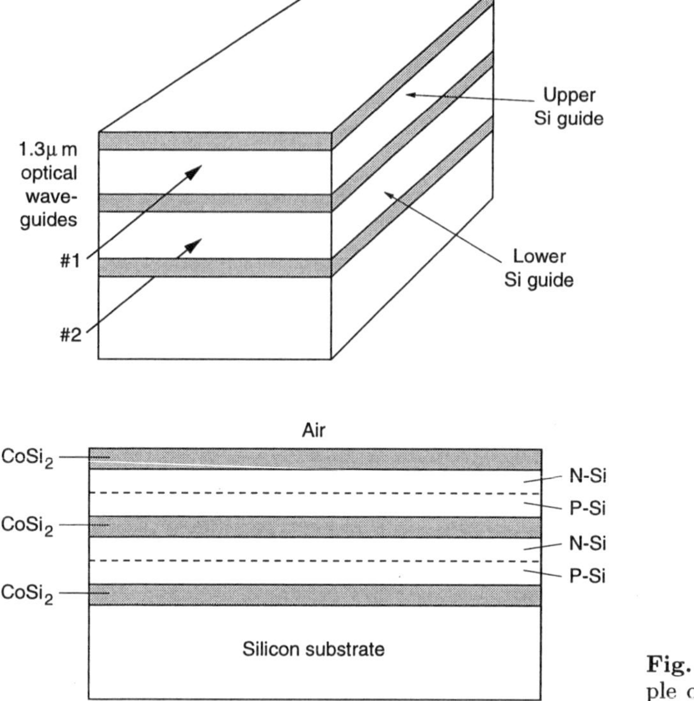

Fig. 10.15. Example of a stack of two Si/$CoSi_2$ waveguides [426]

20 µm each in this structure was approximately 4 dB/cm. The optical isolation of slightly more than 10 dB between the stacked waveguides, however, was quite moderate.

10.2 Photonic Bandgap Filter

The confinement of light to small volumes changes the probability of spontaneous emission from atoms. Both enhancement and inhibition can be achieved. The enhancement of the spontaneous emission rate [427] in an optical cavity is of special interest here. New photonic chip architectures and devices like low-threshold microlasers, filters, and signal routers may arise from these achievements. Photonic-bandgap (PBG) materials [428–431] sometimes also called photonic crystals are a means to confine light within a small volume of the order of $(\lambda_0/2\bar{n})^3$, where λ_0 is the vacuum wavelength and $\bar{n}$ is the refractive index of the material. For this purpose, photonic-bandgap materials exploit the scattering of light by periodic arrays of scattering centers.

Here, an advanced photonic-bandgap micro-cavity integrated into an optical waveguide [432] will be described (Fig. 10.16). A single-mode waveguide

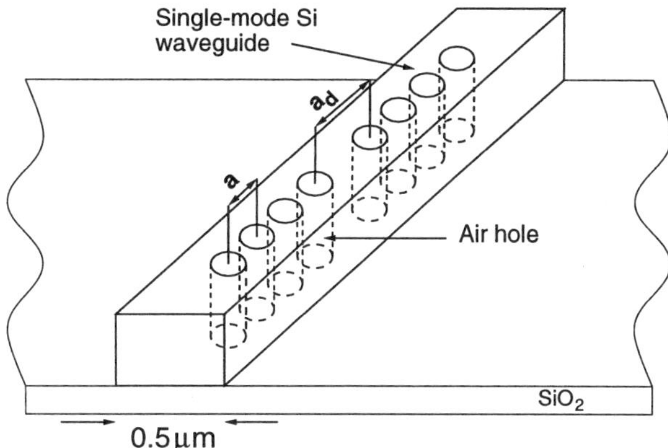

Fig. 10.16. Schematic of a photonic bandgap waveguide [432]

was fabricated in a SOI layer. A periodic array of air holes with a spatial period a in the waveguide limits the wave vector to π/a. This allowed range of wave vectors is similar to the Brillouin zone used in solid-state physics. In addition, the periodic array of holes, which actually is the PBG structure, has the effect of folding the dispersion relation of the strip waveguide and of splitting the lowest-order mode to form two allowable guided modes [433]. There is an energy gap, which is called photonic bandgap, between the two modes. Photons with energies in this energy range cannot pass the waveguide with the periodic hole structure. The size of the photonic bandgap is determined by the dielectric contrast of the periodic structure. The dielectric contrast for the device shown in Fig. 10.16 is defined by the Si waveguide ($\epsilon_{\mathrm{Si}} = 11.9$) and the air holes ($\epsilon_{\mathrm{air}} = 1$) etched into the waveguide. The large contrast results in a gap which is 27 % of the mid-gap energy. With a mid-gap energy of 0.80 eV ($\lambda = 1.54\,\mu\mathrm{m}$), the photonic bandgap is about 0.2 eV or 0.4 μm in width.

States in the photonic bandgap can be formed when a defect is included in the PBG structure. These states are analogous to energy levels within the semiconductor bandgap, which are introduced by impurities in the crystal. The PBG becomes transparent for photons with an energy equal to the energy of the state in the photonic bandgap. Defects in PBG structures can be obtained by incorporating breaks in the periodicity of the PBG device. The distance a_{d} of the air holes in the center of the PBG structure is increased by a factor of 1.5 to 0.63 μm, for instance, in order to introduce a state with a wavelength of $\lambda = 1.54\,\mu\mathrm{m}$ [432].

The device shown in Fig. 10.16 was fabricated using a Unibond SOI wafer with a 0.2 μm thick single-crystal Si layer on a 1.0 μm thick SiO_2 layer. The minimum feature size present in the waveguide micro-cavity occurs between the edges of the holes with a diameter of 0.2 μm and the edge of the waveguide with a width of 0.47 μm. The minimum feature size, therefore, is 0.135 μm, which was beyond the limits of optical lithography when the device was fabri-

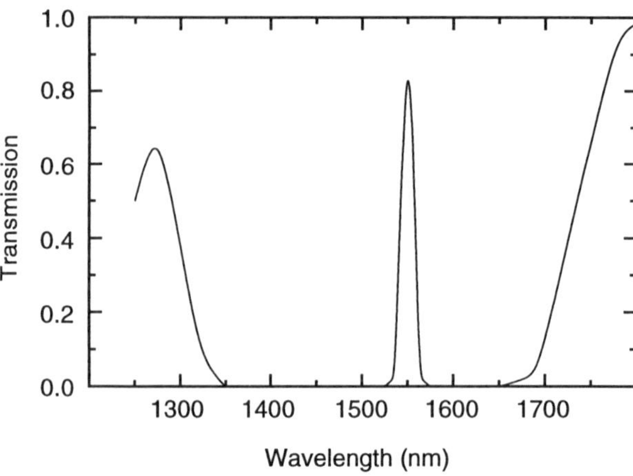

Fig. 10.17. Transmission for a PBG micro-cavity with four holes on each side of the micro-cavity [432]

cated. X-ray lithography with a Cu_L source at 1.3 nm was used to achieve pattern transfer at these small dimensions. The mask was fabricated by electron-beam lithography [434]. The 0.135 μm PBG feature was easily resolved with this approach.

The transmission spectrum of the PBG device (Fig. 10.17) was measured with an optical multichannel analyzer. A quality factor $Q = \lambda/\triangle\lambda$ of 265 can be determined from the center wavelength and the spectral width of the transmission peak shown in Fig. 10.17. Q is a measure for the ratio of the optical energy stored in the micro-cavity to the cycle-average power radiated out of the cavity. High quality factors can be exploited to increase the efficiency of LEDs. The spontaneous emission rate of a resonant-cavity light-emitting diode can be enhanced by a factor η_{mc} over the rate without cavity. The expression for η_{mc} is [435]:

$$\eta_{\mathrm{mc}} = \frac{Q}{4\pi V}\left(\frac{c}{\bar{n}\nu}\right)^3 \tag{10.2}$$

In this equation, V denotes the modal volume, c the speed of light, and ν the optical transmission frequency. Here, the modal volume V is $(5 \times \lambda/(2\bar{n}))^3 = 0.055\,\mu\mathrm{m}^2$. For the PBG device shown in Fig. 10.16, η_{mc} can be calculated to be 35. This large spontaneous emission enhancement could, for instance, lead to efficient Er-doped Si LEDs. An additional advantage of PBG waveguide micro-cavities over conventional stacked Bragg reflector mirror micro-cavities is the large bandgap. With a 400 nm stop-band, it is possible to have only a single resonant mode in the gain bandwidth of a semiconductor laser. In contrast to vertically integrated resonant cavity devices (VCSELs), the PBG configuration used in [432] allows a light-emitting cavity to be directly coupled into a waveguide, reducing coupling losses considerably.

It should also be mentioned that with high-dielectric contrast materials, waveguide cross-sectional dimensions can be reduced by more than an order of

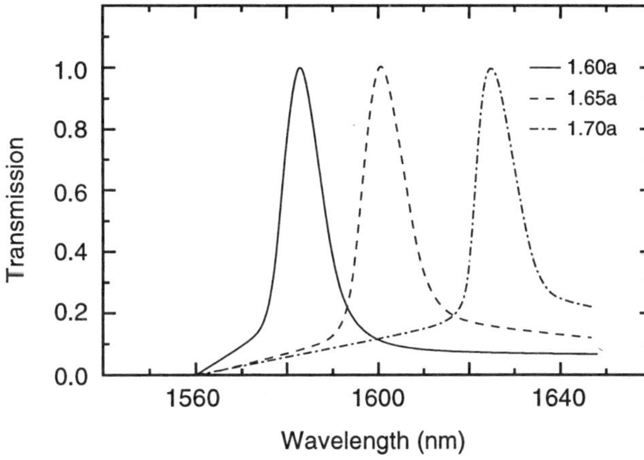

Fig. 10.18. Transmission characteristics of PBG waveguide micro-cavities for three different defect lengths. The resonance peak shifts to longer wavelengths with increasing defect length [432]

magnitude compared to low-index-difference waveguides. Densely integrated waveguide structures for optical communication systems become possible. This high-dielectric contrast waveguide technology could lead to an integrated optics approach compatible in materials and dimensions with sub-micrometer VLSI and ULSI.

The resonance wavelength of the PBG waveguide micro-cavity can be altered by changing the size a_d of the defect. Figure 10.18 shows the transmission spectra for $a_d = 1.60\,a$, $1.65\,a$, and $1.70\,a$. As the defect length increases, the resonance wavelength shifts to larger values. The results shown in Fig. 10.18 demonstrate that wavelength selective PBG devices can be realized for wavelength-division-multiplexed (WDM) optical data transmission.

10.3 Optical Amplifier

The spontaneous emission characteristics from Er in a Si/SiO_2 cavity consisting of two Si/SiO_2 distributed Bragg reflectors and an Er-implanted active SiO_2 region was investigated in [436]. The schematic structure of the Si/SiO_2 cavity is shown in Fig. 10.19. This structure may be interesting for optical amplifiers integrated on Si chips. The cavity wavelength is in resonance with the $1.54\,\mu m$ emission wavelength of the 4f electronic transition of Er^{3+} in SiO_2. As a consequence, the emission intensity is enhanced and the spectral purity is improved compared to a non-cavity structure.

The cavity consists of a bottom distributed Bragg reflector, the SiO_2 active region, and a top distributed Bragg reflector. The bottom and top reflectors consist of 4 and 2.5 pairs of Si ($\lambda/4 = 115\,nm$) and SiO_2 ($\lambda/4 = 270\,nm$) quarter-wavelength layers, respectively. The reflectivities were 99.8 and 98.5 %, respectively. The active region has a thickness of 540 nm. An area of $1\,cm^2$ has been implanted with Er at a dose of $7.7 \times 10^{15}\,cm^{-2}$ and an energy of

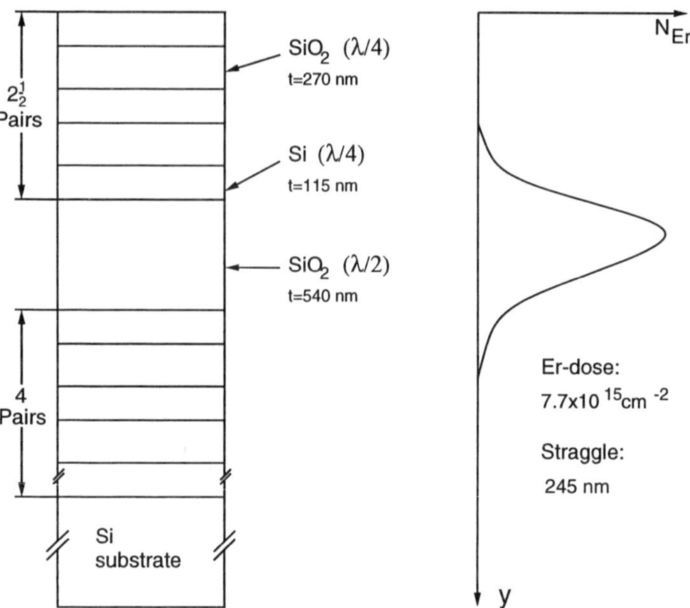

Fig. 10.19. Optical amplifier [436]

3.55 MeV. The projected range of this implant is 1.15 μm. The maximum of the Er profile, therefore, occurred in the center of the SiO_2 active region. The projected straggle of this implant is 245 nm. A post-implantation anneal at 700–900 °C has been carried out for 30 min. The photoluminescence measurements have been carried out at room temperature using a Ti-sapphire laser tuned to the excitation wavelength 980 nm of Er. Without a cavity, the Er-emission lines are inhomogeneously broadened by the random local environment of Er^{3+} ions in the SiO_2 host and the Er emission occurs typically between 1500 and 1600 nm with two distinct maxima at 1535 and 1550 nm. The measured reflectivity spectrum of the cavity, however, contains only one narrow tip at $\lambda = 1.54\,\mu m$ indicating the resonance mode of the cavity. The full width at half maximum (FWHM) of the tip is only $\Delta\lambda = 5$ nm. The cavity quality factor Q, therefore, is 310.

Reference samples without the top mirror showed a 50 times weaker photoluminescence (PL) than the cavity samples. The reference samples also had the two distinct peaks at 1535 and 1550 nm. The spectral width (FWHM) of the single PL line of the cavity sample was only 5 meV (≈ 10 nm).

The spontaneous emission from the resonant cavity is strongly directed along the optical axis of the cavity, which is desirable in optical fiber applications for high coupling efficiency. Since the resonance wavelength depends linearly on the layer thickness, a highly accurate control of the growth process is required in order to obtain the desired resonance wavelength.

10.4 Examples of Integrated Optical Systems

10.4.1 Optical Interconnects

For practical purposes, the cost/performance benefit does not justify the use of optical interconnect technology within digital computer systems at a clock frequency much below 1 GHz [437]. Even when this clock frequency can be handled by optoelectronic devices, packaging costs of optoelectronic devices have to be reduced in order to achieve a breakthrough in optical interconnect technology. For the light emitters this means that edge-emitting lasers, despite their successful use for over two decades in point-to-point long-haul telecommunication links, are not a viable array source to meet the cost and performance requirement for short-haul data communication for instance inside computer systems. Unlike an edge-emitting laser, a vertical cavity surface emitting laser (VCSEL) does not require facet cleavage, so the packaging cost is significantly reduced. The most important feature of VCSELs, in addition to threshold currents below 10 mA, modulation currents below 1 mA, driver voltages below 2 V, and bandwidths exceeding 1 GHz, is that they are fabricated using planar processes similar to those used in CMOS technology. This is a key factor that allows hybrid CMOS VCSEL integration to achieve a cost/performance advantage surpassing all other device technologies [437].

A high-performance, high-density, and low-cost VCSEL array packaging technology being compatible with the electronic assembly process is the final requirement to bring VCSELs and optical interconnect technology inside every computer. Therefore, it is necessary to use a common packaging substrate and die attachment method. However, die attachment methods such as flip-chip, commonly used for electronic devices, require a solder reflow, and therfore impose additional thermal and mechanical constraints for the optical devices. The extra alignment step required for the optical device further increases the complexity to adopt these packaging methods in manufacturing.

The planar hybrid-integration electronic-module packaging (Fig. 10.20) allows CMOS VCSEL integration as well as high-density optical interconnect technology and offers, therefore, an attractive packaging solution for light-emitting array modules. The CMOS VCSEL integration thereby is realized on the package level. The planar hybrid-integration electronic-module packaging includes the fabrication of a polymer waveguide [438] on the planar surface of the module. The polymer waveguide cross section has larger dimensions than monolithically integrated waveguides and allows, therefore, a high coupling efficiency for the optical fiber into the waveguide. Optical receivers with a high efficiency can be realized as a consequence using the planar hybrid-integration electronic-module packaging in addition to integrated CMOS VCSEL modules.

This hybrid-integration approach allows the mixing of different types of device technologies, like silicon and GaAs for instance, to be packaged on a common substrate, either ceramic or plastic. In essence, this packaging

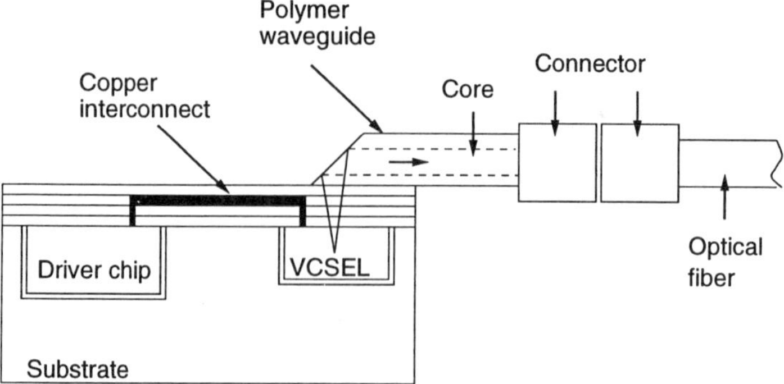

Fig. 10.20. Schematic cross section of a hybrid-integrated laser driver and VCSEL module with a planar polymer waveguide directly placed and attached to a VC-SEL module without air gap. A 45° facet is fabricated at the end of the polymer waveguide for coupling the VCSEL output into the waveguide [437]

technology extends the monolithic wafer-scale electrical interconnects to the first-level packaging by hybrid integration and, at the same time, adds optical interconnects. All interconnects are fabricated using planar processes, and multi-chip modules can be built in batch. This hybrid interconnect technology has been successfully implemented with single-chip and multi-chip module packaging for digital, analog, microwave, and optoelectronic applications in military and commercial systems [437]. High-density, high-speed, and small-size module packaging approaches required for board and backplane applications using the planar hybrid-integration electronic-module packaging technology have been demonstrated [437].

Let us describe the optoelectronic application for the hybrid-integration approach in the following. The hybrid integration technology has been used to fabricate optical transmitter and receiver array modules. In the transmitter module, VCSEL array, drivers, and passive components were placed on a substrate with trenches, and electrical lines were fabricated using a planar thin film process on a polyimide layer laminated over the chips and devices. The electrical lines were fabricated using a copper metallization process that includes resist patterning, sputtering, electroplating, and etching. This process allows the impedance control of the electrical lines to minimize reflection noises. Low parasitic capacitance is achieved by avoiding bondpads and wire-bonding with thin-film-deposited electrical interconnects. As the polyimide is thicker than the oxide between silicon substrate and bondpads the parasitic capacitances are reduced. The hybrid integration technology allows to attach polymer waveguides with waveguide-to-fiber ribbon connectors to mounted light emitters and receivers. An optical link consisting of light-emitter module, fiber ribbon, and receiver was tested successfully at 1 GHz [439].

Low-loss silica-based optical fibers led to optical long-haul telecommunication. Low-loss planar polymer waveguides, together with VCSEL arrays, could push optical interconnects on a circuit-board and backplane level. Recently, several promising polymeric materials have been developed with intrinsic fiber losses of 0.02 to 0.2 dB/cm in the near infrared spectral range at 850 nm [437]. An acrylate-based polymer has shown an excellent layer quality and thickness control allowing both multimode and single-mode planar waveguides to be fabricated by using a lamination and photolithography process. Laser direct writing has also been suggested for polymer waveguide definition. On a backplane level, planar optical waveguides with a 50 μm core and a 100 μm pitch, for instance, gave an increase in the interconnect density by a factor of five compared to electrical interconnects. Preliminary test results have achieved a channel bandwidth close to 1 Gb/s over a total backplane distance of 288 mm [437].

One possibility to combine different technologies, like Si VLSI/ULSI, III-V microwave components, and III-V optoelectronic functions, are multi-chip modules (MCMs). The number of interconnects in these MCMs steadily increases posing problems because of the necessary area for bondpads and increasing interconnect lengths. According to studies, the digital clock rate on electrical interconnects is limited to about 1 GHz for a 2 inch square module [440]. Three-dimensional stacking exploiting optical interconnects between the chip layers like the Z-axis photonic interconnect (ZAPI), therefore, was suggested [441]. Figure 10.21 shows such a two-layer vertical MCM stack. The photonic dies are distributed among electronic signal processing and memory chips that are electrically interconnected on each level, such that all electrical lines are kept as short as possible. Arrays of photonic signals pass vertically through optical via holes in the MCM substrate to provide interconnect paths between layers. These high-aspect-ratio via holes were laser drilled into the 625 μm thick silicon MCM substrate and had a diameter of 50 μm. This vertical optical interconnect scheme requires much less area than one bondpad on the first-layer module, one bondpad on the second-layer module, and the bond wire between them. With the vertical optical interconnect scheme, the first-layer module does not have to be larger in area than the second-layer module as would be necessary for the wire bonded electrical connection. Silicon is a good choice for the MCM substrate because of its high thermal conductivity, because it is inexpensive, and because of the possiblity of using micromechanical structuring techniques. Laser drilling of large arrays has disadvantages in that the structural integrity and thermal capability of the MCM could be compromised. Therefore, it would be desirable to use long wavelength (> 1200 nm) sources with Si MCMs, because Si is transparent at these wavelengths.

As shown in the cross section of Fig. 10.21 and in the exploded view of Fig. 10.22, the photonic dies are mounted in a mixed mode, where the electrical contacts are made on the bottom side of the dies on the first bottom layer of the stack. On this first layer of the stack, the laser and photoreceiver arrays are flip-chip mounted using bump bonds for electrical connection. On

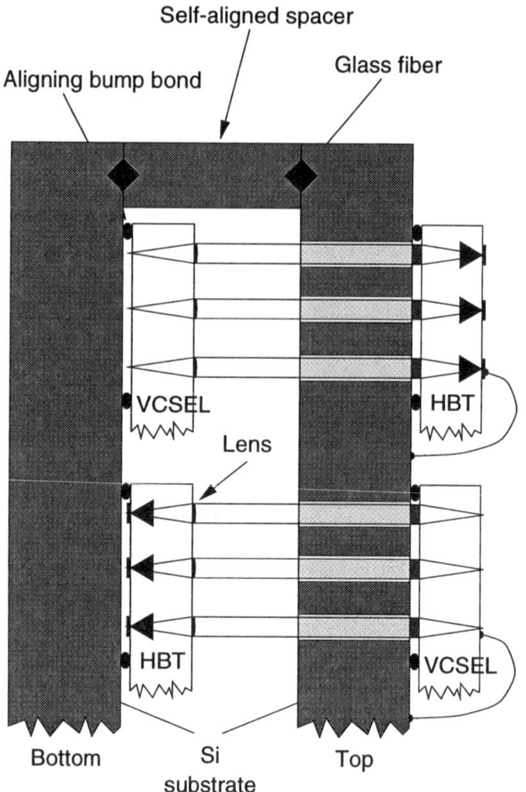

Fig. 10.21. Cross section of a two-layer MCM stack with Z-axis photonic interconnects [441]

the second top layer of the stack, electrical connections to the Si substrate are made by bond-wires and the bump bonds applied to the back of the dies are thus only used for mechanical alignment. Electrical connections are made here by wire bonding. The two layers are aligned using V-groove etching into the two Si substrates and putting pieces of optical fibers into the grooves.

The laser-drilled silicon MCM substrates that formed the demonstrator stack were assembled for testing as in the exploded view of Fig. 10.22. Here, the CMOS drivers, buffers, decoupling capacitors, and photonic devices are assembled onto the two-layer stack, which fits into a large cavity pin grid array package.

With VCSELs emitting at 980 nm and CMOS drivers, a bit rate of 100 Mb/s was reported [441]. Arrays of InGaAs/InP PIN photodiodes were used because of this wavelength. These PIN photodiode arrays were integrated together with InGaAs/InP HBT amplifier arrays. These photoreceiver arrays operated in excess of 200 Mb/s. The power consumption per channel was 45 mW at a supply voltage of 3.3 V considerably exceeding the desired value of 10 mW, as would be necessary for a power consumption limit of 20 W for 2000 channels. The design described in [441], however, was not op-

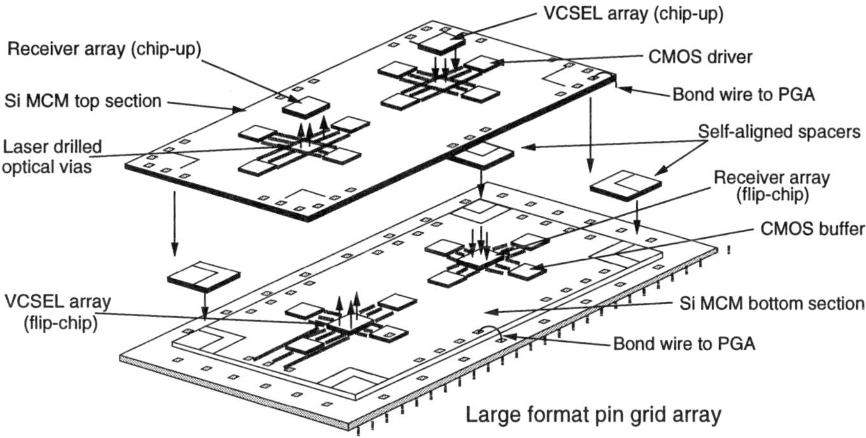

Fig. 10.22. Exploded view of a two-layer MCM stack with Z-axis photonic interconnects showing the VCSELs, photoreceivers, and laser drivers [441]

timized. Appropriate PIN HBT receivers allow much higher data rates. The CMOS drivers were commercial standard components and the VCSELs, for instance, had a power conversion efficiency of only 4 % at an optical output power of 1.0 mW requiring a drive current of 12 mA. With submicrometer CMOS drivers and low-threshold VCSELs, higher data rates and a lower power consumption per channel are possible.

10.4.2 Optical Transceiver

An optical transceiver (see Fig. 10.23 using a laser diode with $\lambda = 1.3\,\mu\mathrm{m}$ and a III/V PIN photodiode for receiving incoming light with $\lambda = 1.55\,\mu\mathrm{m}$ realized as a fiber-embedded lightwave circuit has been proposed [442]. This transceiver enables full duplex bi-directional transmission using a 1.3/1.55 μm scheme. The laser diode has been surface mounted on the Si substrate with the driver circuit in order to minimize interconnect length and line inductance. The Si substrate has been used to adjust the optical fiber passively by exploiting the technique of etching a V-shaped groove.

The receiver with a surface mounted III/V PIN photodiode and a transimpedance amplifier in another Si chip is mounted on the substrate of the emitter chip. The flip-chip mounting eliminates bondpad capacitances and minimizes interconnect length between photodiode and amplifier. A trench has been etched into the receiver Si substrate for fiber alignment. A wavelength division multiplex (WDM) filter placed in a slit in the receiver substrate is transparent for the 1.3 μm laser light and reflects the incoming 1.55 μm light into the photodiode.

For a data rate of 622 Mb/s, the transceiver with an outside placed preamplifier in a different package and in a distance of 9.6 mm only had a −3dB

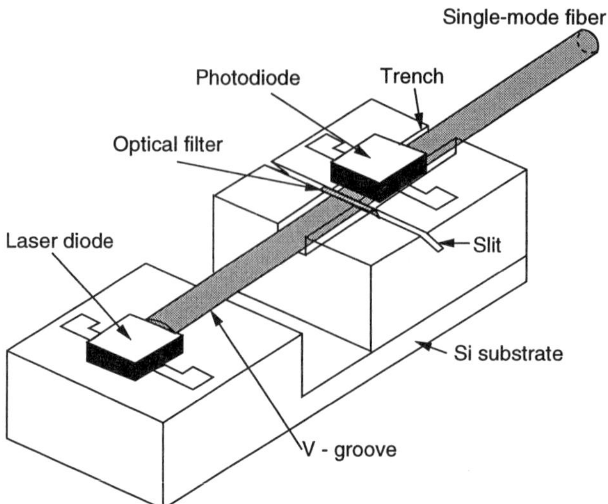

Fig. 10.23. Example of a laser to fiber coupling including fiber to photodiode coupling [442]

bandwidth of 335 MHz. The Nyquist bandwidth of 2/3 of the data rate, i. e. about 420 MHz, could be exceeded with the preamplifier on the Si substrate for fiber and photodiode mounting (see Fig. 10.23) by achieving a bandwidth of 460 MHz [442].

10.4.3 Micro-Spectrometer

A single-chip CMOS micro-spectrometer using Fabry–Perot resonators has been demonstrated [443]. The resonators have fixed cavities resulting in a better optical quality and higher long-term stability than those of tunable cavities [444, 445]. The single-chip micro-spectrometer contains an array of 16 Fabry–Perot resonators in order to cover a large spectral range from 360 to 500 nm with a rather high resolution of FWHM = 18 nm.

Figure 10.24 shows the cross section of the basic sensor structure. The photodetectors have been integrated together with readout, multiplexing, and sensor bus circuits in a conventional 1.6 μm polysilicon-gate CMOS process. The photodetector is a vertical PNP device, where the deep junction is formed by the P-epitaxial layer and the N well, and the shallow junction is formed by the P^+-source/drain implant and the N well. Such a device is called a double photodiode in Sects. 3.5.1 and 3.7. Both junctions are used for photodetection to achieve a high responsivity in a broad spectral range. The two junctions in parallel also result in a higher capacitance for charge storage. The sensors have been arranged in a 4×4 array of square photodiodes with a light sensitive area of $500 \times 500\ \mu m^2$ each. The responsivity of the photodiodes has been reported to be 0.21 A/W for $\lambda = 600$ nm. Dark currents

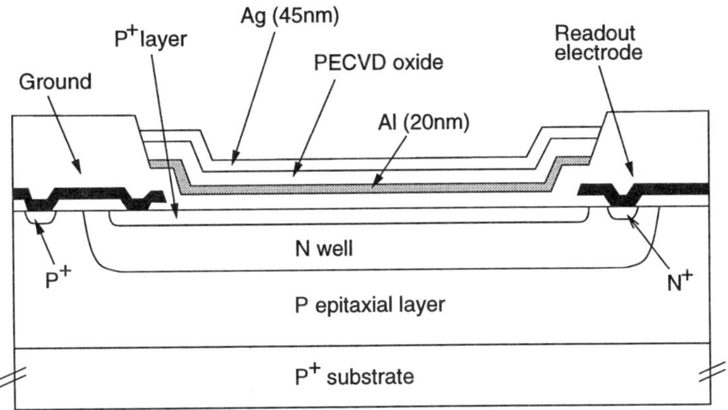

Fig. 10.24. Photodiode with Fabry–Perot resonant cavity [443]

of typically 30 fA corresponding to 12 pA/cm^2 have been measured for both junctions in parallel with a reverse bias of 5 V [443].

On top of the photosensors, Fabry–Perot resonant cavities have been fabricated by postprocessing after completion of the CMOS process. This postprocessing consists of deposition of an Al/SiO_2/Ag layer stack on top of each photodiode. The Al layer is used for compatibility and the Ag layer is the last step. The thickness of the PECVD SiO_2 layer, enclosed between the two semitransparent metallic mirrors determines the transmission peak wavelength. The initially deposited SiO_2 layer has been thinned using four etching steps with four different masks. This has resulted in 16 different resonance cavity lengths and, therefore, in 16 different peak wavelengths. Each of the 16 detectors is sensitive in only one narrow spectral band with a FWHM of 18 nm. Due to some overlapping, a spectral range from 360 to 500 nm has been covered in total by the 16 detectors of the micro-spectrometer. For the 45 nm Ag and 20 nm Al layers used, a maximum transmittance of about 15 % has been measured. A maximum value of about 0.013 A/W for the responsivity has been found for the photodiodes with the Fabry–Perot cavity on top. Only low-temperature CMOS postprocessing steps have been used for the fabrication of the Fabry–Perot resonators. The cavity length can be tuned, to cover a different spectral range, by adjusting the times of the SiO_2 etching during postprocessing, using the same masks.

A multiplexer allows to connect one of the photodiodes to a photocurrent-to-frequency-converter comparator circuit [446]. The photocurrent proportional to the light intensity determines the output frequency f_o according to $f_o = S_{if} \cdot I_{ph}$ with $S_{if} = 278$ kHz/μA, where the responsivity of the photodiode and the transmission of the Fabry–Perot filter enter in the photocurrent I_{ph} and, therefore, have to be accurately reproduced. In the CMOS micro-spectrometer chip an integrated smart sensor bus (ISS-bus) [447] has been integrated. Therefore, only four external connections are needed: VDD (+5 V), ground, clock input, and bi-directional data line, since this line can be used as an input to address the photodiodes and for transmission of the frequency output.

10.4.4 Optical Pressure Sensor

Integrated optics combined with micromechanics and monolithic integration of photodetectors and electronic circuits allows electro-optical signal processing. In [421] for instance, an optical pressure sensor was introduced (Fig. 10.25). This micromechanical optical pressure sensor utilizes the fact that the light path will be lengthened by stretching or bending of the waveguide crossing the membrane. Tension or compression of the membrane, on the other hand, changes the refractive index of the waveguiding films corresponding to their density. Both effects result in a phase change, which can be measured, for instance, by a Mach–Zehnder interferometer. According to [421], the waveguide should not cross the membrane in the center of the trench but in the compressed region of the membrane for maximum phase change (Fig. 10.25). In this case both the refractive index and the optical path increase.

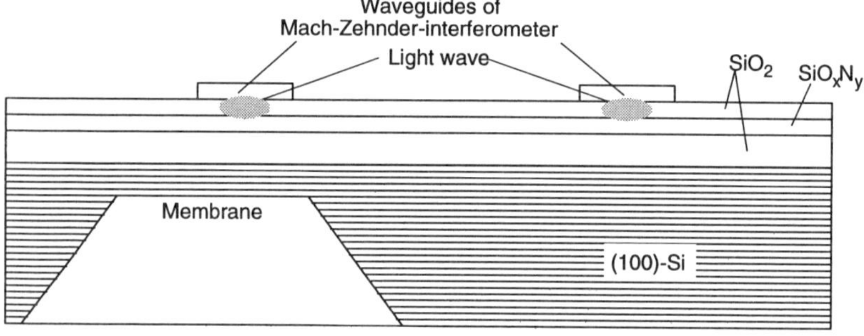

Fig. 10.25. Schematic cross section of a pressure sensor with a Mach–Zehnder interferometer [421]

A Mach–Zehnder interferometer consists of a Y-junction. The optical wave incident in one waveguide is split equally into two branches. Both branches can be made subject to different physical influences changing the phase difference of the waves in both branches. This phase change can be detected by bringing both branches together in a second Y-junction and observing the intensity of the resulting interference signal. In a pressure sensor, one branch of the Mach–Zehnder interferometer is the reference path lying over the silicon substrate, the other branch lies over a thin Si membrane. The waveguide used in [421] is a strip-loaded $SiO_2/SiO_xN_y/SiO_2$ waveguide. V-grooves for the mounting of optical fibers and the membrane were fabricated by methods of silicon micromechanics. This micromachining of the silicon substrate is based on anisotropic etching of (100)-oriented silicon in KOH (potassium hydroxide). Wet thermally grown silicon dioxide, for instance, can serve as the etching mask. The anisotropic etching behavior leaving the (111) direction results in trench walls with an inclination of 53°. The membrane thickness is adjusted via etch time or better by an etch-stop layer. The depth of V-grooves for embedding optical fibers is simply adjusted by the mask opening.

10.4.5 Optical Distance Sensor

Distance sensors based on an interferometric measurement were also suggested (Fig. 10.26). From a fiber coherent laser light is coupled into one end of a Mach–Zehnder interferometer. Each branch of the interferometer is part of a 3dB coupler. The 3dB coupler splits the incident optical power from one incoming branch into two equal parts in the two out-going branches. These couplers are important elements for mixing and phase detection. Here, one half of the original wave leaves the other end of the interferometer, passes a lens, is reflected by the object and received again by the lens. This signal splits and interferes with the waves coming back from the outer branches of the 3dB couplers where they have been reflected by aluminum mirrors. The resulting interference patterns are detected by photodiodes integrated into the silicon.

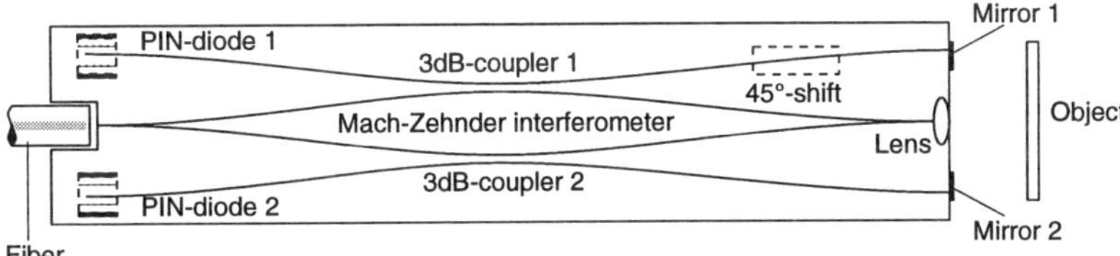

Fig. 10.26. Schematic diagram of a distance sensor [421]

A shift of the object corresponds to the number of minima multiplied by $\bar{n}_{\text{eff}}\lambda/2$. In addition to the measurement of short distances between the sensor and the object, i. e. of distance changes, however, the direction of movement can be detected by the distance sensor shown in Fig. 10.26. This is possible because the phase difference between the mirror branches amounts to $\pi/4$. Consequently both signals are orthogonal, allowing the direction of object movement to be detected [421].

10.4.6 Optical Pickup Devices

A monolithic optical-disk pickup device capable of detecting readout, focus error, and tracking error signals was investigated in [448]. A film waveguide, a grating beam splitter, a focusing grating coupler, and photodiodes were integrated on a Si substrate (see Fig. 10.27) resulting in an integrated-optic disc pickup device (IODPU). Light from an external laser diode was coupled into the waveguide. This light passes through the beam splitter and is coupled out of the waveguide in the focusing grating coupler, whereby a spherical wave is formed and focused on the optical disk. Depending on the stored information more or less light is reflected and coupled into the waveguide by the focusing grating coupler (FGC). Within the waveguide, this reflected light is focused onto the two photodetector pairs by the FGC and the twin grating focusing beam splitter (TGFBS).

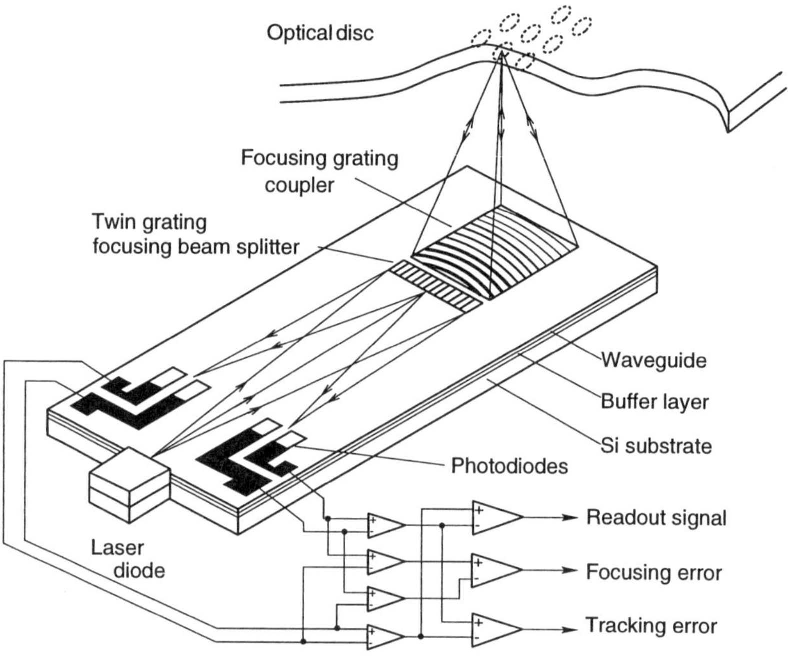

Fig. 10.27. Integrated optical-disc pickup device [448]

The readout signal (stored information) is the sum of the four photocurrents $(I_1 + I_2 + I_3 + I_4)$. The focus error is determined by $(I_1 - I_2 - (I_3 - I_4))$ and the tracking error by $(I_1 + I_2) - (I_3 + I_4)$. The interested reader will find more details for this working principle called the Foucault or push-pull method [449] in [448].

The (100)-oriented N-type substrate was thermally oxidized to grow the SiO_2 buffer layer with a thickness of 1.86 μm and the PN photodiode array was fabricated. The Corning #7059 glass waveguide layer with a thickness of 0.95 μm was then deposited. Electron-beam lithography was used to write the grating pattern of the FGC and TGFBS into a positive resist coated on the waveguide. The pattern was transferred to a 0.035 μm thick SiN layer on the FGC and TGFBS by reactive ion etching to form the grooves of the grating.

The dimensions of different components in the IODPU for $\lambda = 790$ nm will be given in the following. The overall area of the IODPU shown in Fig. 10.27 was $\approx 10 \times 2.3\,\text{mm}^2$. The operating groove depth of the FGC of 0.035 μm resulted in a 60 % theoretical output coupling efficiency. The diffraction limited focus spot size was calculated to be 1.4 μm, which, of course, would be too large for digital video disk systems. The 3dB spot width was measured to be 3.5 μm. The broadening of the spot was due to fabrication errors in the FGC pattern [448]. The TGFBS had the same structure as the FGC and the grooves formed an angle of 7.8 ° with the direction of light incidence from the laser diode to focus the light in the middle of the photodiode pairs. The length of the grooves was 95 μm. The P^+/N-substrate photodiodes had a size of $150 \times 50\,\mu\text{m}^2$. The oxidized SiO_2 buffer layer was tapered (see Fig. 10.27) from the guiding area to the detecting area, and the guided wave was led into the photodiodes without serious losses.

A similar approach was investigated in [450] in order to read optical cards and optical memories. Figure 10.28 shows this integrated optical parallel pickup device. The laser is guided to the linearly focusing grating coupler and directed upward onto the read-only optical card. According to the stored information in the digital marks more or less light is reflected from the focal line on the optical card. The focusing grating coupler directs the reflected

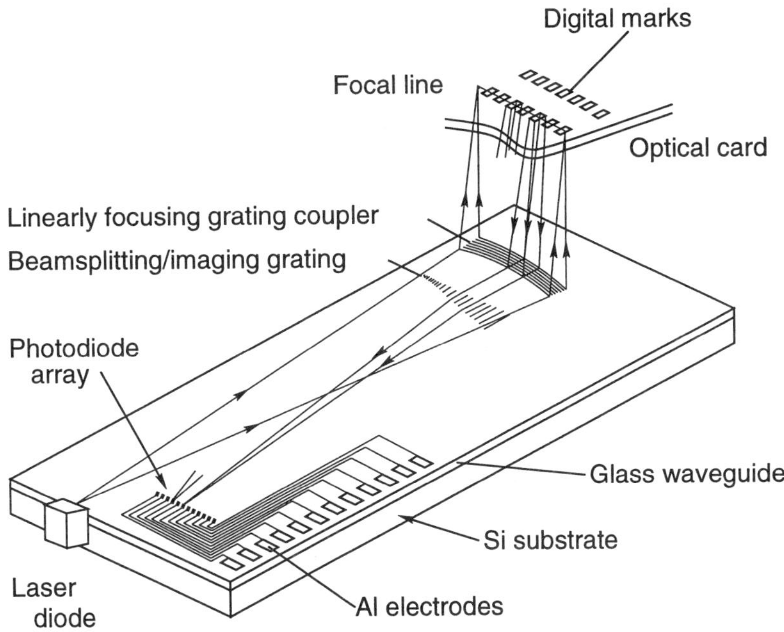

Fig. 10.28. Integrated optical parallel pickup device [450]

light into the photodiode belonging to each digital mark. In such a way, the stored information can be read in parallel. The device was designed to detect signals from 9 μm sized optical reflection marks with a pitch of 18 μm. The chip size was $10 \times 17\,\mathrm{mm}^2$ [451].

11. Design of Integrated Circuits

Most of the optoelectronic integrated circuits (OEICs) described in this work are analog circuits. We, therefore, will restrict ourselves to the design of analog integrated circuits. The exclusion of the design of digital integrated circuits, furthermore, seems to be justified because there are many books describing this topic (for instance [452–454]). The design of analog integrated circuits is, for instance, described in [455]. It will be, therefore, sufficient to give here an overview on the aspects being most relevant for the design of analog OEICs. Since photocurrents have to be converted to signal voltages in most applications, the transimpedance amplifier will be discussed in some detail, especially with respect to its implementation in OEICs.

11.1 Circuit Simulators and Transistor Models

Circuit simulation is useful for the development of integrated circuits in order to avoid long time periods for fabrication and high development costs. Actually, circuit simulation is absolutely necessary for the development of integrated circuits in advanced submicrometer silicon technologies, because analytical models and calculations do not cover short channel effects and, therefore, are not accurate enough. In addition, most of the circuits contain too many devices for analytical investigations.

Circuit simulators implement transistor models covering their temperature dependence as well as short-channel effects and guarantee the correct calculation of the operating point, which is the prerequisite for the simulation of frequency and transient responses, because the amplification of the circuit depends on the operating point of the circuit, which determines the voltages across all the transistor terminals and the currents flowing in all the devices. Therefore, the transconductance g_{m} of each transistor, which determines the amplification, and the junction capacitances in each transistor, which determine the frequency and transient responses, depend on the operating point of the circuit.

There are several widespread circuit simulation programs: for instance HSPICE originally developed by the University of California at Berkeley and ACCUSIM as part of the MENTOR design framework as well as SPECTRE-S as part of the CADENCE design framework. ACCUSIM and SPECTRE-S are part of the professional design frameworks being used by semiconductor chip manufacturers. Foundries and manufacturers of application specific integrated circuits (ASICs) also support these frameworks. The manufacturers of ASICs make the transistor models and parameters available to customers and design houses. One of the newest and widely used transistor models covering the small signal and large signal behavior for MOSFETs is BSIM3.3 [456]. This model was used here for the development of CMOS OEICs with ACCUSIM within the MENTOR framework and for the development of BiCMOS OEICs with SPECTRE-S within the CADENCE framework.

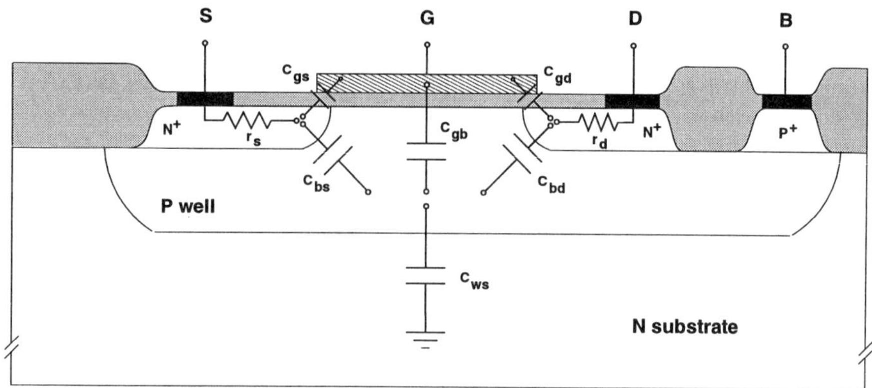

Fig. 11.1. Integrated-circuit MOS transistor structure including parasitic elements

The modeling of MOSFETs dates back to [457, 458]. The models consider parasitic capacitors and resistors within the MOS transistors shown in Fig. 11.1. The parasitic elements are included in the small-signal equivalent circuit (Fig. 11.2) underlying the transistor models implemented in modern circuit simulators. A small-signal model implicitly assumes a linear behavior. The parameters of the small-signal model are usually designated by lower case letters.

The conductances g_{bd} and g_{bs} are the conductances of the bulk-to-drain and bulk-to-source junctions. These conductances are normally very small, since these junctions are reverse-biased. The channel transconductances g_{m}, g_{mb}, and g_{ds} are defined as

$$g_{\mathrm{m}} = \frac{\partial i_{\mathrm{d}}}{\partial v_{\mathrm{gs}}},\ g_{\mathrm{mb}} = \frac{\partial i_{\mathrm{d}}}{\partial v_{\mathrm{bs}}},\ g_{\mathrm{ds}} = \frac{\partial i_{\mathrm{d}}}{\partial v_{\mathrm{ds}}} \tag{11.1}$$

at the quiescent point, i. e. at the operating point. The transconductance g_{m} gives rise to the desired amplification of a MOSFET. The so-called backgate effect, arising from g_{mb}, however, reduces the amplification when source and bulk, i. e. well, cannot be connected as in the case when PMOS input transistors would be used in a difference amplifier on an N-type substrate,

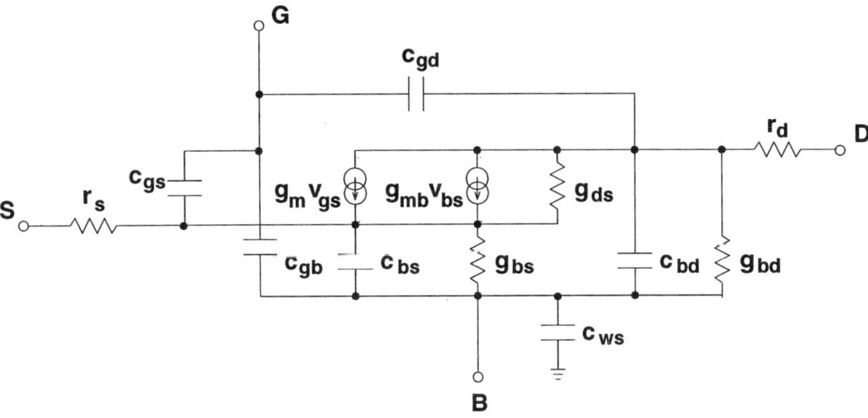

Fig. 11.2. Complete MOS transistor small-signal equivalent circuit [459]

for instance. The conductance g_{ds} considers the channel length modulation effect for the transistor in saturation.

For bipolar transistors in HSPICE, level 1 can be used, for instance. SPECTRE-S, which was used in the development of BiCMOS OEICs, implements several high-current effects in addition to the large-signal model of Gummel and Poon. The simpler Ebers–Moll large-signal model is obtained, when certain parameters are left unspecified. The complete large-signal and small-signal models consider parasitic capacitors and resistors within the bipolar transistors shown in Fig. 11.3.

The complete small-signal equivalent circuit of a bipolar transistor considering the parasitic capacitors and resistors is shown in Fig. 11.4.

The parameter c_π not shown in Fig. 11.3 is the sum of the base-charging capacitance $c_{\mathrm{b}} = \tau_{\mathrm{F}} g_{\mathrm{m}}$ and the emitter-base depletion layer capacitance c_{je} [460]. For the diffusion transistor with a constant base doping concentration, the base transit time τ_{F} is equal to $Q_{\mathrm{E}}/I_{\mathrm{C}} = W_{\mathrm{B}}^2/(2D_{\mathrm{N}})$, where Q_{E} is the minority-carrier charge in the base, W_{B} is the base width, and D_{N} is the minority-carrier diffusion coefficient. The base-collector capacitance c_μ is the sum of $c_{\mu 1}$ and $c_{\mu 2}$ shown in Fig. 11.3. The collector-substrate capacitance c_{cs} is the sum of c_{cs1}, c_{cs2}, and c_{cs3}. The parameters r_π and r_μ are obtained easily:

$$r_\pi = \frac{\beta}{g_{\mathrm{m}}}, \tag{11.2}$$

and

$$r_\mu = \frac{\Delta V_{\mathrm{CE}}}{\Delta I_{\mathrm{B}}} = \frac{\Delta V_{\mathrm{CE}}}{\Delta I_{\mathrm{C}}}\frac{\Delta I_{\mathrm{C}}}{\Delta I_{\mathrm{B}}} = r_{\mathrm{o}}\beta, \tag{11.3}$$

where r_{o} is the small-signal output resistance. The collector series resistance r_{c} is the sum of r_{c1}, r_{c2}, and r_{c3} (compare with Fig. 11.3). The emitter series

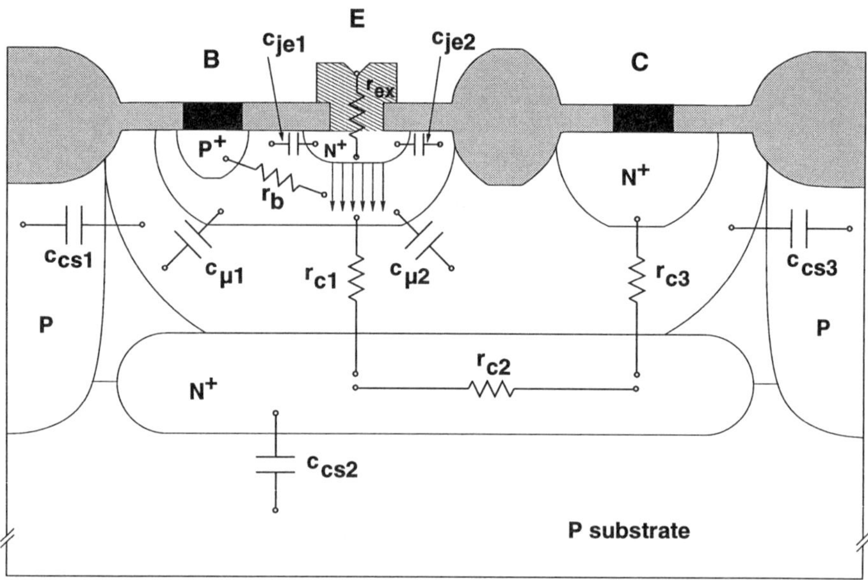

Fig. 11.3. Integrated-circuit NPN bipolar transistor structure including parasitic elements [460]

resistance is represented by r_{ex}. The base series resistance r_{b} is one of the most important parameters.

The values for the parameters shown in Fig. 11.4 are available in the parameter file from the manufacturers of ASICs for several predesigned transistors.

In addition to typical parameter values, parameter values for the so-called slow and fast cases are also available. These extreme parameter values have to be considered in order to guarantee the correct function of a circuit within the complete specified range for the fabrication process. Furthermore, the

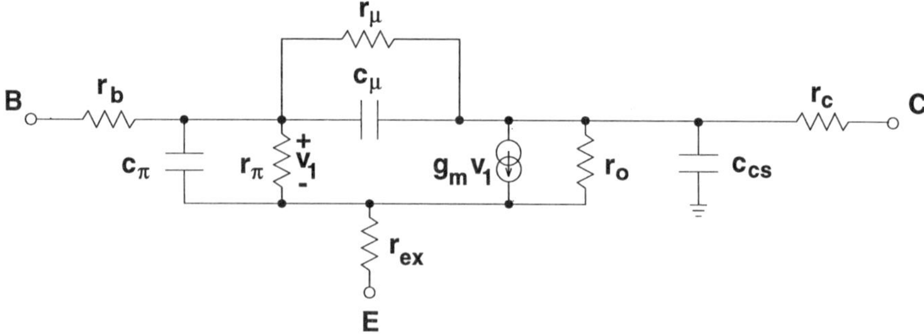

Fig. 11.4. Complete bipolar transistor small-signal equivalent circuit in a common-emitter circuit [460]

influence of the temperature on the performance of a circuit has to be investigated within the specified operating temperature range. For MOS circuits, the bandwidth usually reduces with increasing temperature and increases for temperatures lower than room temperature. Simulations have to be made in order to check whether the bandwidth is sufficient for the highest specified operating temperature and whether the circuit is immune against oscillations for the lowest specified operating temperature. The worst cases for MOS circuits are the combination of the highest specified operating temperature with the slow MOS transistor parameter values and the combination of the lowest specified operating temperature with the fast MOS transistor parameter values.

The current in bipolar transistors increases with temperature resulting in an increase of the circuit bandwidth with temperature. Therefore, the slow bipolar transistor parameter values have to be combined with the lowest specified operating temperature and the fast bipolar transistor parameter values have to be combined with the highest specified operating temperature in order to consider the worst cases for bipolar circuits.

In addition to the differences in the worst case behavior between MOSFETs and bipolar transistors, there are fundamental differences in the amplification and frequency behavior of MOS and bipolar transistors as well as analog MOS and bipolar circuits, which result from the fundamental difference of the output characteristics of the bipolar transistor $I_C(U_{CE}, U_{BE})$ in the forward active region and of the MOS transistor $I_D(U_{DS}, U_{GS})$ in the saturation region:

$$I_C = I_S \left(1 + \frac{U_{CE}}{U_{Ea}}\right) \exp[U_{BE}/(k_B T/q)], \tag{11.4}$$

$$I_D = \frac{1}{2}\mu_n C'_{ox} \frac{W}{L}(1 + \lambda_{ch} U_{DS})(U_{GS} - U_{Th})^2. \tag{11.5}$$

These equations consider the Early effect, i. e. the increase of I_C with U_{CE} due to a decrease in the base width, by the incorporation of the Early voltage U_{Ea} and the channel length modulation effect, i. e. the increase in I_D because of the decrease in the effective channel length with an increase in the drain space-charge region width due to an increase in U_{DS} by the incorporation of λ_{ch}.

The differences between bipolar and MOS analog circuits are caused by the exponential dependence of the collector current I_C on the base voltage U_{BE} and by the quadratic dependence of the drain current I_D on the gate voltage U_{GS}. The transconductance of a bipolar transistor, therefore, is

$$g_m = \frac{\partial I_C}{\partial U_{BE}} = \frac{I_C}{k_B T/q}, \tag{11.6}$$

whereas the transconductance of a MOSFET is

$$g_{\mathrm{m}} = \frac{\partial I_{\mathrm{D}}}{\partial U_{\mathrm{GS}}} = \frac{2I_{\mathrm{D}}}{U_{\mathrm{GS}} - U_{\mathrm{Th}}}. \tag{11.7}$$

When we assume that $I_{\mathrm{C}} = I_{\mathrm{D}}$ and that $U_{\mathrm{GS}} - U_{\mathrm{Th}} = 0.5\,\mathrm{V}$, then the transconductance of the bipolar transistor is 10 times larger than that of the MOSFET, because $k_{\mathrm{B}}T/q = 25\,\mathrm{mV}$. This means that bipolar amplifiers have a much larger amplification factor than MOS amplifiers, since the amplification factor is proportional to the transconductance for most types of amplifiers. Capacitive loads can be charged more rapidly by bipolar than by MOS transistors and the bipolar transistor is said to have the better driver capability. The bandwidth of bipolar amplifiers as a consequence is larger than the bandwidth of MOS amplifiers. For many operational amplifiers, as an example, the gain-bandwidth-product is proportional to the transconductance of the input transistor [455].

The design frameworks offer schematic entry for the definition of circuit diagrams, i. e. the circuit diagram can be drawn by placing device symbols and connecting them. Within the CADENCE framework, the schematic entry program is called COMPOSER and within the MENTOR framework there is the schematic entry program DESIGN ARCHITECT. It is, therefore, not necessary for a designer to write a netlist. The circuit simulator derives the netlist from the schematic. In such a way the number of possible errors is minimized.

11.2 Layout and Verification Tools

When the required specifications are met with the circuit diagram according to circuit simulations, the next step is the layout of the integrated circuit. The frameworks contain the layout tools necessary for this task. The chip manufacturers also make the process definition and design rule files available to the customer or design house. These files are necessary for the design rule check of the layout in order to detect violations and to be able to correct them.

When the design rule check does not report any errors, there is still the possibility of deviations from the circuit diagram. Therefore, tools are included within the frameworks for the comparison of layout and schematic. These layout-versus-schematic (LVS) tools require the design rule file. LVS is one part of the design verification. A further verification tool is included within each framework, which enables the extraction of parasitic capacitances and resistances of interconnects from the layout. These parasitics were not included in the circuit diagram for prelayout circuit simulations. They can, however, deteriorate the behavior of the circuit. Therefore, the extraction tools allow us to add these parasitics to the circuit diagram or to include them in a netlist. In such a way the so-called postlayout simulations become possible. Usually, the bandwidth of the circuit calculated by postlayout simulations will be lower than initially determined by the prelayout simulations.

Since the OEICs described here are optimized (full custom) designs, no library cells could be used for the amplifiers, i. e. all mask levels had to be designed, and the layout was done mostly manually.

11.3 Design of OEICs

Figure 11.5 shows the flowchart for the development of OEICs. Starting from the technological flowchart of a bipolar, CMOS, or BiCMOS process, the PN junction of a diode or the two PN junctions of a transistor have to be selected for a photodetector. After this selection, the topology and the doping profiles of the photodetector can be calculated using a one- or two-dimensional process simulation program considering the technological flowchart. Such programs are, for instance, SUPREM3 [461] and TSUPREM4 [462]. This topology and these doping profiles can serve as an input for a device simulator, which offers a model for photogeneration and optical transmission of isolation and passivation stacks. Such a device simulator is, for instance, MEDICI [5]. For the device simulations, the operating conditions like photodetector bias, operating temperature, and optical wavelength have to be considered. With the help of the device simulations, rise and fall times of the detector photocurrent, bandwidth, quantum efficiency, and capacitance C_D (compare Fig. 2.12) can be extracted. For the example of the integration of PIN photodiodes in a CMOS process, the results of the device simulations were used to develop process modules for the implementation of shallow junctions in order to increase the internal quantum efficiency of the PIN photodiodes and for the implementation of an antireflection coating in order to increase the optical quantum efficiency. The device simulations also showed that the series resistance of the PIN photodiodes, which would result from the use of surface substrate contacts, can be eliminated by a back contact.

The most important aspect in the design flowchart of OEICs is the combination of photodetector device simulation and circuit simulation. Experience shows that the use of C_D and of a ramp with the rise time or fall time for the photocurrent leads to good results for transient circuit simulations (Fig. 11.6). For AC simulations of the frequency response, it is also sufficient to use the photodetector capacitance C_D. It should, however, be noted that a DC offset current has to be added to the AC current source, because negative photocurrents are not possible. C_D has to be added to the schematic of the amplifier and appropriate current sources have to be defined for the circuit simulations.

For some photodetectors, C_D alone may not be sufficient and it may be necessary to include the series resistor R_S (see Figs. 2.12 and 11.6)). It should be mentioned that this model may not be sufficient, when diffusion of photogenerated carriers is present in the photodetector. For this case, it may be suggested to store the photocurrent values calculated by the device simulator

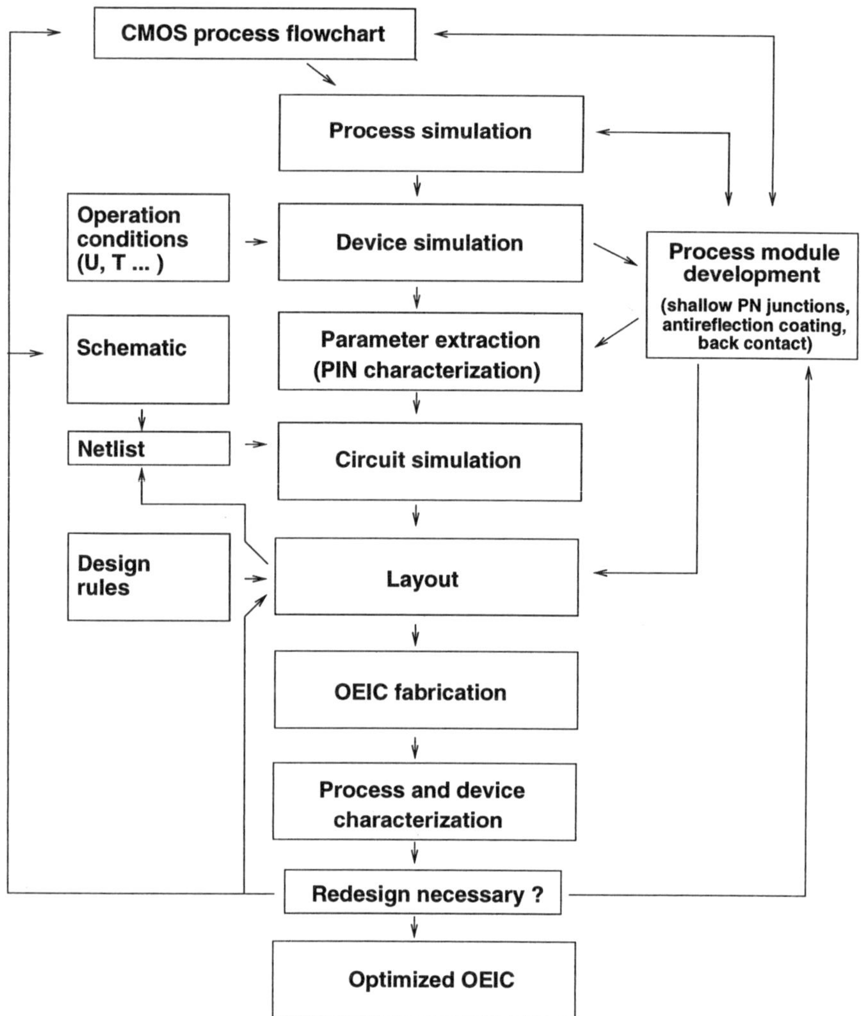

Fig. 11.5. Design flowchart for the development of OEICs. Here the integration of PIN photodiodes in CMOS OEICs is taken as an example

as a function of time in a file and to read this $I_{\mathrm{ph}}(t)$ file by the circuit simulator in order to simulate the transient response of the OEIC. The method being most convenient concerning the combination of device and circuit simulation would be to simulate the OEIC, i. e. the photodetector together with the amplifier, using a circuit simulator on the device level. This method, however, would be very CPU time intensive when all transistors in the amplifier have to be considered by device simulations like the photodetector.

Let us return to the design flowchart of OEICs (Fig. 11.5). Possibly the schematic of the amplifier has to be changed in order to meet the speci-

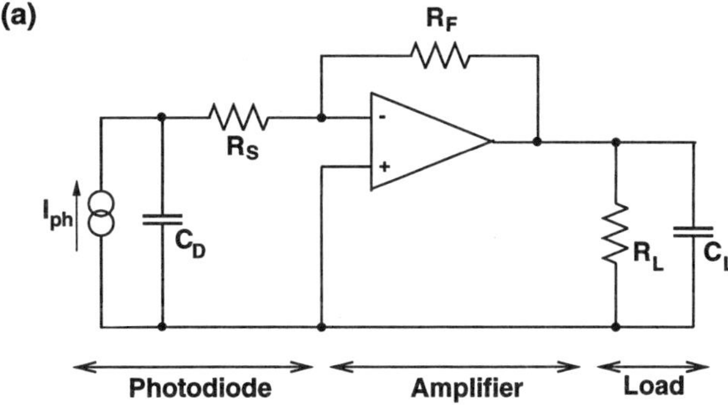

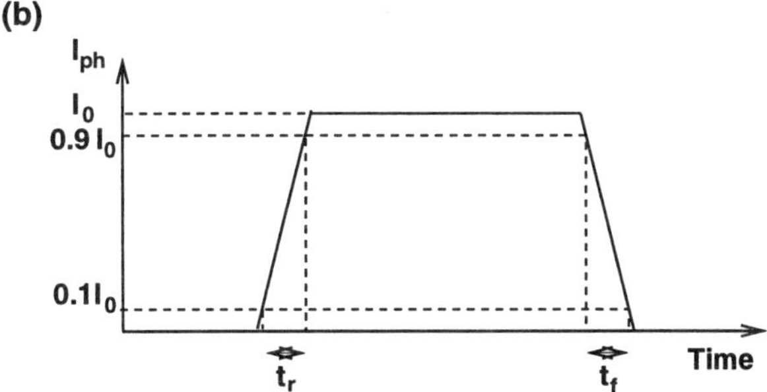

Fig. 11.6. (**a**) Equivalent circuit of a photodiode for the circuit simulation of an OEIC. (**b**) Photocurrent versus time for the consideration of rise and fall times obtained by device simulations or by measurements

fications of the OEICs. When these prelayout simulations predict that the specifications can be fulfilled, the layout can be drawn. For the integration of photodetectors, it is, of course, necessary to obey the design rules of the process and to add as few new design rules as possible for new process modules like the antireflection coating for instance. Due to the reduction of the doping concentration in the N^- epitaxial layer for the speed enhancement of the CMOS-integrated PIN photodiodes, the minimum distance of analog P-type wells had to be increased in order to avoid reach-through currents between P-type wells with different potentials. When the layout is finished, a design rule check is performed. After the correction of possible errors, the layout is compared with the netlist (layout versus schematic) for the detection of possible errors. Finally, when no errors are present or after they are corrected, parasitic capacitors of metal and polysilicon interconnects as well as series resistances of these interconnects are extracted and added to the netlist. For metal interconnects usually only the capacitances are extracted, because their resistances are negligible. Now, the postlayout simulation can be performed. When the specifications for the OEIC are no longer met, an iteration becomes necessary. Changes in the layout, in the netlist, or in the schematic may be necessary. Usually, it will be rather unlikely that the detector has to be modified. When all specifications are fulfilled, the so-called chip finishing is done, i. e. alignment masks and process control modules are added and arranged together with the chip layout on the reticles or on a wafer-size area. Then the masks are fabricated and the chips can be processed.

The OEICs produced have to be characterized. Information obtained by the process monitoring with the help of the process control modules will be used for the interpretation of the OEIC performance. The measured results have to be compared with the specifications for the OEIC. When all knowledge and know-how concerning the design of non-optoelectronic analog integrated circuits has been considered, the design of OEICs should be successful after the first OEIC fabrication as is most desirable and as is usually achieved in industrial chip design. When certain specifications are not met, a correction of the layout, of the schematic, of the process modules, or of the steps for photodetector integration in the technological flowchart may have to be made.

In the example of the PIN CMOS integration, the process modules, and the technological flowchart were developed perfectly at the first attempt. A redesign was only necessary to increase the sensitivity of the OEICs, i. e. of the active load resistors connected to the photodiodes, and to improve the group delay flatness in the specified frequency range.

11.4 Transimpedance Amplifier

A current/voltage (I/U) converter is also called a transimpedance circuit, because an output voltage being proportional to the input current is generated similarly as at an impedance. Each resistor is a current/voltage converter, with its output impedance being equal to its resistor value. In an active current/voltage converter, however, the output impedance is determined by an operational amplifier and, therefore, is independent of the proportionality factor between I and U.

A real current source can be split into an ideal current source with an impedance equal to infinity and into a parallel impedance $Z_1 = R_1 \| C_1$. Without a series resistance, the equivalent circuit of a photodiode (Fig. 2.12) is equal to the equivalent circuit of the real current source in Fig. 11.7, when $R_{\mathrm{D}} = R_1$, $C_{\mathrm{D}} = C_1$, and $R_{\mathrm{s}} = 0$. In fact, the circuit in Fig. 11.7 is the most common amplifier used together with photodiodes. The implementation of an operational transimpedance amplifier configuration is also very interesting for the optoelectronic integration especially when low offset voltages compared to a reference voltage are required. The properties of this configuration, therefore, will be discussed in detail here.

The frequency response function $G(\Omega)$ can be derived thus:

$$U_{\mathrm{i}} + I_2 Z_{\mathrm{F}} + U_{\mathrm{o}} = 0, \tag{11.8}$$

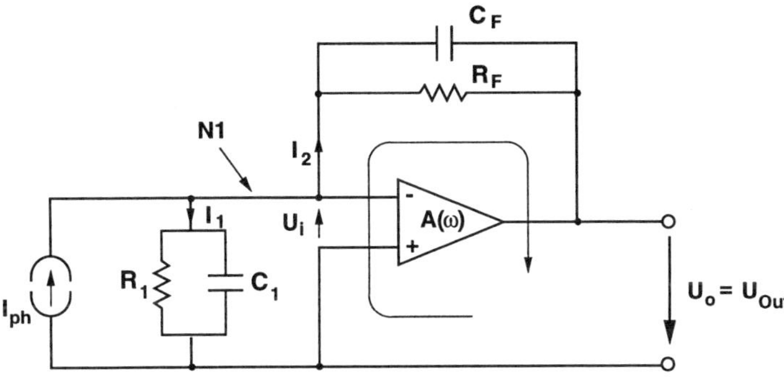

Fig. 11.7. Operational amplifier in transimpedance configuration for I/U conversion of a photocurrent

$$I_{\mathrm{ph}} - I_1 - I_2 = 0 \tag{11.9}$$

with $Z_1 = R_1 \| C_1$, $Z_{\mathrm{F}} = R_{\mathrm{F}} \| C_{\mathrm{F}}$, $U_{\mathrm{i}} = U_{\mathrm{o}}/A(\omega)$, and $I_1 = U_{\mathrm{i}}/Z_1 = U_{\mathrm{o}}/(A(\omega) Z_1)$ where $A(\omega)$ is the frequency dependent amplification factor of the operational amplifier. Inserting U_{i} into (11.8), results in

$$Z_{\mathrm{F}} I_2 + \left(1 + \frac{1}{A(\omega)}\right) U_{\mathrm{o}} = 0. \tag{11.10}$$

Inserting I_1 into (11.9), multiplying both sides of the equation by Z_{F}, and adding to (11.10) leads to the transfer function, which is the *transimpedance* and possesses the dimension Ω:

$$\frac{U_{\mathrm{o}}}{I_{\mathrm{ph}}} = G(\omega) = -\frac{Z_{\mathrm{F}}}{1 + (1/A(\omega))(1 + (Z_{\mathrm{F}}/Z_1))}. \tag{11.11}$$

When we insert the frequency-dependent amplification of the operational amplifier $A(\omega)$ with the transit frequency ω_{T}

$$A(\omega) = \frac{A_0}{1 + \mathrm{i} A_0 \omega/\omega_{\mathrm{T}}} \tag{11.12}$$

into (11.11), we obtain with $A_0 >> 1$:

$$\frac{U_{\mathrm{o}}}{I_{\mathrm{ph}}} = G(\omega) = -\frac{Z_{\mathrm{F}}}{1 + (Z_{\mathrm{F}}/A_0 Z_1) + \mathrm{i}(\omega/\omega_{\mathrm{T}})(1 + (Z_{\mathrm{F}}/Z_1))}. \tag{11.13}$$

After the insertion of

$$Z_1 = R_1 \| C_{\mathrm{D}} = \frac{R_1}{1 + \mathrm{i}\omega R_1 C_{\mathrm{D}}} \tag{11.14}$$

and

$$Z_F = R_F \| C_F = \frac{R_F}{1 + i\omega R_F C_F}, \tag{11.15}$$

the transimpedance is:

$$G(\omega) = -\frac{R_F}{1 - (\omega^2/\omega_T)R_F(C_F + C_D) + i\omega[(1/\omega_T) + R_F(1/(R_1\omega_T) + C_F + C_D/A_0)]}. \tag{11.16}$$

For an ideal operational amplifier with $A_0 = \infty$ and $\omega_T = \infty$, (11.16) can be simplified considerably:

$$G(\omega) = -\frac{R_F}{1 + i\omega R_F C_F}. \tag{11.17}$$

This ideal transfer function obviously is that of a first order low pass filter. The −3dB frequency, therefore, is $f_{3dB} = 1/(2\pi R_F C_F)$. This is not surprising for the ideal operational amplifier, when we look at Fig. 11.7. This assumption of the ideal operational amplifier, however, is too simple an approximation of the real behavior of transimpedance amplifiers. The gain-peaking which occurs often in practice, for instance, is not included in the ideal transfer function. Especially for photodiodes with large capacitance C_D, the ω^2 term in the denominator of (11.16) must not be neglected. For a more thorough analysis, we therefore introduce the following abbreviations: $a = (1/\omega_T) + R_F[(1/R_D\omega_T) + C_F + (C_D/A_0)]$ and $b = (R_F/\omega_T)(C_F + C_D)$. The transfer function for the real operational amplifier then is:

$$\frac{U_o}{I_{ph}} = G(\omega) = -\frac{R_F}{1 - \omega^2 b + i\omega a}. \tag{11.18}$$

The magnitude of the transfer function can be determined from (11.18) to be:

$$\frac{|U_o|}{|I_{ph}|} = |G(\omega)| = \frac{R_F}{\sqrt{(1 - \omega^2 b)^2 + (\omega a)^2}}. \tag{11.19}$$

Equation (11.19) can have a maximum, i.e. a gain-peak (GP), when $(1 - \omega^2 b) = 0$, i.e. when $\omega_{GP} = \sqrt{1/b}$. The frequency for which the gain-peaking occurs then is

$$f_{GP} = \frac{\omega_{GP}}{2\pi} = \frac{1}{2\pi\sqrt{b}} = \frac{1}{2\pi}\sqrt{\frac{\omega_T}{R_F(C_F + C_D)}}. \tag{11.20}$$

If the gain-peaking is not desired, it can be eliminated for a photodiode as a current source by increasing C_F, because in this case C_F has a stronger effect on a than on b. This elimination of the gain-peaking, however, is at the expense of a reduced −3dB bandwidth (see Fig. 11.8).

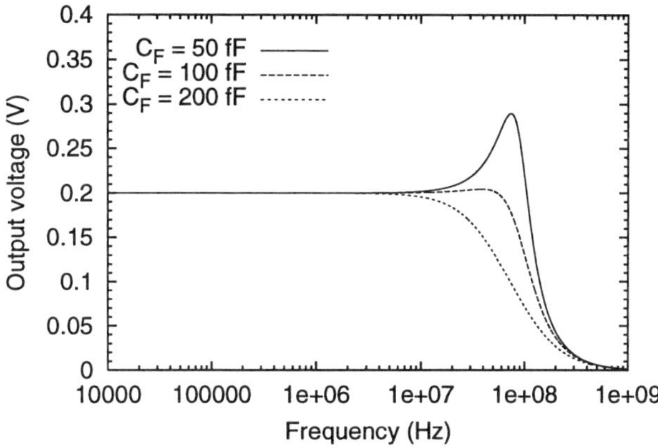

Fig. 11.8. Frequency response of an operational amplifier in transimpedance configuration for three different values of the feedback capacitance C_{F} ($I_{\mathrm{ph}} = 10\,\mu\mathrm{A}$, $R_{\mathrm{F}} = 20\,\mathrm{k}\Omega$, $R_1 = 10^9\,\Omega$, $\omega_{\mathrm{T}} = 6.28\,\mathrm{GHz}$ ($f_{\mathrm{T}} = 1\,\mathrm{GHz}$), $C_{\mathrm{D}} = 1\,\mathrm{pF}$, and $A_0 = 100$) calculated with (11.19)

For $C_{\mathrm{F}} = 50\,\mathrm{fF}$ (or less) a large gain-peak occurs at about 70 MHz. With $C_{\mathrm{F}} = 100\,\mathrm{fF}$, the gain-peak is already almost suppressed. Increasing C_{F} to 200 fF suppresses the gain-peak completely. The -3dB bandwidths for the three cases are about 120 MHz, 90 MHz, and 50 MHz, respectively. The case with $C_{\mathrm{F}} = 200\,\mathrm{fF}$ can practically be considered as over-compensated.

The influence of the depletion capacitance C_{D} of a photodiode on the frequency response is shown in Fig. 11.9. The gain-peak vanishes, when the capacitance of the photodiode is reduced from 1 pF to 0.1 pF. At the same time, the -3dB bandwidth increases from about 120 to 180 MHz. We can conclude that a PIN photodiode with a capacitance of the order of 100 fF is advantageous compared to a double photodiode with a capacitance of the order of 1 pF. In order to understand the advantage of a transimpedance amplifier compared to a voltage follower configuration (compare with Fig. 2.11) with the same sensitivity, we calculate the bandwidth of the load resistor and of the photodiode capacitance $f_{\mathrm{g}} = 1/(2\pi R_{\mathrm{F}} C_{\mathrm{D}})$. For $R_{\mathrm{F}} = 20\,\mathrm{k}\Omega$ and $C_{\mathrm{D}} = 100\,\mathrm{fF}$, we obtain the bandwidth $f_{\mathrm{g}} = 79.6\,\mathrm{MHz}$ for the voltage follower configuration. This bandwidth of the voltage follower is much lower than the bandwidth of about 180 MHz for the transimpedance configuration with $R_{\mathrm{F}} = 20\,\mathrm{k}\Omega$ and $C_{\mathrm{D}} = 100\,\mathrm{fF}$. This example clearly demonstrates the advantage of the transimpedance amplifier. This advantage is due to the amplification factor $A(\omega)$, which reduces the effective photodiode capacitance in the transimpedance configuration with the help of the operational amplifier. Let us now return to the theory of the transimpedance amplifier.

The bandwidth can be determined from the condition

$$|G(\omega)| = G(\omega = 0)/\sqrt{2} = R_{\mathrm{F}}/\sqrt{2}, \tag{11.21}$$

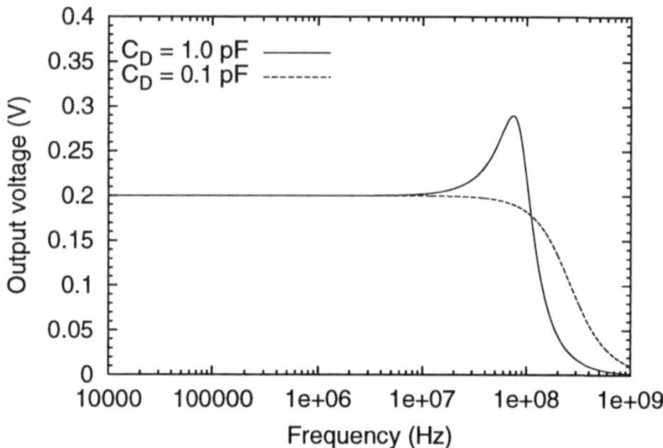

Fig. 11.9. Frequency response of an operational amplifier in transimpedance configuration for two different values of the photodiode capacitance C_D ($I_{ph} = 10\,\mu A$, $R_F = 20\,k\Omega$, $R_1 = 10^9\,\Omega$, $\omega_T = 6.28\,GHz$ ($f_T = 1\,GHz$), $C_F = 50\,fF$, and $A_0 = 100$) calculated with (11.19)

which leads to

$$(1-\omega_g^2 b)^2 + (\omega_g a)^2 = 2. \tag{11.22}$$

The single positive, real solution of this equation is [463]

$$f_g = \frac{\omega_g}{2\pi} = \frac{1}{2\pi}\sqrt{\frac{1}{2b^2}(2b-a^2) + \sqrt{\frac{1}{4b^4}(2b-a^2)^2 + \frac{1}{b^2}}}. \tag{11.23}$$

The bandwidth of the transimpedance configuration is not proportional to $1/R_F$. Therefore, the frequency responses of the transimpedance configuration with $R_F = 20\,k\Omega$ and $R_F = 40\,k\Omega$ were calculated and compared in Fig. 11.10 in order to show the influence of R_F on the bandwidth. For $R_F = 40\,k\Omega$ the bandwidth reduces to about 100 MHz compared to 180 MHz for $R_F = 20\,k\Omega$.

The phase, in degrees, can be obtained in general from

$$\Phi(\omega) = \frac{180°}{\pi}\arctan\left(\frac{\mathrm{Im}(G(\omega))}{\mathrm{Re}(G(\omega))}\right). \tag{11.24}$$

For the operational amplifier in transimpedance configuration, we obtain

$$\Phi(\omega) = -\frac{180°}{\pi}\arctan\left(\frac{\omega a}{1-\omega^2 b}\right) + 180°. \tag{11.25}$$

In order to account for the inverted output signal, 180° are added for all frequencies. The pole in the argument of the arctan for $\omega = 1/\sqrt{b} = \omega_g$

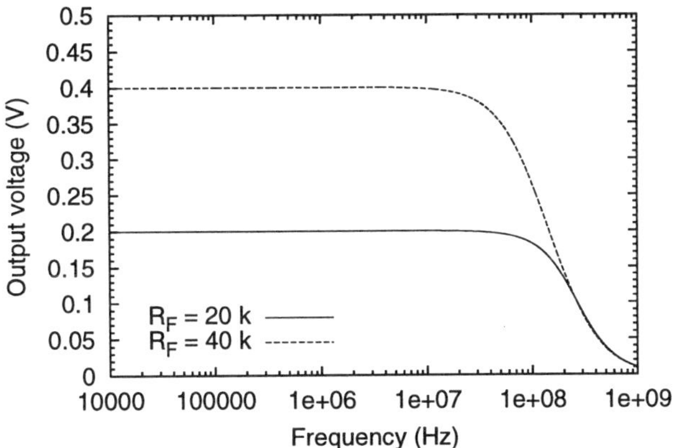

Fig. 11.10. Frequency response of an operational amplifier in transimpedance configuration for two different values of the feedback resistor R_F ($I_{ph} = 10\,\mu A$, $R_I = 10^9\,\Omega$, $\omega_T = 6.28\,GHz$ ($f_T = 1\,GHz$), $C_D = 100\,fF$, $C_F = 50\,fF$, and $A_0 = 100$) calculated with (11.19)

causes a 180° step in the phase due to the periodicity of the arctan function. Fig. 11.11 shows the phase response of the operational amplifier in transimpedance configuration for the same parameters as in Fig. 11.8.

The influence of the photodiode capacitance on the phase response can be seen in Fig. 11.12. The phase does not change as much for a lower photodiode capacitance than for a larger photodiode capacitance when the frequency

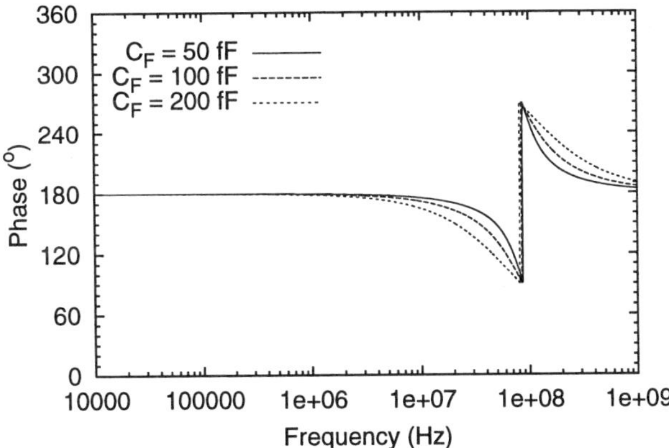

Fig. 11.11. Phase response of an operational amplifier in transimpedance configuration for three different values of the feedback capacitance C_F ($I_{ph} = 10\,\mu A$, $R_F = 20\,k\Omega$, $R_I = 10^9\,\Omega$, $\omega_T = 6.28\,GHz$ ($f_T = 1\,GHz$), $C_D = 1\,pF$, and $A_0 = 100$) calculated with (11.25)

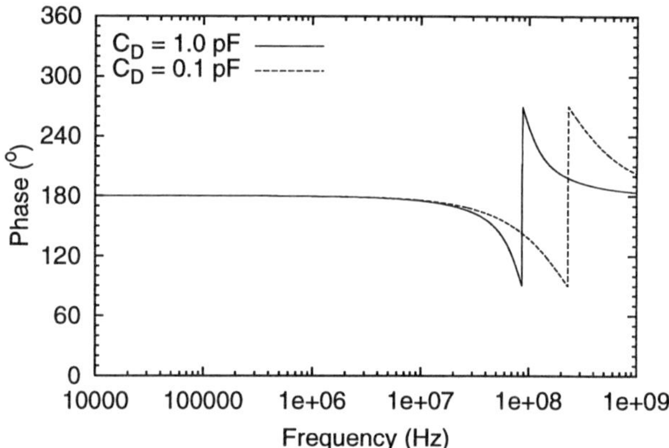

Fig. 11.12. Phase response of an operational amplifier in transimpedance configuration for two different values of the photodiode capacitance C_{D} ($I_{\mathrm{ph}} = 10\,\mu\mathrm{A}$, $R_{\mathrm{F}} = 20\,\mathrm{k}\Omega$, $R_1 = 10^9\,\Omega$, $\omega_{\mathrm{T}} = 6.28\,\mathrm{GHz}$ ($f_{\mathrm{T}} = 1\,\mathrm{GHz}$), $C_{\mathrm{F}} = 50\,\mathrm{fF}$, and $A_0 = 100$) calculated with (11.25)

increases. A PIN photodiode again is advantageous compared to a double photodiode.

For certain applications of amplifiers, the group delay has to be constant in a wide frequency range. We, therefore, will discuss the group delay t_{GD}, which we define here as $t_{\mathrm{GD}} = -\pi/180° \cdot \partial\Phi(\omega)/\partial\omega$, for the transimpedance amplifier. First we introduce the substitution $z = \omega a/(1 - \omega^2 b)$. Knowing that $\partial \arctan(z)/\partial z = 1/(1 + z^2)$, we obtain from (11.25):

$$t_{\mathrm{GD}} = \frac{a(1 + b\omega^2)}{(1 - b\omega^2)^2 + (a\omega)^2}. \tag{11.26}$$

Figure 11.13 shows the group delay for the three cases depicted in Fig. 11.8. There is a peak in the group delay at f_{gp} for $C_{\mathrm{F}} = 50\,\mathrm{fF}$ and $C_{\mathrm{F}} = 100\,\mathrm{fF}$. This peak is suppressed with $C_{\mathrm{F}} = 200\,\mathrm{fF}$ leading to a decrease already at rather low frequencies. The optimum C_{F} value for a good group delay behavior lies between 100 and 200 fF.

The influence of the photodiode capacitance on the group delay is illustrated in Fig. 11.14. For the large capacitance of a double photodiode of the order of 1 pF, a large peak results in the group delay. For the low capacitance of a PIN photodiode of the order of about 100 fF, a constant group delay is obtained for frequencies of up to 100 MHz. The PIN photodiode again is advantageous compared to the double photodiode.

In the following, we will derive an estimate for the maximum bandwidth of an integrated transimpedance amplifier resulting from R_{F} and its parasitic capacitance. An ideal operational amplifier is implicitly assumed. Furthermore it is presumed that C_{F} is not needed for compensation, i. e. that the parasitic capacitance of R_{F} is sufficient for compensation.

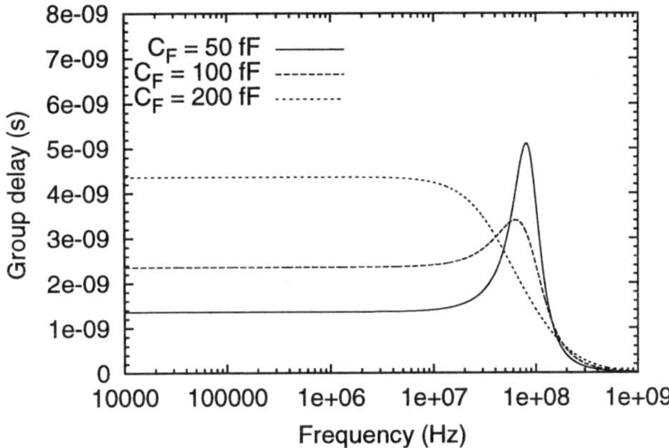

Fig. 11.13. Group delay of an operational amplifier in transimpedance configuration for three different values of the feedback capacitance C_F ($I_{ph} = 10\,\mu A$, $R_F = 20\,k\Omega$, $R_1 = 10^9\,\Omega$, $\omega_T = 6.28\,GHz$ ($f_T = 1\,GHz$), $C_D = 1\,pF$, and $A_0 = 100$) calculated with (11.26)

In analog integrated circuits, polysilicon resistors are used. These polysilicon resistors are fabricated on top of an oxide layer (field oxide, FOX) on the silicon substrate. It is therefore necessary to consider the polysilicon resistor like a chain of RC elements (see Fig. 11.15). The feedback resistor R_F, therefore, has to be split in $R_1 = 0.5R_F/N$, $R_i = R_F/N$ ($i = 2 \ldots N$), $R_{N+1} = 0.5R_F/N$ and the total parasitic capacitance of R_F has to be split in $C_i = C_{R_F}/N$, where $C_{R_F} = A_{R_F}/(\epsilon_{SiO_2}\epsilon_0 t_{FOX})$ with the

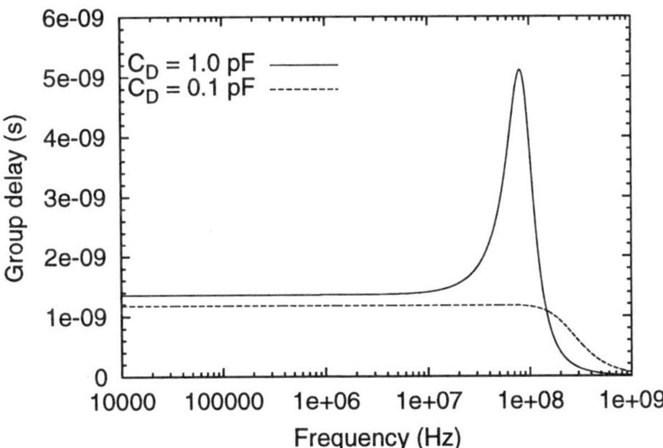

Fig. 11.14. Group delay of an operational amplifier in transimpedance configuration for two different values of the photodiode capacitance C_D ($I_{ph} = 10\,\mu A$, $R_F = 20\,k\Omega$, $R_1 = 10^9\,\Omega$, $\omega_T = 6.28\,GHz$ ($f_T = 1\,GHz$), $C_F = 50\,fF$, and $A_0 = 100$) calculated with (11.26)

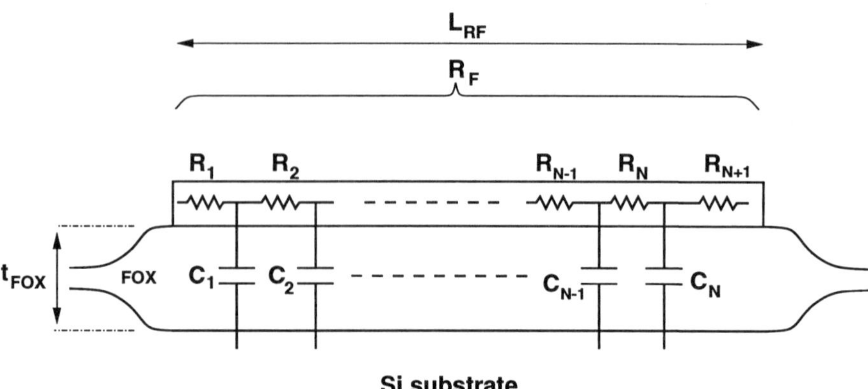

Fig. 11.15. R-C chain of an integrated polysilicon resistor

area $A_{\mathrm{R_F}} = L_{\mathrm{R_F}} \cdot W_{\mathrm{R_F}}$ of R_{F}. The value of R_{F} can be calculated from the sheet resistance $R_{\square}^{\mathrm{Poly}}$ and the length $L_{\mathrm{R_F}}$ and the width $W_{\mathrm{R_F}}$ of R_{F}: $R_{\mathrm{F}} = R_{\square}^{\mathrm{Poly}} \cdot L_{\mathrm{R_F}} / W_{\mathrm{R_F}}$ assuming a linear polysilicon resistor. When R_{F} is meandered, for the corner squares, $R_{\square}^{\mathrm{Poly}} \cdot 0.56$ has to be considered and $A_{\mathrm{R_F}}$ has to be determined in a more complicated way. For simplicity, we, however, will assume a linear polysilicon resistor.

For the integrated resistor, the parasitic capacitance increases when the resistor value has to be increased (for $W_{\mathrm{R_F}}$ = constant, $L_{\mathrm{R_F}} \propto R_{\mathrm{F}}$ and $A_{\mathrm{R_F}} \propto R_{\mathrm{F}}$) in order to achieve a larger sensitivity S of the transimpedance amplifier. Let us, for simplicity, assume the time constant $\tau_{\mathrm{R_F}} = R_{\mathrm{F}} \cdot C_{\mathrm{R_F}}$ for the polysilicon resistor. Then, when $R_{\mathrm{F}} \propto S$ is increased, $C_{\mathrm{R_F}}$ increases proportional to R_{F} and to S, resulting in $\tau_{\mathrm{R_F}} \propto S^2$. This leads to the conclusion that the bandwidth is proportional to $1/S^2$ for an integrated polysilicon resistor in a bipolar, CMOS, or BiCMOS transimpedance amplifier.

In order to illustrate this derived dependence of the bandwidth on S, circuit simulations have been performed with an R-C chain including $N = 10$ elements for $R_{\square}^{\mathrm{Poly}} = 100\,\Omega_{\square}$ and $1000\,\Omega_{\square}$ with $R_{\mathrm{F}} = 10\,\mathrm{k\Omega}$, $20\,\mathrm{k\Omega}$, and $40\,\mathrm{k\Omega}$ for a polysilicon width $W_{\mathrm{R_F}} = 2\,\mu\mathrm{m}$ and $C_{\mathrm{R_F}}/A_{\mathrm{R_F}} = 0.1\,\mathrm{fF}/\mu\mathrm{m}^2$. The fringe

Table 11.1. Simulated bandwidths of R-C chain for an integrated polysilicon resistor

$R_{\square}$ ($\Omega_{\square}$)	$f_{3\mathrm{dB}}(R_{\mathrm{F}} = 10\,\mathrm{k\Omega})$ (MHz)	$f_{3\mathrm{dB}}(R_{\mathrm{F}} = 20\,\mathrm{k\Omega})$ (MHz)	$f_{3\mathrm{dB}}(R_{\mathrm{F}} = 40\,\mathrm{k\Omega})$ (MHz)
100	980	245	61
1000	9800	2450	610

capacitance of the polysilicon resistor has been neglected in this example. The -3dB bandwidths obtained are listed in Table 11.1.

These results show that for $R_{\square}^{\mathrm{Poly}} = 100\,\Omega_{\square}$, polysilicon resistors with values of $10\,\mathrm{k}\Omega$ and $20\,\mathrm{k}\Omega$ will not limit the bandwidths of integrated transimpedance amplifiers. For a value of $40\,\mathrm{k}\Omega$, however, the polysilicon resistor reduces the bandwidth to 61 MHz i. e. below the value of 90 MHz, which was obtained for a BiCMOS OEIC for optical storage systems [464]. For $R_{\square}^{\mathrm{Poly}} = 1000\,\Omega_{\square}$, finally, polysilicon resistors with values of up to $40\,\mathrm{k}\Omega$ possess larger -3dB bandwidths than common integrated high-speed operational amplifiers and will, therefore, not limit the bandwidths of integrated transimpedance amplifiers.

12. Examples of Optoelectronic Integrated Circuits

In this chapter the full variety of receiver OEICs in digital and analog techniques will be introduced. Examples of optical receivers range from low-power synchronous digital circuits for massively parallel optical interconnects and three-dimensional optical memories to Gb/s fiber receivers. Low-offset analog OEICs for two-dimensional optical memory systems like CD-ROM and digital-versatile-disk (DVD) will be described as well as image sensors. Hybrid integrated laser drivers are included as examples of optical emitters.

12.1 Digital CMOS Circuits

In this section, the properties of digital CMOS OEICs are described because of their potential application in massively parallel optical interconnects and volume holographic optical memories. Synchronous circuits are appropriate for such purposes. Furthermore, an asynchronous photoreceiver for application in the testing of digital CMOS circuits on the wafer level deserves to be described here.

12.1.1 Synchronous Circuits

For the application in massively parallel optical interconnects between VLSI chips a novel monolithic optoelectronic receiver system was presented in a standard 0.7 μm N-well CMOS technology [465]. The circuit, which requires two clock signals, *reset* (RST) and *store* (STORE), and therefore is a synchronous circuit, is shown in Fig. 12.1.

The heart of the synchronous receiver is a CMOS bistable flip-flop, which acts as a sense-amplifier. The flip-flop is formed by two CMOS inverters, M2/M4 and M3/M5. The input of each inverter is connected to the output of the other inverter. In such a way a dual state can be stored. In order to use the flip-flop as a light receiver, two photodiodes were added to the drains of the P-channel transistors M2 and M3. In fact the diodes were not separated from the transistors. The diodes were obtained by extending the drains to an area of $15 \times 15\,\mu m^2$. The P^+ drain island and the N well, therefore, form the photodiode (Fig. 12.2).

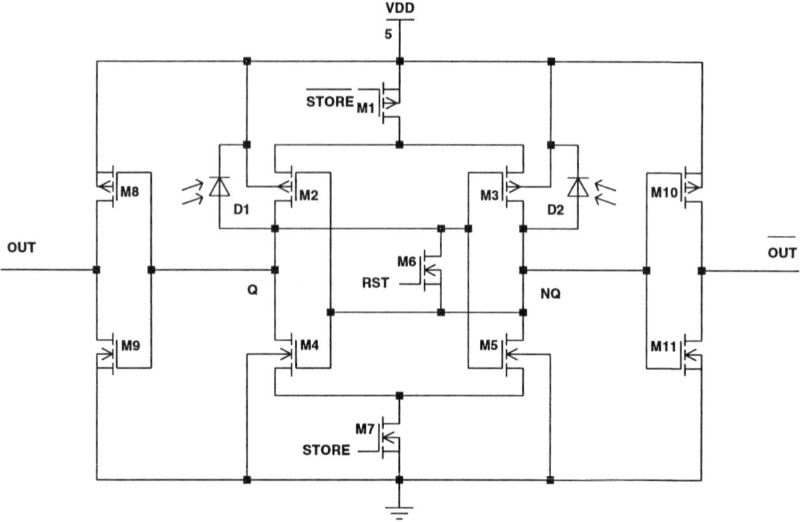

Fig. 12.1. CMOS synchronous photoreceiver circuit [465]

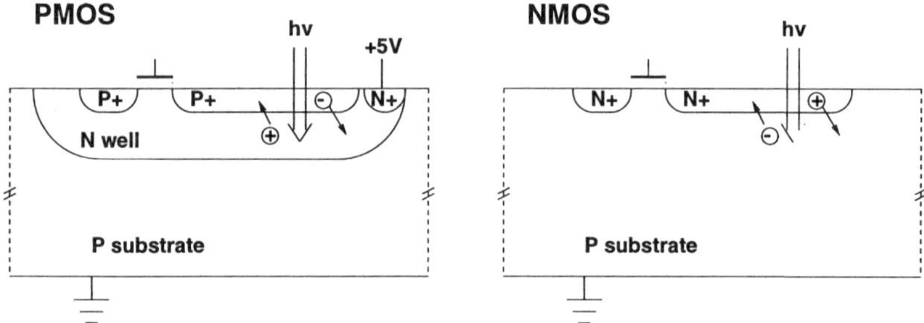

Fig. 12.2. Cross section of MOS transistors with enlarged drain areas as photodiodes [465]

The timing of the synchronous receiver is shown in Fig. 12.3. Before the flip-flop can receive, i. e. store, a new bit, a reset signal has to be applied to M6. During this reset signal, the flip-flop is deactivated by a *low* voltage level at the gate of M7 and by a *high* voltage level at the gate of M1. M6 is conducting during *reset* and forces the nodes Q and NQ to the same potential. After the reset phase, M6 is switched off and light can fall into one of the diodes, let us say into D1. Electron–hole pairs are generated at the P^+N junction between the drain of M2 and the N well. The N well is biased at VDD = 5 V. The potential of the drain can be assumed to be at a floating level of approximately VDD/2. The P^+N photodiode is biased in the reverse direction and the photogenerated carriers are separated in the electric field region of the PN junction effectively. The photogenerated electrons drift into the N well and the photogenerated holes drift into the P^+ region. There is also a slower contribution of diffusing minority carriers to the photocurrent, because the light penetrates also in deeper field-free regions of the N well. As a consequence of the photocurrents, the potential of node Q becomes more positive than that of node NQ. When the supply voltages are applied to the flip-flop via M1 and M7, the sense-amplifier flip-flop begins to work. A more positive input signal of the inverter M3/M5 results in a more negative output signal at node NQ, which causes a higher output signal of the inverter M2/M4 at node Q. This amplifying ring process continues until digital levels, i. e. VDD and ground levels are obtained at the two output nodes Q and NQ of the flip-flop, respectively. It was reported that this final stable state was reached after approximately 3 ns from the beginning of the light incidence [465].

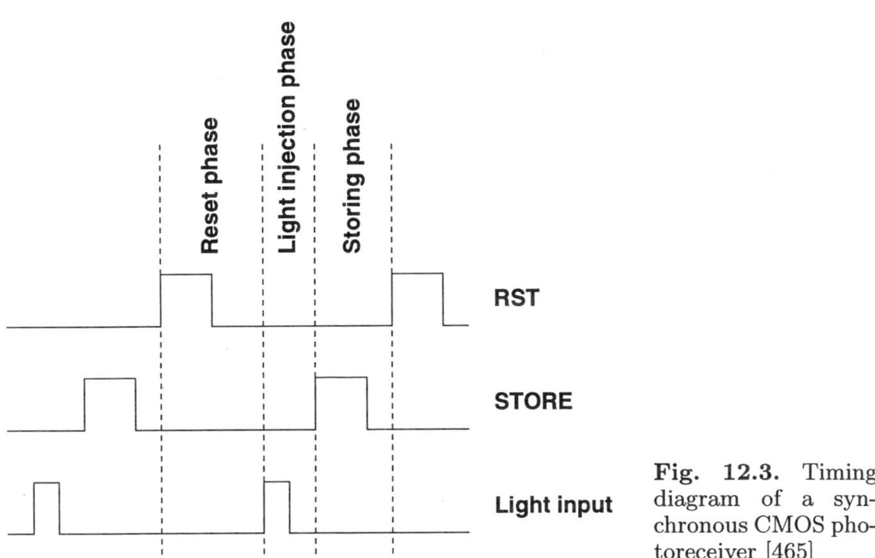

Fig. 12.3. Timing diagram of a synchronous CMOS photoreceiver [465]

A minimum light input energy of 176 fJ, which corresponds to an optical power of 59 μW within a time interval of 3 ns, was needed for an optical wavelength of 830 nm in order to obtain a correct decision of the sense-amplifier flip-flop. This minimum light energy causes a voltage change of 264 mV at node Q [465]. This voltage change is amplified by the flip-flop to a digital level. With the PMOS version described above a bit rate of 120 Mb/s was obtained. In the NMOS version, i. e. using the N^+-drain to P-substrate diodes of the transistors M4 and M5 as photodiodes (see Figs. 12.1 and 12.2), a maximum bit rate of 180 Mb/s was reported. This higher speed of the NMOS version of the synchronous receiver can be explained by the faster diffusion of electrons in the P substrate. The PMOS version is slower because holes are the minority carriers in the N well and holes are diffusing slower than electrons due to the lower hole mobility.

For massively parallel optical interconnects it is necessary to minimize the die area of the receivers. The receiver circuit shown in Fig. 12.1 occupied an area of $55 \times 24\,\mu m^2$ without the output buffers. The synchronous receiver with the sense-amplifier flip-flop is especially interesting for application in massively parallel optical interconnects due to its very small area consumption.

The function of this receiver is called synchronous because the clock signals *reset* and *store* are needed. These signals have to be transmitted in additional optical fibers, for which asynchronous receivers are needed, or have to be supplied electrically.

In the following, another application of sense-amplifier flip-flops, where the reset and store signals are readily available, will be described. Sense-amplifier flip-flops can readily be used for the read process of page-oriented optical memories (POMs), like volume holographic storage systems [466]. Such optical memories need highly parallel read circuits. A small size of each detector and each amplifier is therefore essential. The clock signals for reset and store (or latch) of the sense-amplifier flip-flop are readily available in optical memories and these signals do not have to be extracted for this application. The POMs offer the potential for high capacity, random access times from 10 to 100 μs, and page sizes up to 1 Mb, yielding data transfer rates of 10–100 Gb/s. Maximum output data rates of 250 Mb/s for CCD arrays [125] are insufficient for POM systems. Pixel circuits, like that in Fig. 12.4 with sense-amplifier flip-flops, combine a large speed, a high gain, and a small size. They can be embedded in each pixel yielding active receivers in a highly parallel arrangement. The circuit of a pixel in a photodetector array for a page-oriented optical memory (Fig. 12.4) combines two flip-flops (latches) in order to achieve a high gain.

The operation of the circuit is controlled by reset $\overline{\mathrm{RST}}$ and latch $\overline{\mathrm{LAT}}$. The P latch with transistors P1 and P2 is isolated from the N latch with N1 and N2 via P3 and P4 for $\overline{\mathrm{LAT}} = 1$. The pull-down devices N3 and N4 are conducting and the N latch is reset for $\overline{\mathrm{LAT}} = 1$. Next $\overline{\mathrm{RST}}$ is set to 0. During this reset, the photosensitive inputs A and B are shorted through P5 and $V_\mathrm{A} = V_\mathrm{B} = V_\mathrm{RST} \approx VDD + V_\mathrm{Th,PMOS}$. When $\overline{\mathrm{RST}}$ is taken high again, parasitic capacitances hold nodes A and B at V_RST putting the P latch in a metastable state. A photocurrent I_ph at one of the differential inputs causes the corresponding input voltage to drop. Positive feedback in the P latch increases the differential voltage. When $\overline{\mathrm{LAT}}$ is changed to 0, this differential voltage is amplified by the N latch and the signal is acquired at Q and $\overline{\mathrm{Q}}$. P1 and P2 should be small in size for a high sensitivity to small photocurrents. The static power consumption of the circuit in Fig. 12.4 is determined by leakage currents. The dynamic power consumption mainly depends on the capacitance of the photodiode. In [467], the N-well to P-substrate photodiode

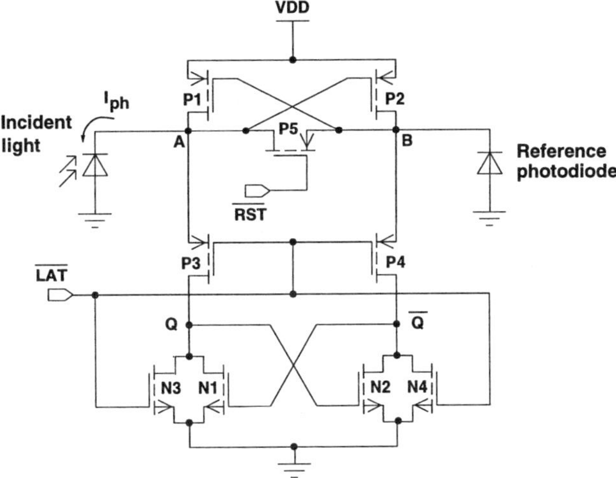

Fig. 12.4. Photoreceiver circuit for smart photodetector array [467]

in a 0.35 μm CMOS process was used with a photosensitive N-well area of $50 \times 50\,\mu m^2$. The capacitance of this photodiode was approximately 1 pF. The size of the photoreceiver circuit in Fig. 12.4 without the photodiodes was $20 \times 22\,\mu m^2$. An optical power of 2.5 μW for $\lambda = 839$ nm was necessary to toggle the receiver. With an assumed photodiode responsivity of 0.3 A/W, a switching energy of 150 fJ was estimated [467]. A single-pixel data rate of 245 Mb/s was determined for the circuit in Fig. 12.4. With the circuitry necessary for error correction, a maximum pixel number of approximately 26 700 per 0.35 μm CMOS chip was projected in [467]. The bit rate per chip for corrected data was estimated to be 102 Gb/s stemming from the high parallelism.

12.1.2 Asynchronous Circuits

Compact and fast photoreceivers with on-chip photodiodes in standard CMOS technology have been developed as optical inputs for testing of digital circuits [468]. Novel CMOS circuits work at increased speed and they cannot be tested on the wafer level at their working speed with conventional electric needle contacts. In order to overcome this limitation, optical inputs can be used. The light is coupled into photodiodes on the chips via optical fibers being adjusted on a wafer prober (Fig. 12.5). The outputs of the chip having lower frequency were contacted with electric needles or capacitively with a special probe sensor. The output signals are checked for correctness by the comparator in a test equipment (Fig. 12.5).

In a 1.5 μm CMOS technology an input frequency of 250 MHz was obtained for an optical wavelength of 635 nm with an optical receiver requiring a die area of only $70 \times 70\,\mu m^2$. Maximum input frequencies of 243 and 187 MHz

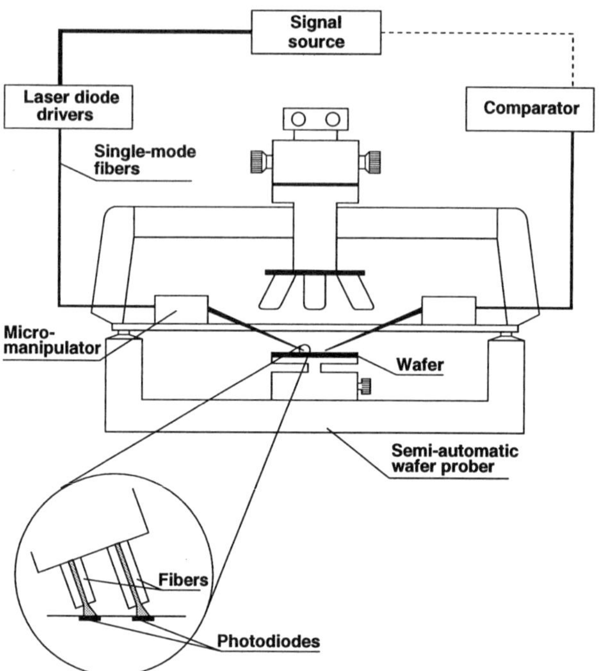

Fig. 12.5. Test equipment using optical inputs on a wafer level [469]

were reported for the wavelengths of 685 and 787 nm, respectively. The N^+-source/drain to P-substrate diode in the N-well CMOS process was used for the photodiodes with an area of $20 \times 20\,\mu m^2$. The speed of the circuit test on the wafer was increased several times compared to the conventional electrical input technique using this optical input technique. This performance could be obtained with a current comparator circuit with a statically supplied reference photodiode. The circuit diagram of this photoreceiver is shown in Fig. 12.6.

The transistors M1, M2 and M3, M4, respectively, form current mirrors, which amplify the photocurrents of the signal and reference photodiodes. The transient response has its optimum for a width ratio W2/W1 of 2 to 2.5 [468]. The amplified photocurrents are compared to each other by the transistors M2 and M4. The advantage of this configuration is its good sensitivity to small differences in the photocurrents due to the high impedance node N3. Due to the high drain resistances a small difference in the currents results in a large voltage change. The advantage of this photoreceiver clearly is that the photocurrent of the N^+P signal photodiode does not have to drop much to obtain a logical *zero* at the output node N3 of the current comparator, when the light is switched off for a short period, i. e. a large contribution of the slow diffusion current to the photocurrent is possible. It has to be mentioned, however, that large photocurrents and large optical powers are necessary to obtain fast current mirror amplifiers. The optical power used for

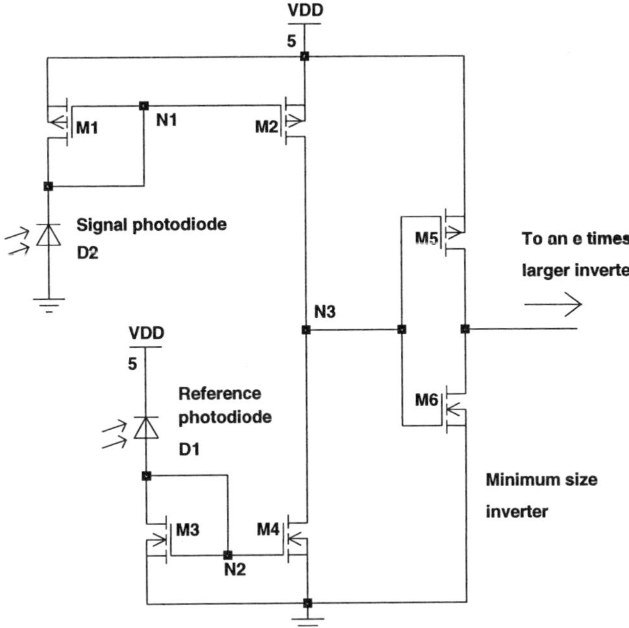

Fig. 12.6. CMOS current comparator photoreceiver circuit [468]

chip testing was of the order of 1 mW [468]. Another advantage of this current comparator circuit is its robustness for the application in automatic chip testers. Compared to the synchronous receiver shown in Fig. 12.1, the current comparator circuit shown in Fig. 12.6 works asynchronously and, therefore, does not need overhead circuitry for timing. The speed of the current comparator circuit in a 1.5 µm CMOS technology is comparable to the speed of the synchronous receiver in a 0.7 µm CMOS technology for the optical wavelength of 787 nm. The speed of the current comparator circuit, therefore, can be improved considerably by using a technology with a smaller gate length. The signal photodiode feeding an N-channel current mirror and the reference photodiode feeding the P-channel current mirror probably would allow a further increase in the input frequency.

12.2 Digital BiCMOS Circuits

The principle of a current comparator circuit shown in Fig. 12.6 for a CMOS photoreceiver was also applied to a BiCMOS photoreceiver for the high speed testing of frequency dividers on the wafer level [469]. Figure 12.7 shows the BiCMOS version of a current comparator. In the 1.2 µm BiCMOS process fast NPN transistors were available, which preferably were used in the sig-

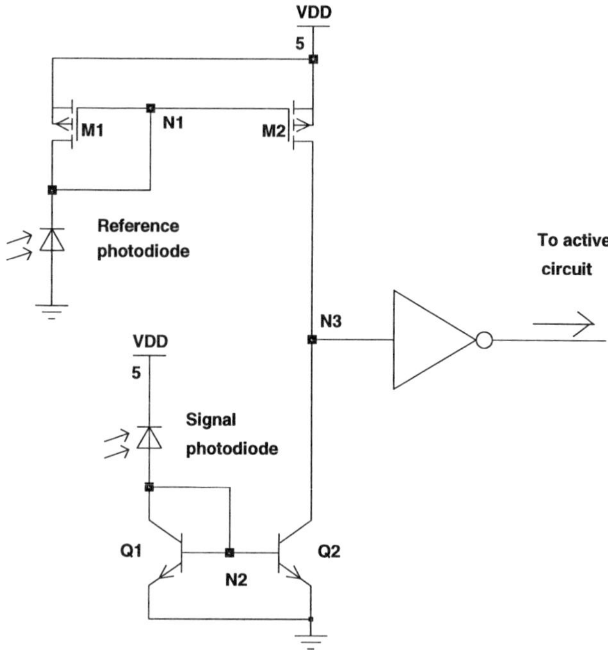

Fig. 12.7. BiCMOS current comparator photoreceiver circuit [469]

nal current mirror. The slower P-channel current mirrors are kept for the reference path.

The circuit shown in Fig. 12.7 was used for a high speed test of a BiCMOS frequency divider on the wafer level. N^+ to P-substrate and P^+ to N-well photodiodes with areas of $20 \times 20\,\mu m^2$ were used. A frequency of 800 MHz was successfully fed into the BiCMOS frequency divider optically with a wavelength of 635 nm via a single-mode fiber. The optical output power of the commercially available semiconductor laser used in [469] was of the order of 1 mW. The area consumption of an optical input was reported to be less than $70 \times 70\,\mu m^2$.

12.3 Laser Driver Circuits

A simple two-transistor CMOS driver (Fig. 12.8) based on a current-shunting principle, which provides a low-area, tunable power circuit with a measured small-signal bandwidth of 2 GHz in 0.5 μm CMOS technology, has been implemented in a flip-chip bonded CMOS VCSEL chip [346].

The PMOS transistor is used to supply an adjustable current through the laser, and the NMOS transistor is used to quickly shunt the current into and out of the VCSEL for digital operation. The multimode VCSELs

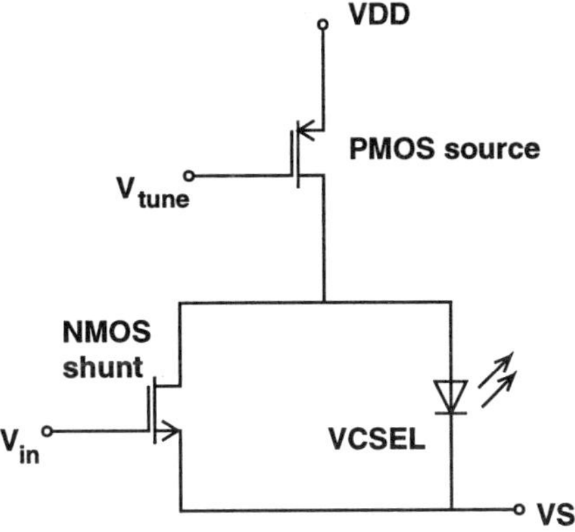

Fig. 12.8. CMOS shunt laser driver circuit [346]

could be operated at 1.25 Gb/s from below the laser threshold in this digital operation. In this case, the eye pattern indicated a turn-on delay of the lasers of about 180 ps. For a bit error rate of less than 10^{-11} at 1.25 Gb/s, the total power consumption of one driver and laser was about 17.5 mW at an optical power of −6.9 dBm. The NMOS transistor used as a small signal modulator, reducing the laser current only by a small amount and especially not below the threshold current, allowed to verify a bandwidth of the order of 2 GHz. This low-power high-speed operation demonstrates the utility and potential of the flip-chip bonding technique for optical interconnect technology.

Another hybrid CMOS-VCSEL transmitter has been introduced [470]. An array of 8 GaAs-AlAs lasers with an InGaAs triple-quantum-well active region was wire-bonded to driver circuits in a 1.0 μm CMOS technology. The threshold current I_{th} of the lasers was 2.7 mA and the peak optical output power exceeded 1 mW without heat sinking. Typical turn-on voltages $V_{\mathrm{ld}}(I_{\mathrm{th}})$ were (3 ± 0.5) V. The driver circuit is shown in Fig. 12.9.

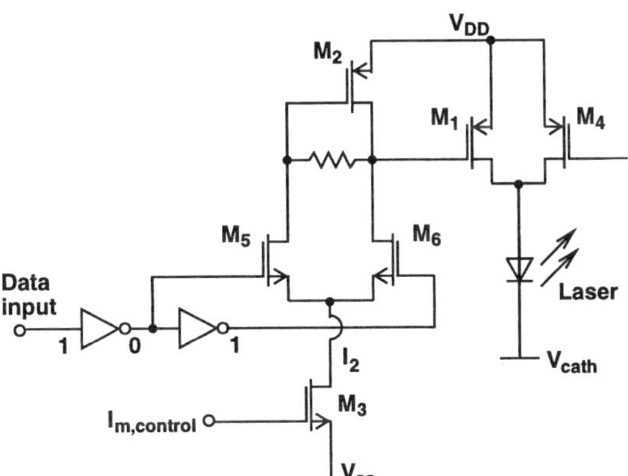

Fig. 12.9. CMOS laser driver circuit [470]

Since the laser array had one common cathode, only the lasers' anode terminals are independent and the modulation current I_m is controlled by the PMOS transistor M_1. The PMOS transistor M_4 regulates the laser bias current, which has to be somewhat larger than I_{th} in order to allow fast laser modulation. By minimizing stored charge in the current steering transistors, fast rise and fall times with low jitter and skew can be achieved. The choice of the width of M_1, however, is a trade-off between low parasitic capacitance for small device width and a reduced output impedance for a large width. A device width of 500 μm has been chosen as a good compromise for $I_m = 15$–25 mA in [470].

The NMOS devices M_5 and M_6 together with the inverter between their gates perform a single-ended to differential conversion. The NMOS device M_3 controls the modulation current I_m. The PMOS transistor M_2 serves as a current mirror reference in the '1' state and is a self-biased inverter in the '0' state. In such a way U_{gs} of M_1 is not zero but approximately equal to the threshold voltage in the '0' state, when the laser is off, and a high modulation speed can be obtained. In the '1' state, I_2 is mirrored (amplified by a certain factor) through M_2 to M_1 and into the laser. With a 50 Ω resistor load instead of the laser and with $I_m = 25$ mA the rise and fall times were about 0.5 ns and the eye pattern was wide open at 622 Mb/s. When a chip with 8 CMOS drivers was incorporated into one package with a VCSEL array flip-chip mounted onto a BeO substrate, the bond-wire inductances somewhat degenerated the slew rate [470]. A data rate of 622 Mb/s with a bit error rate of less than 10^{-9}, however, still has been obtained.

The cell size of one driver circuit with output pads was $175 \times 300\,\mu m^2$. The average power dissipation of one channel was 137 mW for an optical output power of 1 mW with $I_m = 20$ mA, where the laser consumed 75 mW (55 %), M_1 dissipated 45 mW (33 %), and its driver used 17 mW (12 %). For an array with many CMOS-VCSEL channels, this large power dissipation generates too much heat and better designs are necessary.

A laser driver circuit consisting of two N-channel MOS transistors (see Fig. 12.10) is advantageous compared to the shunt driver circuit in Fig. 12.8 with respect to power consumption. Here, M_2 is used to bias the laser above threshold ('0')and M_1 increases the laser current in order to obtain a high

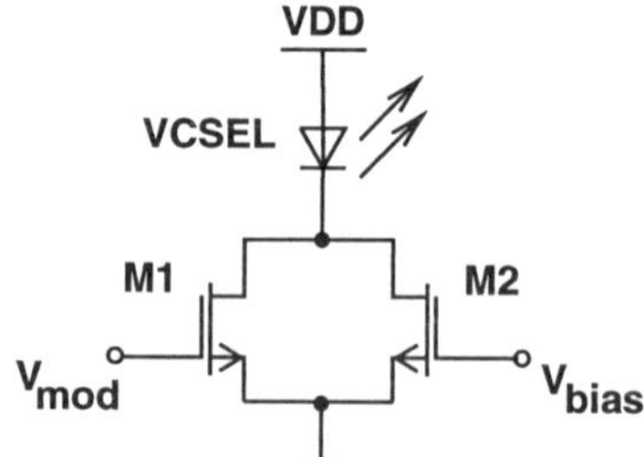

Fig. 12.10. Laser driver circuit with two NMOS transistors [471]

optical output power ('1'). Here, the current flowing continuously is lower than in the shunt driver circuit shown in Fig. 12.8. The circuit in Fig. 12.10 was implemented in [471] to modulate flip-chip bonded InGaAs quantum well VCSELs with an I_{th} of 6.4 mA.

The N-channel MOSFETs had a gate length of 0.8 μm and a gate width of 30 μm. With V_{mod} between 3.2 V and 5 V the optical output power could be controlled between 0.05 and 1.3 mW, when VDD = 8.5 V had been chosen. A possible modulation rate of 2 Gb/s for the circuit in Fig. 12.10 has been estimated [471].

12.4 Analog Circuits

In this section, a bipolar amplifier circuit, for an optical flame detection system, with a very high transimpedance of 1.1×10^9 V/A will be described. Another bipolar amplifier for audio CD systems is also described because of a so-called T-type feedback circuitry.

Two pixel circuits of a-Si:H CMOS image sensors with reduced cross-talk and a very high dynamic sensitivity range, respectively, and a fingerprint detector using lateral bipolar transistors will follow. Then CMOS and BiCMOS circuits for optical storage systems are described.

Finally, fiber receiver circuits in bipolar SiGe, NMOS, BiCMOS, and CMOS technology are explained. A comparison of the performance of optical silicon receivers compactly represents the state of the art.

12.4.1 Bipolar Circuits

The detector of a UV-sensitive OEIC [47] was already described in Sect. 3.3. The electronic circuit of the UV sensor system for flame detection where the photocurrent was only 20 pA to 1 nA will be explained here.

Due to the small photocurrents a large amplification with a large transimpedance of 10^9 V/A was necessary in order to obtain signal voltages of up to 1 V. Fortunately, the system did not require accurate control of the gain and, therefore, the current amplification factors of the NPN and PNP transistors of the complementary bipolar process could be fully exploited without a feedback circuitry. The circuit diagram of the bipolar amplifier is shown in Fig. 12.11.

The transistors Q_{13} and Q_{15} both in common base configuration together with their current sources Q_{12} and Q_{14}, respectively, provide a low-impedance input for the photocurrent ($R_{\text{in}} = 1/g_{\text{m},13}$). The UV photodiode is biased at zero volts. The anodic UV photocurrent I_{UV} (compare Fig. 3.6) flows into the emitter of Q_{13} and is injected into the base of Q_{16}. Transistor Q_{16} amplifies the photocurrent by its current gain factor β_{16}. This amplified current forms the base current of Q_{18}. At the collector of Q_{18}, the photocurrent is amplified

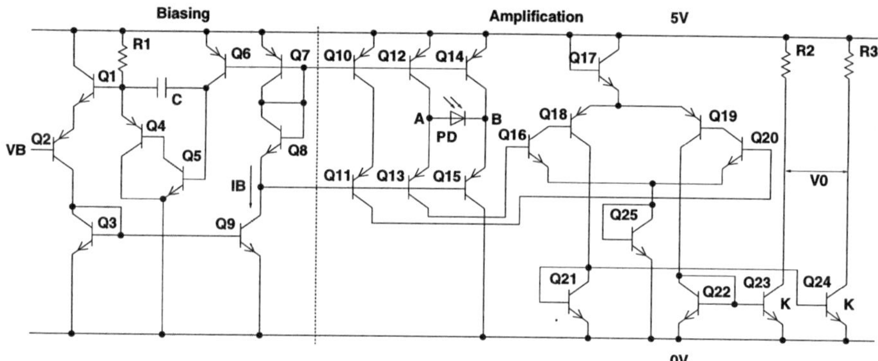

Fig. 12.11. Circuit diagram of a UV sensitive OEIC [47]

by the value of the product $\beta_{16} \times \beta_{18}$. The collector current of Q_{18} is mirrored into Q_{24} by Q_{21} and finally transformed into a voltage by R_3.

The infrared-dependent cathodic current $I_{UV} + I_{IR}$ (see Sect. 3.3) is shunted to V_{CC} by Q_{14}. The transistors Q_{10} and Q_{11} supply a reference bias current I_B to Q_{20}, which is the input of the second branch of the differential amplifier with Q_{16}, Q_{18}, Q_{19}, and Q_{20}. A reference current amplified by β_{19} and β_{20} is mirrored via Q_{22} and Q_{23} to R_2. The voltage V_o across the output terminals of the circuit is, therefore, given by

$$V_o = \beta_{16}\beta_{18}KR_3(I_{UV} + I_B) - \beta_{19}\beta_{20}KR_2I_B. \tag{12.1}$$

When perfect device matching ($Q_{16} = Q_{20}$, $Q_{18} = Q_{19}$, and $R_2 = R_3$) can be assumed, the output voltage V_o does not depend on the bias current I_B:

$$V_o = \beta_{16}\beta_{18}KR_3I_{UV}. \tag{12.2}$$

For the complementary bipolar process, β is approximately 140. R_3 was designed to be $28\,\mathrm{k\Omega}$, and the mirror factor K was 2. This resulted in a transimpedance of 1.1×10^9 V/A.

Although the output voltage is independent of I_B, I_B has to be very well controlled for correct operation and it has to be very small, because of the large transimpedance value. The bias section Q_1 to Q_9 replicates the gain stage (Q_4 corresponds to Q_{18}, Q_5 corresponds to Q_{16}, and $R_1 = R_3$) and feeds back the proper bias current via Q_6, Q_{10}, and Q_{12}. I_B can be adjusted accurately in the nA range with the reference voltage V_B. For V_B ranging from 0 V to 3.5 V, I_B is adjustable between 15 nA and 0 nA. The compensation capacitor $C = 12\,\mathrm{pF}$ is needed for the stability of the feedback loop.

The response time of the sensor of less than 100 ms was determined by the capacitance of the UV photodiode and the small photocurrents. The input-equivalent noise was smaller than $3.7\,\mathrm{pA}/\sqrt{\mathrm{Hz}}$ at 1 Hz. The power consumption of the UV-OEIC was 4 mW at 5 V. The total chip area was 4 mm^2 of which the UV photodiode occupied 1 mm^2.

After this example of an open loop very-high-gain amplifier, an amplifier with a T-type feedback network will be explained. Such a bipolar preamplifier OEIC for compact disk (CD) systems was described in [472]. The amplifier provides a high transimpedance gain, i. e. a low photocurrent is converted to a large voltage, and a relatively large bandwidth. In order to achieve these properties, the feedback via the T-type network shown in Fig. 12.12 is applied.

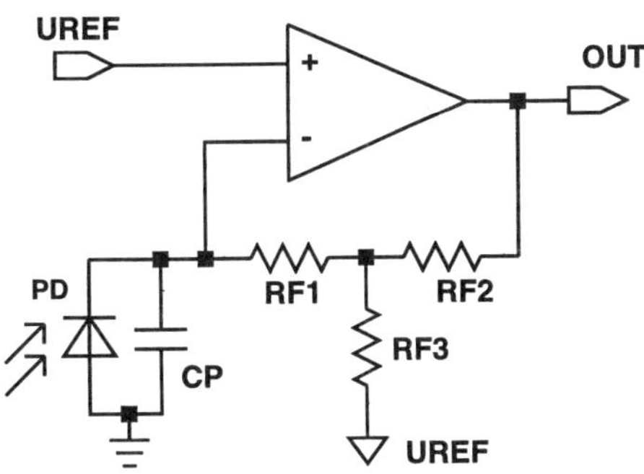

Fig. 12.12. T-type closed-loop configuration for high transimpedance gain [473]

The T-type network consists of the resistors R_{F1}, R_{F2}, and R_{F3}, whereby the values of R_{F1} and R_{F2} were chosen to be equal in [472]. The T-type network is an appropriate measure to simulate a high resistance R_F with low resistor values. The effective value R_F is

$$R_F = (R_{F1}R_{F2} + R_{F1}R_{F3} + R_{F2}R_{F3})/R_{F3} \tag{12.3}$$

and the output voltage V_o is obtained

$$V_o = -I_{ph}R_F. \tag{12.4}$$

In such a way it is possible to construct high-gain amplifiers without a high-resistivity polysilicon process module. Emitter polysilicon or gate polysilicon in a (Bi)CMOS process is sufficient. The T-type feedback can also be considered as a means to avoid large RC time constants of large resistance values, requiring a large polysilicon area with a large parasitic capacitance. Effective resistance values of 82 kΩ for a maximum photocurrent of 1.2 μA and of 328 kΩ for a photocurrent of 0.3 μA were realized in [472], whereby the true resistors had much lower values and the die area could be kept low at 400 × 350 μm^2. It should be mentioned, however, that the input offset voltage of the operational amplifier is also amplified [473].

The circuit is shown in Fig. 12.13. Only NPN transistors are used in the OEIC for a compact disk (CD) system in order to achieve a large bandwidth.

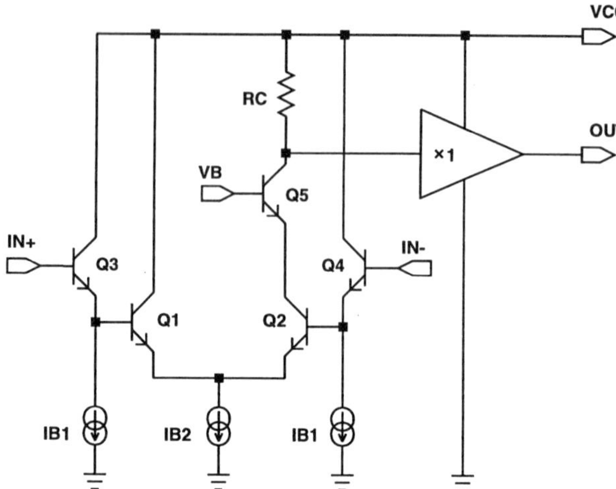

Fig. 12.13. Simplified circuit diagram of a bipolar OEIC for CD systems [472]

The transit frequency of the NPN transistors was 1.5 GHz. The operational amplifier has the voltage followers Q3 and Q4 in front of the common-emitter difference amplifier stage Q1/Q2 to obtain a low input current, whereby a large offset voltage due to the voltage drop across the T-type network caused by the base current of Q2 can be avoided.

The transistors Q2 and Q5 form a cascode stage to avoid a large Miller capacitance of Q2. High performance PNP transistors were not available and instead of a PNP current mirror load, the resistor R_C is used. Another voltage follower is used in order to obtain a low output impedance and to isolate the collector of Q5 from the load capacitance. The low-frequency open-loop gain was estimated to be $A_0 = 0.5 g_{m2} R_C = 40$. A compensation network was reported to be necessary, however, it was not described in [472]. The photodiodes used in the CD-OEIC also were not described.

The complete CD-OEIC contained fast channels with a bandwidth of 16 MHz for a maximum photocurrent of 1.2 μA and slower channels for a photocurrent of 0.3 μA. For the fast channels, rise and fall times of 22 ns with a settling time to 0.1 % of 200 ns were reported. The systematic offset voltage due to the base current of Q4 was smaller than 4 mV. The power consumption for one fast channel was 9.8 mW at 5 V.

12.4.2 CMOS Imagers

In Sect. 3.5.9, the concept of Thin Film on ASIC (TFA) was described. As an alternative to a self-structuring technology for the a-Si photodetectors in image sensors, an electronic circuit within each pixel was presented which reduces cross-talk between neighboring pixels [205]. The TFA sensor overcoming the coupling effect of neighboring photodetectors in an unstructured a-Si:H film by electronic means was called AIDA (Analog Image Detector

Array). A circuit inside each pixel (in c-Si) provides here a constant rear electrode potential for the photodetectors, thereby eliminating lateral currents between neighboring photodetectors in the a-Si:H film. The photodetectors are used in a constant voltage mode. The circuit diagram of a pixel is given in Fig. 12.14. Each pixel consists of an a-Si:H photodetector, eight MOSFETs and one integration capacitance C_{int}. C_{int} is discharged by the photocurrent. The inverter M2, M3 and the source follower feedback M1 keep the cathode voltage of the detector constant. M4 limits the power consumption of the inverter. M5 restricts the minimum voltage across the integration capacitance to 1.2 V in order to always keep the constant voltage circuit working. The integrated voltage on C_{int} is read out via M7 in source follower configuration and via the switch M8. The reset operation is performed as M6 recharges C_{int} after readout. The effective integration time of the pixel is the time period between two reset pulses, because readout is nondestructive and is performed at the end of the integration period.

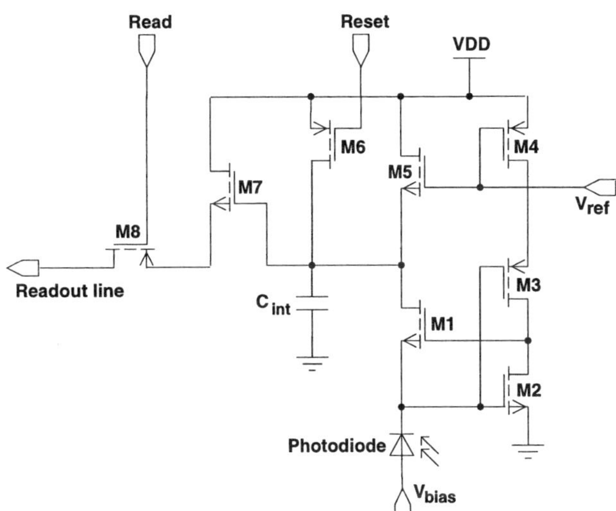

Fig. 12.14. Circuit diagram of a pixel in AIDA for operating the photodiode in a constant voltage mode [205]

The integration time may be varied according to the brightness of the scene. By this means, the sensitivity is controlled for all pixels globally. The AIDA sensor consists of 128 × 128 pixels with a size of 25 × 25 μm^2 each. The dynamic range of the sensor amounts to 60 dB for an integration time of 20 ms. The dynamic range can be extended significantly by means of the sensitivity control. The sensor was tested for illumination levels as high as 80 000 Lux. No blooming effects or image lag were observed [205].

For applications of an image sensor in a vehicle guidance system, for instance, which requires a very high degree of safety, the global sensitivity control is not appropriate. A number of pixels with excessive illumination may be saturated, whereas only slightly illuminated pixels may generate signal voltages below the noise and dark current levels. As pixels with logarithmic output characteristics exhibited a seriously increased sensitivity to temperature changes and fixed pattern noise, a Locally Adaptive Sensor in TFA-technology (TFA-LAS) was suggested [205]. The TFA-LAS allows to control the integration time and, therefore, the sensitivity of each pixel individually [474]. Figure 12.15 gives the complete pixel circuitry realized in the c-Si below each a-Si:H pixel photodetector.

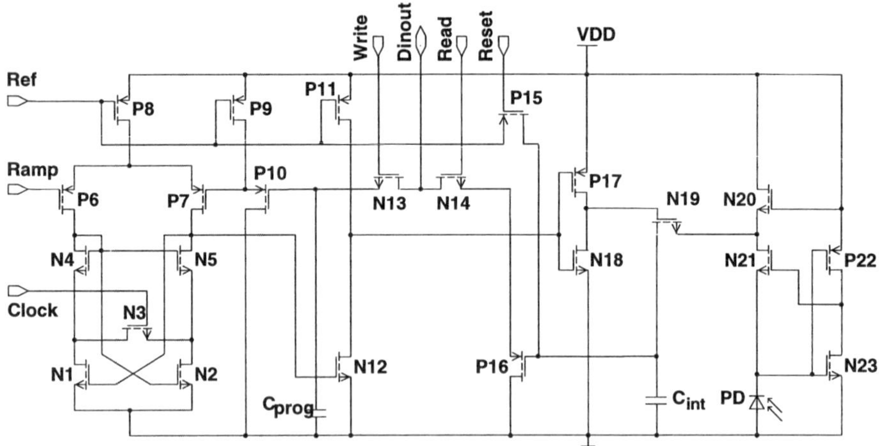

Fig. 12.15. Pixel circuit diagram of a locally adaptive image sensor [205]

The photocurrent is integrated into the MOS capacitance C_{int} while the rear electrode of the photodiode (cathode) is kept at a constant potential (compare Fig. 12.14). The integration time value is calculated externally for each illumination period and programmed into the pixel as an analog voltage V_{prog}, which is stored on the capacitor C_{prog}. V_{prog} standing at *Dinout*, is switched via N13 to C_{prog} for this purpose. This voltage on C_{prog} is compared to a linear voltage ramp generated by the peripheral electronics during the integration phase. The comparator consisting of the transistors N1–N5 and P6–P8 starts the integration as soon as the voltage ramp rises above V_{prog} and stops it at the falling edge of the ramp. A standard 0.7 μm CMOS ASIC technology implementing the TFA-LAS pixel schematic of Fig. 12.15 enabled a dynamic illumination range of more than 100 dB throughout the complete pixel array at any time. This corresponds to the outstanding dynamic range of c-Si PIN detectors, which is over 100 dB at an illumination intensity of 1000 Lux. For the TFA-LAS, the voltage range for the pixel signal amounts to 54 dB. The remaining dynamic range of 46 dB is included in the integration time and, therefore, in the programming voltage. The combination of both signal and programming voltage thus gives the information on the illumination level of a specific pixel. The TFA-LAS consisted of 64 × 64 pixels with a size of 40 × 50 μm^2 each.

Another interesting CMOS image sensor in bulk silicon should be mentioned here. It combines a lateral bipolar phototransistor array and a current comparator for digitizing the image [218]. The lateral bipolar phototransistor described in Sect. 3.5.10, Fig. 3.79, was used as a photodetector in a fingerprint detection chip fabricated in standard CMOS technology. The fingerprint detection chip contained an 64 × 64 array of the lateral phototransistors.

The phototransistors in the array were scanned by connecting one at a time to a current comparator using MOS transistors as selection switches. The circuit of the current comparator is shown in Fig. 12.16.

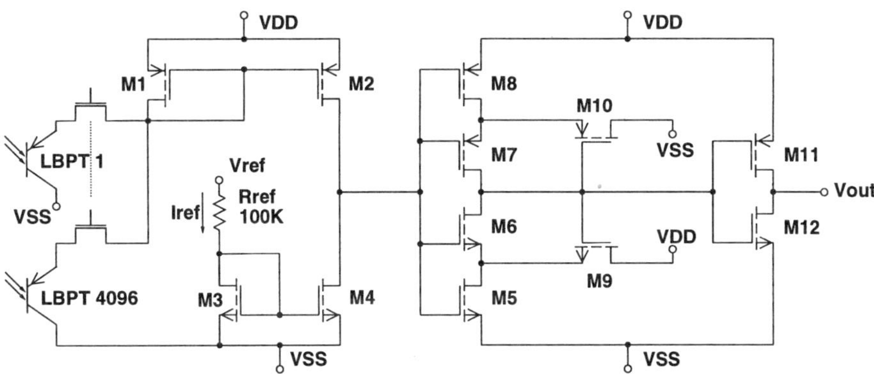

Fig. 12.16. Current comparator circuit of a fingerprint detector [218]

A threshold must be established so that pixel current values may be compared in order to give a black/white image. The most convenient point to place the threshold is the average of all pixel currents in order to obtain a global average. It also may be advantageous to select only m pixels from the entire array to give a local average from a sub-array. The average of the pixel currents is then obtained by dividing the measured current by m. In the fingerprint detection system, for instance, four rows of the 64 × 64 array were used and a 256 : 1 current mirror was used to give the average pixel current. The current mirror gate voltage is stored on a 10 pF capacitor. From this capacitor the reference voltage V_{ref} is derived setting the reference current for the current comparator. During the following 4096 clock cycles, all pixels are selected one at a time and their photocurrents are digitized to '1' or '0' by the current comparator. During each clock cycle, one selection MOS switch is activated and the pixel current is pulled through the PMOS current mirror M1 and M2 (Fig. 12.16). If this pixel current is greater than the average current, which was set up through M3 and M4, V_{out} goes high; otherwise V_{out}

goes low. The transistors M5–M12 compose a noninverting Schmitt trigger, which imposes a hysteresis and, in turn, stable digitization.

The field of application of such an image detection system is not limited to a fingerprint detection system, of course. Other types of pictures can also be digitized.

12.4.3 CMOS Circuits for Optical Storage Systems

Compact disk (CD), CD-ROM, and Digital-Video-Disk (DVD) are optical storage (OS) systems with a storage capacity of the order of 1 to 10 GByte [475]. The stored information is read with a focused laser beam. Depending on the stored state of 0 or 1, more or less light intensity is reflected into the read circuit, which we will call OS-OEIC. Special arrangements of 6 to 8 photodiodes are implemented in OS-OEICs in order to obtain the signals for tracking and for focusing in addition to the RF signal, which contains the stored information (see Fig. 12.17). In contrast to the name *digital*-video-disc-system, the DVD-OEIC is a purely *analog* front-end circuit, which is a key-device for the whole DVD-system. OS-OEICs for CD and CD-ROM are also analog circuits. The data rate and therefore the speed of the optical storage system are determined by the OS-OEIC.

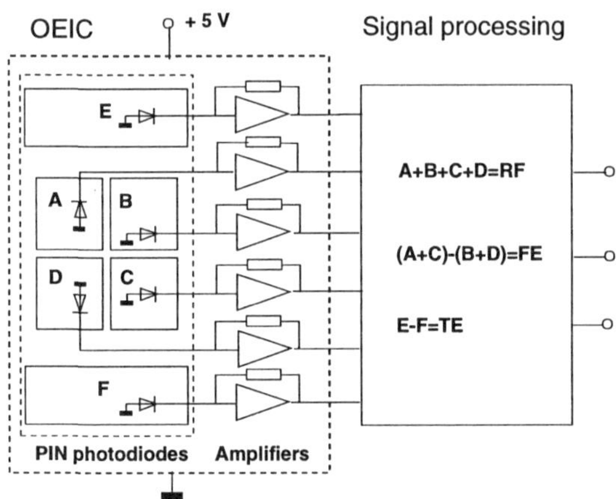

Fig. 12.17. OEIC for compact disc, CD-ROM and DVD applications

In OS systems, accordingly, the demand for fast OEICs is steadily increasing, especially in the red spectral range ($\lambda \approx 635$–650 nm). The integration of both the optical devices and the electronic circuits on the same chip leads to a smaller die area, to lower manufacturing costs and to faster systems. Furthermore, the reliability and the immunity against electromagnetic interference of OEICs are enhanced compared to a two-package solution with a photodiode package and an amplifier package. In comparison to a two-chip solution with a photodetector chip being wire-bonded to an amplifier chip in one package, the die area consumption of a monolithic OS-OEIC is smaller, because the area of a photodiode in an OS system is smaller than the area of a bondpad.

For fast OS systems, integrated PN photodiodes are not sufficient, because only a bandwidth of 10–15 MHz is achievable [67,69]. Although a −3dB bandwidth of 32 MHz can be derived from the frequency response of a P^+ to N-substrate photodiode shown in Fig. 3.15, the photocurrent already begins to decrease at a frequency of 2 MHz due to slow diffusion of photogenerated carriers. Hence, PIN photodiodes are required. It was shown that for the integration of a PIN photodiode in a standard twin-well CMOS process (1 μm), which uses epitaxial wafers, little modification is necessary. Compared to the published approaches in [38] and [59] with standard-buried-collector (SBC) based bipolar OEICs, less additional process complexity is required [79] as was already pointed out in Sects. 3.3.2 and 3.5.2. The cross section of the CMOS-OEIC was already shown in Fig. 3.33.

In order to obtain fast integrated PIN photodiodes, the standard doping concentration of the epitaxial layer of $1 \times 10^{15}\,\mathrm{cm}^{-3}$ has to be reduced to approximately $5 \times 10^{13}\,\mathrm{cm}^{-3}$ [79]. This reduction of the doping level does not influence the electrical parameters of the CMOS devices, since the MOSFETs are placed in wells and, therefore, the model parameters for circuit simulation did not have to be modified for the design of the OEICs [85]. ESD (electrostatic discharge) and latch-up immunity could be obtained by appropriate layouts.

OS-OEICs have to fulfill a stringent requirement concerning the output offset voltage. This requirement implies that only operational amplifiers can be used. Other amplifier types and especially amplifiers without feedback, which could be fast, do not guarantee that the output signals refer to the same reference voltage for a dark detector field within several millivolts.

The PIN photodiodes were integrated together with operational amplifiers in a voltage follower configuration (see Fig. 12.18), since a transimpedance amplifier, which is normally used for small photocurrents, was not advantageous in a digital CMOS process [85]. In order to compare the performance of a monolithically integrated and a wire-bonded circuit, which corresponds to a two-chip solution with photodiode chip and amplifier chip in one package, two test OEICs were developed. In the first case, the PIN photodiode was directly connected to the input of the CMOS operational amplifiers, whereas in the second case the PIN photodiode is connected via two bondpads with the input of the CMOS operational amplifier (see Fig. 12.18). This is not really a wire-bonded circuit, but it was possible to demonstrate the benefits of the monolithically integrated circuit [85]. A discrete circuit with an external photodiode has an even poorer performance than the wire-bonded two-chip circuit, due to package pin capacitances.

The voltage follower was realized by a two-stage CMOS operational amplifier, which was compensated ($R_\mathrm{C}, C_\mathrm{C}$) to ground (see Fig. 12.19). This

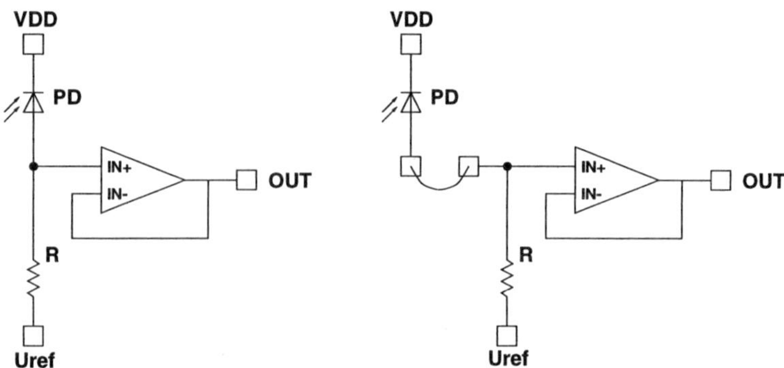

Fig. 12.18. Circuits of monolithic and wire-bonded CMOS DVD OEICs [85]

compensation technique was implemented instead of the Miller compensation because of the digital CMOS process. Due to the unavoidable use of the MOS capacitance a larger voltage drop across the capacitance was necessary in order to avoid the dip in the capacitance/voltage (CV) curve and, therefore, to obtain a practicable size (for the layout) and a value, which was more independent of process deviations within the relatively wide specification limits of the digital CMOS process.

The dimensions of the transistors were chosen in order to satisfy the zero-systematic-offset condition, which is given by the following equation [476]:

$$\frac{(W/L)_3}{(W/L)_6} = \frac{(W/L)_4}{(W/L)_6} = \frac{1(W/L)_5}{2(W/L)_7}. \tag{12.5}$$

Due to circuit simulations the open loop gain A_0 of the operational amplifier was larger than 40 dB for a load of 1 kΩ parallel 10 pF for all operating

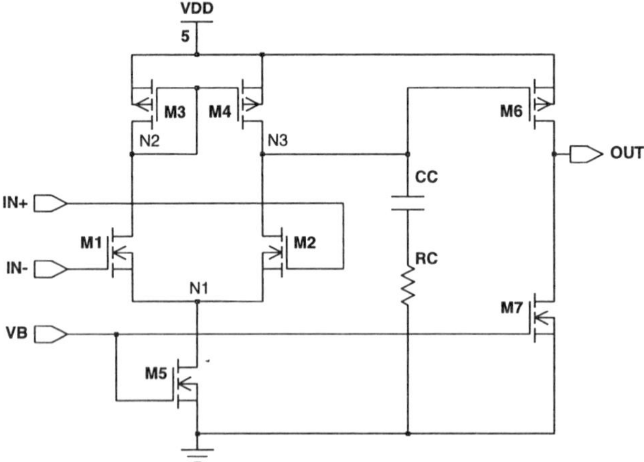

Fig. 12.19. Schematic of the operational CMOS amplifier of an OS-OEIC [85]

temperature and transistor parameter combinations. The phase margin was 46°, 58°, and 68° for the 'fast' case with the lowest temperature, for the 'nominal' case with room temperature, and for the 'slow' case with the highest temperature, respectively.

Samples on different N-substrate wafers (doping concentration in the epitaxial layer: approximately $1 \times 10^{15}\,\mathrm{cm}^{-3}$ and $5 \times 10^{13}\,\mathrm{cm}^{-3}$) were compared. The monolithic test circuit consisted of a PIN photodiode ($\approx 50 \times 50\,\mu\mathrm{m}^2$) with an integrated load resistor ($\approx 20\,\mathrm{k\Omega}$), which was realized by an N-MOSFET, since no analog high polysilicon resistors were available, followed by the CMOS operational amplifier in a voltage follower configuration (see Fig. 12.18). The voltage follower was loaded with $C_L = 10\,\mathrm{pF}$ and $R_L = 1\,\mathrm{k\Omega}$ (nominal values) for the measurements. The 'wire-bonded' test circuit consisted of the same modules plus two additional bondpads between PIN photodiode and operational amplifier. ESD-protection circuitry was implemented on all the chips in order to guarantee ESD immunity.

In the following, measured results will be presented. The supply voltage of the OEICs was 5 V, $U_{ref} = 2.5\,\mathrm{V}$, and the wavelength of the laser light was $\lambda = 638\,\mathrm{nm}$. The laser light was coupled into the photodiodes of the OEICs on the wafer prober via a single-mode fiber. A network analyzer HP8751A was used for the modulation of the laser and for the frequency response measurements of the OEICs.

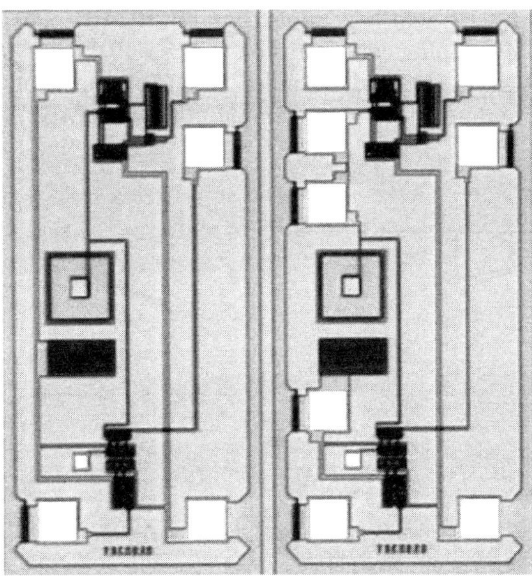

Fig. 12.20. Microphotograph of a test chip for the comparison of monolithic and wire-bonded CMOS receiver OEICs. The circuit on the *left* represents the monolithic OEIC, whereas the circuit on the *right* representing the wire-bonded OEIC includes two bondpads between photodiode and amplifier in the upper part of the chip [85]

Figure 12.20 shows a microphotograph of the test circuits. On the left half of the figure the fully integrated circuit and on the right half the 'wire-bonded' circuit are located. The operational amplifiers are located in the upper half of the chips, whereas the integrated resistors are placed in the lower half. The PIN photodiode with a 50 μm wide metal shield and a guard ring around is located in the middle part of the chips with an additional substrate contact, which, however, was not used. The layout was not area-optimized since it was only a test chip.

Furthermore, an 8-channel OEIC for universal focusing and tracking methods of optical storage systems like DVD was designed [85]. This OEIC consists of 8 channels with PIN photodiodes and voltage followers. Four channels were so-called fast channels for data extraction and focusing and the other four channels were so-called slow channels with a ten times larger sensitivity. The gain of the amplifiers in the fast and slow channels is switchable (high, medium, and low), so that three levels of photocurrent in DVD-ROM and DVD-RAM applications can be detected and amplified. This OEIC was also integrated on N-substrate wafers with different epitaxial layers.

For frequency response measurements on a wafer prober, the output signal of the OEICs was fed via a picoprobe into the network analyzer. Figure 12.21 shows the frequency response of the monolithic OEICs on a standard and on a low doped epitaxial layer and of the 'wire-bonded' OEIC. The −3dB bandwidth of the OEIC with the standard epitaxial layer with a doping concentration of approximately $1 \times 10^{15}\,\mathrm{cm}^{-3}$ is approximately 10 MHz. The best results are achieved with a low doping concentration (i. e. $5 \times 10^{13}\,\mathrm{cm}^{-3}$) in the epitaxial layer ($f_{3\mathrm{dB}} = 19\,\mathrm{MHz}$). The 'wire-bonded' OEIC on the same low doped epitaxial material has a much poorer performance ($f_{3\mathrm{dB}} = 4\,\mathrm{MHz}$).

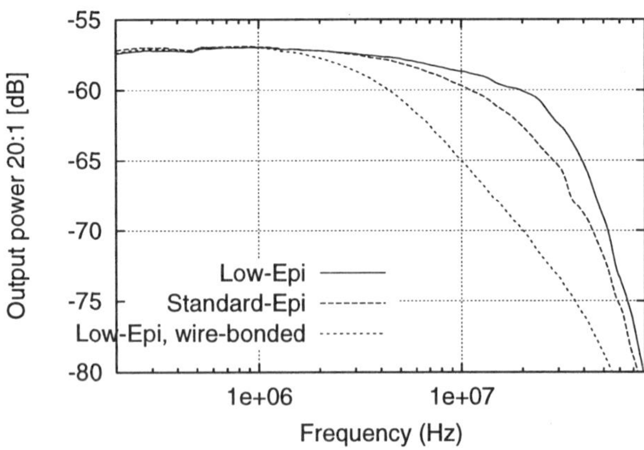

Fig. 12.21. Comparison of measured frequency responses for three different CMOS OEICs [85]

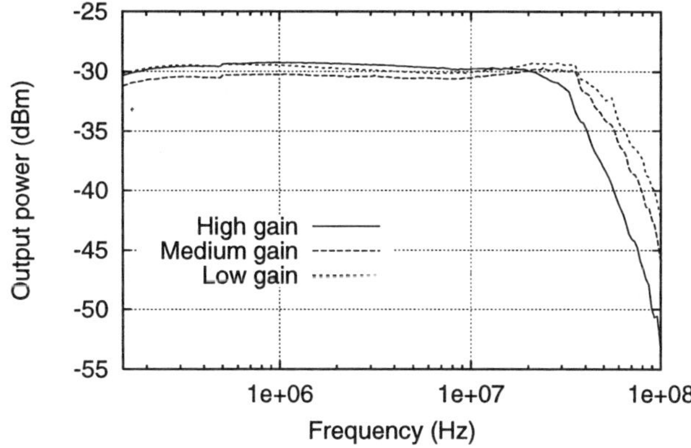

Fig. 12.22. Frequency responses of a fast channel of the CMOS OEIC with different optical input power for the same low-frequency output voltage

For the packaged 8-channel DVD-OEIC with a total power consumption of approximately 70 mW the following results were obtained: the offset voltage was smaller than 10 mV and the sensitivity of the fast channels was 3.3 mV/μW. A value of 5.6 mV/μW was achieved with a special antireflection coating layer. The −3dB bandwidth was measured with an active probe head, whereby the amplifiers were loaded with $C_{\mathrm{L}} = 11.2\,\mathrm{pF}$ and $R_{\mathrm{L}} = 1\,\mathrm{k\Omega}$. Figures 12.22 and 12.23 show the frequency responses of a fast channel and of a slow channel with a ten times larger sensitivity, respectively.

For a fast channel, the −3dB bandwidth was ≈ 33 MHz for the high gain, ≈ 50 MHz for the medium gain, and ≈ 54 MHz for the low gain. These values far exceed the bandwidth of 7.3 MHz of the circuit for 8× speed CD-ROM,

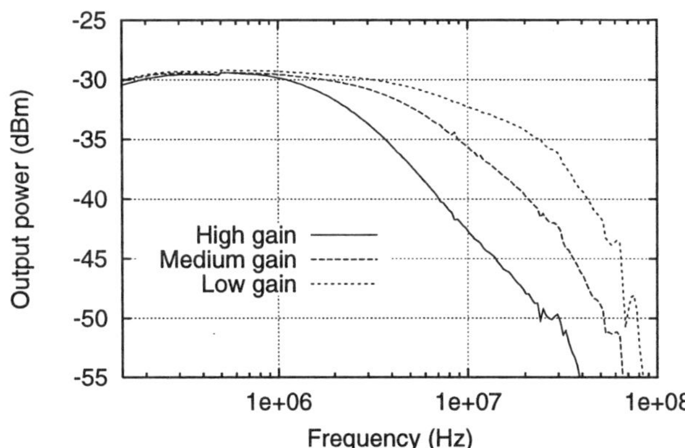

Fig. 12.23. Frequency responses of a slow channel of the CMOS OEIC with different optical input power for the same low-frequency output voltage

which was fabricated in a 0.8 μm CMOS technology and which used off-chip photodiodes [477]. The noise level at 10 MHz with a resolution bandwidth (RBW) of 30 kHz was −89 dBm. The group delay was constant to within ± 2.5 ns for frequencies up to approximately 15 MHz. For a slow channel, the −3dB bandwidth was ≈ 2.2 MHz for the high gain, ≈ 4.8 MHz for the medium gain, and ≈ 9.0 MHz for the low gain.

Let us summarize. For the standard doping level of the epitaxial layer, the first result is the 2.5 times larger bandwidth of the fully integrated circuit compared to the wire-bonded circuit. This enhanced performance is due to the minimized parasitic capacitances between the photodiodes and the amplifiers. The second result is that the performance of the PIN CMOS OEICs is enhanced when they are integrated on substrates with an epitaxial layer, which has a low doping concentration (e. g. $5 \times 10^{13}\,\mathrm{cm}^{-3}$). The depletion layer width of the PIN photodiode, which is reverse-biased, is greater in the case of the epitaxial material with $5 \times 10^{13}\,\mathrm{cm}^{-3}$ than in the case of the standard material. For a doping level of $5 \times 10^{13}\,\mathrm{cm}^{-3}$ in the epitaxial layer, the depletion layer reaches through the whole intrinsic region and the slow diffusion of photogenerated carriers is eliminated. This results in a faster frequency response of the photodiode. With the results achieved, double-speed DVD video systems with PIN CMOS OEICs are possible.

12.4.4 BiCMOS Circuits for Optical Storage Systems

For optical storage (OS) systems with an enhanced data rate, BiCMOS OEICs were developed within the BiCMOS project. These OEICs contain four fast channels (A–D) for data extraction and focus control plus four slower channels (E–H) for tracking control with a ten times larger sensitivity (Fig. 12.24).

No process modifications were necessary in the BiCMOS process implementing a double photodiode shown in Fig. 3.90 [238]. The schematic of one fast channel of an eight-channel OS-BiCMOS-OEIC is shown in Fig. 12.25 [238]. Polysilicon-polysilicon capacitors are available in the 0.8 μm BiCMOS process and a transimpedance amplifier is realized here. The gain is switchable between high (H), medium (M), and low (L). Only NPN transistors are used in the signal path and the resistor loads R1 and R2 are implemented in the difference amplifier in order to achieve a high −3dB bandwidth. The value of 44 MHz was confirmed by measurements for the high gain [239]. The emitter follower Q5 reduces the output impedance of the operational amplifier.

The base currents of Q1 and Q2 in the input stage of the operational amplifier are approximately 5 μA. Such a large input current of the operational amplifier, flowing through the transimpedance resistor R3, leads to a systematic output offset voltage of approximately 0.1 V, when no special measures are taken. These special measures are: The transistors Q3 and Q4 are used to sense the base currents of Q1 and Q2, respectively. The current

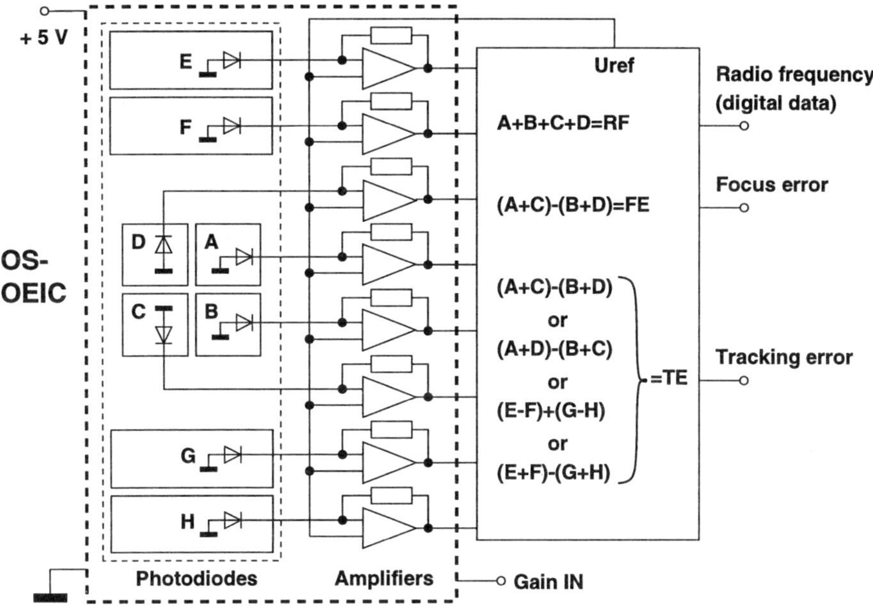

Fig. 12.24. Block diagram of an OS-BiCMOS-OEIC

mirrors M1/M2 and M3/M4 mirror the base currents of Q3 and Q4 into the bases of Q1 and Q2, respectively [478]. With this bias current cancellation, the output offset voltage could be reduced to less than 9 mV [238].

The frequency responses for the three different gain factors of the OS-BiCMOS-OEIC are shown in Fig. 12.26. The −3dB bandwidth for the high gain is 44 MHz, for the medium gain it is 59 MHz, and for the low gain it is 64 MHz. The slightly decreasing frequency responses between 1 MHz and 40 MHz are due to parasitic capacitors in the amplifier and not due to slow carrier diffusion of photogenerated carriers in the double photodiode (DPD). The power consumption of the circuit in Fig. 12.25 is 7 mW at 5.0 V and with an antireflection coating a sensitivity of 10.5 mV/μW is achieved with the highest gain factor. The active die area of this circuit is 0.054 mm^2. An 8-channel OEIC with four of the above described fast amplifiers and four ten times more sensitive MOS amplifiers consumed a power of approximately 40 mW.

At the expense of a higher power consumption, an even higher speed of OEICs for optical storage systems is possible [464]. A high-bandwidth BiCMOS OEIC has been demonstrated, which implemented fast amplifiers with the same schematic shown in Fig. 12.25 but with larger bias currents than in [238]. These fast amplifiers exhibit −3dB bandwidths in excess of 90 MHz (Fig. 12.27).

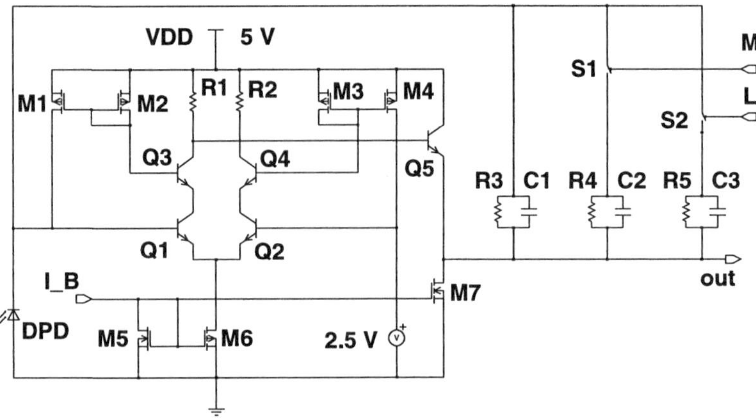

Fig. 12.25. Schematic of the fast channels A–D in an OS-BiCMOS-OEIC [238]

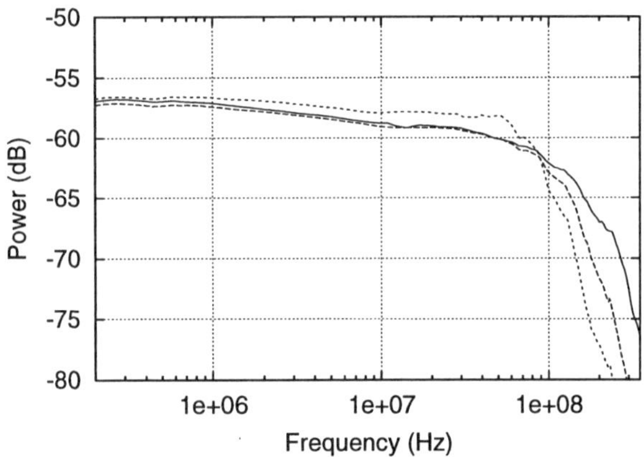

Fig. 12.26. Frequency response of an BiCMOS OEIC for optical storage systems for three different optical input powers, i. e. three different gain factors [238]

An integrated double photodiode (Fig. 12.25) is connected to a transimpedance amplifier using an operational amplifier in order to obtain a low output offset voltage compared to a reference voltage of 2.5 V as is required for applications in optical storage systems. For a universal applicability, the gain is switchable by MOS elements between high (H, R3), medium (M, R4 $\|$ R3) and low (L, R5 $\|$ R4 $\|$ R3) with a ratio of approximately 1/3 each. Polysilicon-polysilicon capacitors are used for frequency compensation with C1, C2, and C3. Here, the bias current cancellation of the input transistors Q1 and Q2 reduces the systematic output offset of approximately 110 mV, which would result from the base current of Q1 across the resistor R3 $\approx 20\,\mathrm{k}\Omega$, to below 11 mV. According to simulations, the low-frequency open-loop gain

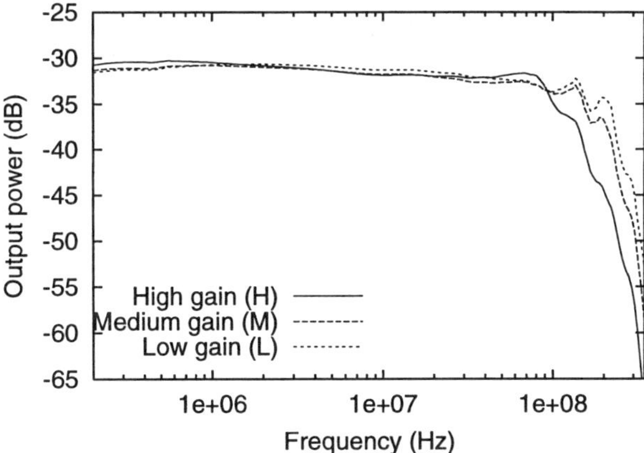

Fig. 12.27. Frequency responses of the fast channels A–D of a high-bandwidth BiCMOS OEIC for optical storage systems measured with three different optical input powers, i. e. three different gain factors [464]

of the operational amplifier is 27 dB and its transit frequency is 870 MHz in the case of a load of 1 kΩ and 10 pF. An OEIC was packaged, mounted together with these load elements on a printed circuit board, and the frequency responses were measured with a probe head having an input capacitance of 1.7 pF. The slight decrease in the frequency responses (Fig. 12.27) between about 5 MHz and 80 MHz is due to parasitic capacitors in the amplifier and is not due to slow diffusion of photogenerated carriers in the DPD. The measured bandwidths exceed a value of 92 MHz. This value is much larger than the bandwidth of 7.3 MHz of the circuit for 8× speed CD-ROMs fabricated in 0.8 μm CMOS technology with off-chip photodiodes [477].

Figure 12.28 shows the schematic of the four sensitive channels E–H with a ten times larger sensitivity for tracking control in the optical storage system. A double photodiode with approximately twice the size of the DPDs in the channels A–D is implemented in the channels E–H. The N-channel MOSFET source followers M1 and M2 are added in front of the bipolar difference amplifier Q1 and Q2 in order to avoid input currents and the resulting output offset voltages across the feedback resistors of about 200 kΩ. A high sensitivity of 100 mV/μW in combination with a low offset voltage can be realized in such a way. The PMOS load elements M3 and M4 are implemented for Q1 and Q2 in the difference amplifier in order to achieve a larger open-loop gain than with resistor load elements. The compensation is split between C1 and C2, C1 and C3, as well as C1 and C4. According to circuit simulations, the low-frequency open-loop gain of the circuit shown in Fig. 12.28 is 36 dB with a transit frequency of 130 MHz.

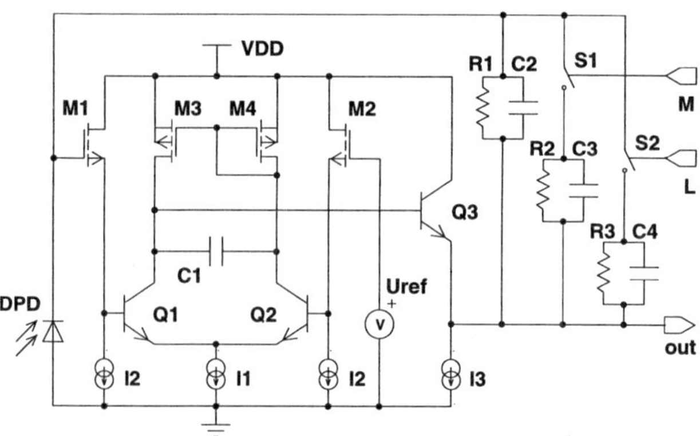

Fig. 12.28. Schematic of the sensitive channels E–H in a high-bandwidth OS-BiCMOS-OEIC [464]

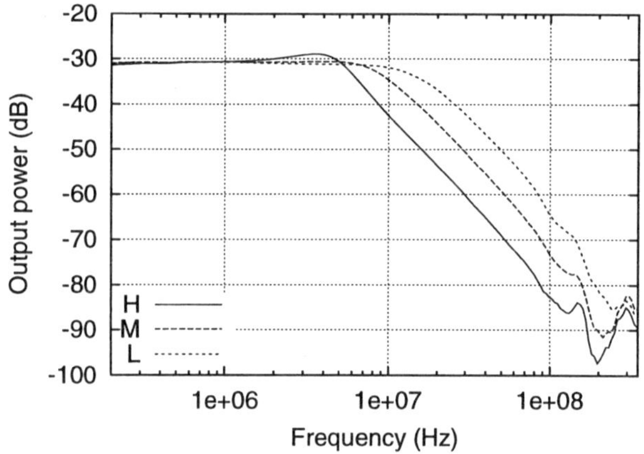

Fig. 12.29. Frequency responses of the sensitive channels E–H of a high-bandwidth BiCMOS OEIC for optical storage systems measured with three different optical input powers, i. e. three different gain factors

The frequency responses of the channels E–H are shown in Fig. 12.29. The values for the −3dB bandwidths are listed in Table 12.1 together with other results.

Each amplifier in the channels A–H covers an active die area of about 0.079 mm^2, and the total die area of the high-bandwidth BiCMOS OEIC amounts to 3.25 mm^2. The power consumption of the high-bandwidth OS-BiCMOS-OEIC is less than 75 mW at 5.0 V. Table 12.1 summarizes further technical data of the fast and the sensitive channels of the high-bandwidth OS-BiCMOS-OEIC.

Table 12.1. Measured results of the high-bandwidth OS-BiCMOS-OEIC

	H	M	L
$f_{-3\,\mathrm{dB}}$ (MHz) A–D	92.0	94.9	95.1
$f_{-3\,\mathrm{dB}}$ (MHz) E–H	5.2	8.5	14.6
Sensitivity (mV/μW) A–D	8.8	2.9	0.9
Sensitivity (mV/μW) E–H	88.1	29.3	9.1
U_{Offset} (mV) A–D	< 10.8	< 9.5	< 9.0
U_{Offset} (mV) E–H	< 7.4	< 6.4	< 6.4
Noise (dBm) @10 MHz with 30 kHz RBW A–D	−81.5	−85.0	−85.2
Noise (dBm) E–H	−66.0	−67.5	−73.5

Figure 12.30 shows the microphotograph of the OS-BiCMOS-OEIC with bandwidths in excess of 90 MHz, which was realized in full custom design. The results demonstrate that it is possible to avoid the slow carrier diffusion problem by exploiting double photodiodes in standard BiCMOS technology.

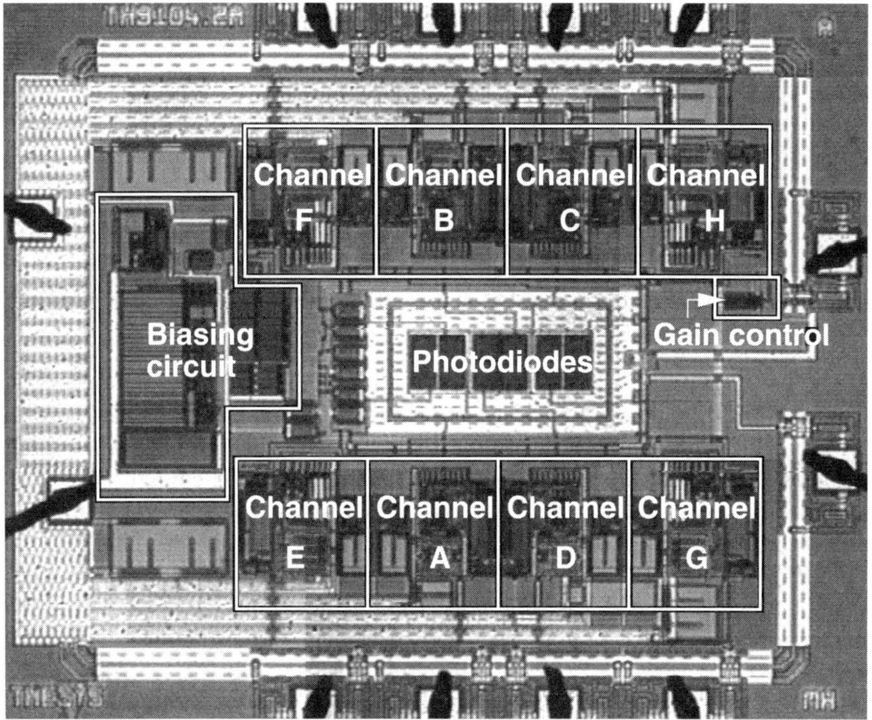

Fig. 12.30. Micrograph of a high-bandwidth BiCMOS DVD OEIC [464]

When a high speed and a higher sensitivity are required in addition to a low output offset voltage for the OS-OEICs, a two-stage optical receiver may be necessary. The circuit principle of such a two-stage amplifier [479] is shown in Fig. 12.31. The circuit consists of a transimpedance amplifier for the photocurrent and a reference I/U converter for offset compensation plus an operational amplifier in subtractor configuration. The subtractor can be used simultaneously as a voltage amplifier with the amplification factor RS2/RS1 enabling a high overall sensitivity of the OEIC. It must be mentioned, however, that offset voltages due to mismatch of the two transimpedance input amplifiers and due to mismatch in the operational amplifier are also amplified.

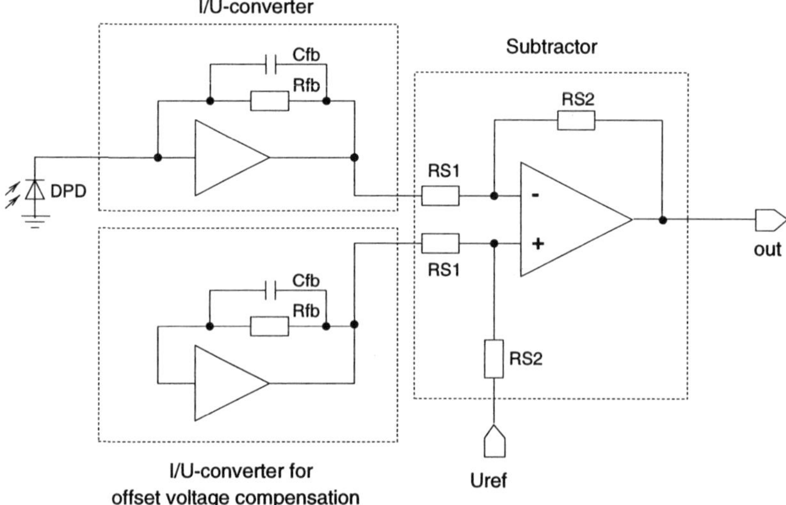

Fig. 12.31. Block diagram of a two-stage optical receiver for one fast channel of a high-speed OS-BiCMOS-OEIC

The circuit diagram of the complete circuit is shown in Fig. 12.32. In order to achieve a high bandwidth, only NPN transistors are used in the signal paths of the preamplifiers. Q1 is used in common-emitter configuration, Q3 is used as emitter follower, and the feedback resistor R_{fb1} together with Q1 and Q3 represent a low input impedance for the photocurrent of the double photodiode (DPD). Thereby, the effect of the DPD capacitance is minimized. The reference voltage V_{ref} is chosen as the emitter potential of Q1 in order to increase the reverse voltage of the DPD to $V_{\mathrm{BE,Q1}} + V_{\mathrm{ref}}$. The values for R_1, R_{fb1}, and C_{fb1} were $3\,\mathrm{k}\Omega$, $27\,\mathrm{k}\Omega$, and $25\,\mathrm{fF}$, respectively. A second emitter follower (Q11) is implemented for level shifting and decoupling of output and feedback path. The second preamplifier consists of transistors Q2, Q4, and Q12 as well as the current mirror with Q6 and Q7 and the feedback

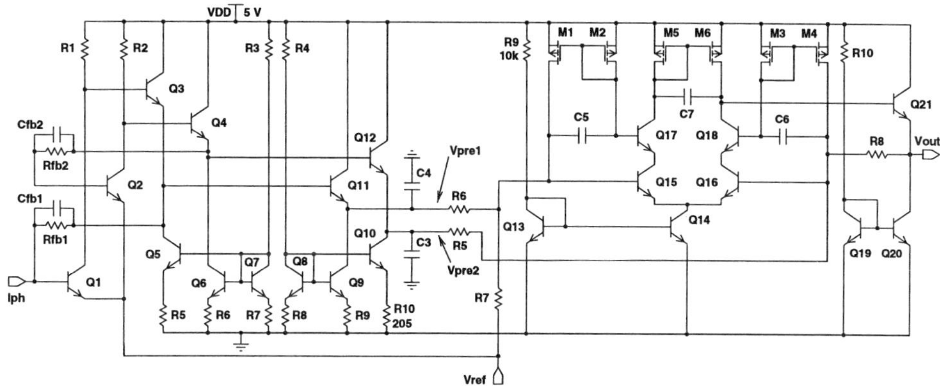

Fig. 12.32. Schematic of high-speed BiCMOS-OEIC for the fast channels A–D in an OS-BiCMOS-OEIC [480]

resistor R_{fb2} plus the compensation capacitor C_{fb2}. At the outputs of the preamplifiers, C3 and C4 are added as further compensation capacitors.

The large signal DC transfer functions of the preamplifiers are given by

$$V_{\text{pre1}} = V_{\text{ref}} + \eta_{\text{tia}} I_{\text{ph}} R_{\text{fb1}} + I_{\text{B,Q1}} R_{\text{fb1}}, \tag{12.6}$$

and

$$V_{\text{pre2}} = V_{\text{ref}} + I_{\text{B,Q2}} R_{\text{fb2}}, \tag{12.7}$$

when we assume $U_{\text{BE,Q1}} = U_{\text{BE,Q11}}$ and $U_{\text{BE,Q2}} = U_{\text{BE,Q12}}$. The efficiency factor η_{tia} of the preamplifier is given by $\eta_{\text{tia}} = R_1\beta/(R_{\text{fb1}}+R_1\beta)$. For $\beta = 100$ and for the resistor values given above the efficiency factor η_{tia} of the transimpedance preamplifier is equal to 0.917. The base current of Q1 (and Q2) causes a voltage of about 0.1 V across R_{fb1} (and R_{fb2}). In order to achieve a low output offset voltage compared to V_{ref}, therefore, the second preamplifier without a photodiode and the subtractor operational amplifier are necessary. Perfect matching of the two preamplifiers ($I_{\text{B,Q1}} = I_{\text{B,Q2}}$, $R_{\text{fb1}} = R_{\text{fb2}}$, $R_1 = R_2$, $\beta_1 = \beta_2$, $U_{\text{BE,Q3}} = U_{\text{BE,Q4}}$, $U_{\text{BE,Q11}} = U_{\text{BE,Q12}}$) is, however, necessary in order to obtain $V_{\text{pre1}} = V_{\text{pre2}}$ for a dark photodiode. This perfect matching of the two preamplifiers requires a careful layout to achieve a low output offset voltage.

The preamplifiers are connected to the subtractor operational amplifier via R5 and R6. The bias current cancellation introduced in Fig. 12.25 is applied to reduce the input currents of the operational amplifier necessary for a low output offset voltage. A PMOS current mirror load with M5 and M6 is used here in order to obtain a higher open loop gain of the operational amplifier. For $R_5 = R_6$ and $R_7 = R_8$, an analysis of the subtractor amplifier yields the transfer function

$$V_{\rm out} = V_{\rm ref} + \frac{R_7 A_{\rm d}(s)}{(R_6 + R_7) + R_6 A_{\rm d}(s)}(V_{\rm pre1} - V_{\rm pre2}) \quad (12.8)$$

and

$$V_{\rm out} \approx V_{\rm ref} + \frac{R_7}{R_6}\eta_{\rm tia} I_{\rm ph} R_{\rm fb1}, \quad (12.9)$$

when we assume a large open loop voltage gain $A_{\rm d}(s)$ of the operational amplifier. A -3dB frequency of 189 MHz was determined by numerical prelayout simulation for the complete amplifier with a load of $R_{\rm L} = 1\,{\rm k}\Omega$ and $C_{\rm L} = 10\,{\rm pF}$. The complete two-stage amplifier was designed for a sensitivity of 10 mV/μW and an offset voltage of less than 10 mV for $R_5 = R_6 = R_7 = R_8$.

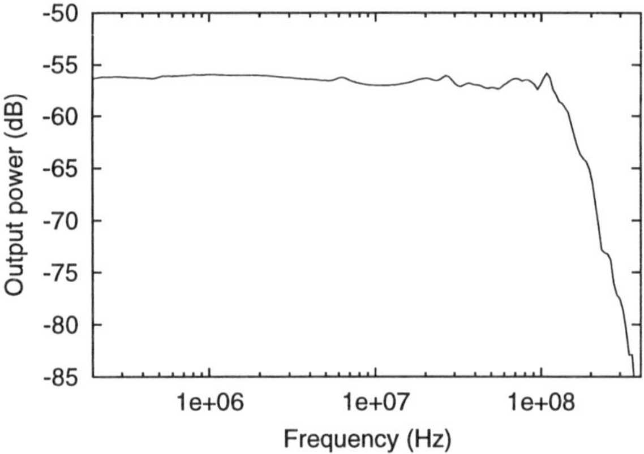

Fig. 12.33. Frequency response of a two-stage optical BiCMOS receiver for the fast channels A–D of a high-speed BiCMOS OEIC for optical storage systems [480]

The OEIC with the DPD and the two-stage amplifier was fabricated in a 0.8 μm BiCMOS technology. The measured frequency response of this two-stage optical receiver is shown in Fig. 12.33. A -3dB frequency of 147.7 MHz is determined from this frequency response. The power consumption of the two-stage optical receiver is 35 mW at a supply voltage of 5 V. The active die area of the two-stage optical receiver is $340 \times 140\,\mu{\rm m}^2$.

12.4.5 Fiber Receivers

Optical multimode fibers for optical data transmission usually possess a core diameter of 50 or 62.5 μm. The core diameter of single-mode fibers for wavelengths shorter than 1.1 μm actually is less than 10 μm. Small-area photodiodes therefore can be used in order to realize a low capacitance at the input of the receiver circuits. The bondpad capacitances of wire-bonded receivers are much larger than the capacitance of PIN photodiodes. Monolithically integrated optical fiber receivers should be the first choice as a consequence. In the following, a bipolar OEIC, two NMOS OEIC, two BiCMOS OEICs and several CMOS OEICs for application as fiber receivers will be described.

Bipolar SiGe Receiver. The superior speed of SiGe HBTs compared to Si bipolar transistors has already been mentioned and the structure of a monolithic SiGe-Si PIN-HBT receiver was described in Chap. 6. Here, the bipolar transimpedance amplifier circuit of this receiver (see Fig. 12.34) will be discussed.

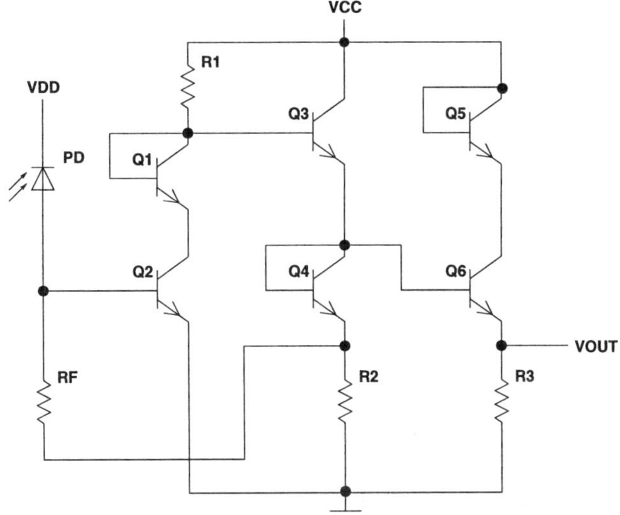

Fig. 12.34. Circuit diagram of a bipolar photoreceiver [314]

The receiver consists of a PIN photodiode, a common emitter gain stage, two emitter follower buffers, and a resistive feedback loop. NiCr thin-film resistors were used in the monolithic SiGe HBT receiver [314]. The transistors Q1, Q4, and Q5 are used as level shifting diodes. Q1 and Q5 reduce U_{CE} of Q2 and Q6, respectively, because the breakdown voltages of high-speed transistors are quite low.

The two voltage sources VDD and VCC were necessary to optimize the PIN transient behavior and the operating point of the amplifier. The value of the feedback resistor R_{F} determines the bandwidth, gain, and noise characteristics of the photoreceiver. The value of R_{F} is usually chosen based on a trade-off between these three parameters. In [314], a value of $640\,\Omega$ was chosen for R_{F} resulting in a transimpedance gain of 52.2 dBΩ. The bandwidth of 1.6 GHz was obtained for the transimpedance amplifier with a f_{T} of 25 GHz for the HBTs with an emitter area of 5×5 µm. The optical bandwidth of 460 MHz of the PIN-HBT receiver was measured for VDD = 9 V and VCC = 6 V. The bandwidth of the receiver was limited by the photodiode and the trade-off mentioned above might be improved with respect to an increased gain, i. e. a larger sensitivity. An input noise spectral density of $8.2\,\text{pA}/\sqrt{\text{Hz}}$ up to 1 GHz caused by shot noise from the base current and thermal noise from the feedback resistor was given. With these values, the photoreceiver sensitivities of −24.3 and −22.8 dBm were estimated for 0.5 and 1 Gb/s, respectively, for a bit error rate (BER) of 10^{-9} and $\lambda = 850\,\text{nm}$.

NMOS Receivers. The next example is an NMOS OEIC. A lateral PIN photodiode was integrated in a 1.0 μm NMOS technology using a nominally undoped substrate, which was actually P-type with $N_A = 6 \times 10^{12}\,\text{cm}^{-3}$ [75, 76]. This lateral PIN photodiode is described in Sect. 3.5.2.

An NMOS transimpedance preamplifier (Fig. 12.35) implementing a depletion transistor at the input was integrated together with a lateral PIN photodiode (see Fig. 3.18). The second and third stages were formed by source followers. The source follower stage M3/M4 with M4 as a constant current source is used to establish the correct operating point across the feedback loop with MF as an active resistor. The source follower stage M5/M6 at the output was sized to match a 50 Ω load impedance. The power dissipation including the output driver was only 3 mW. Accordingly, open-eye operation for bit-rates of only up to 40 Mb/s with a photocurrent of 3 μA for $\lambda = 870\,\text{nm}$ and for a transimpedance of 3 kΩ was reported [75].

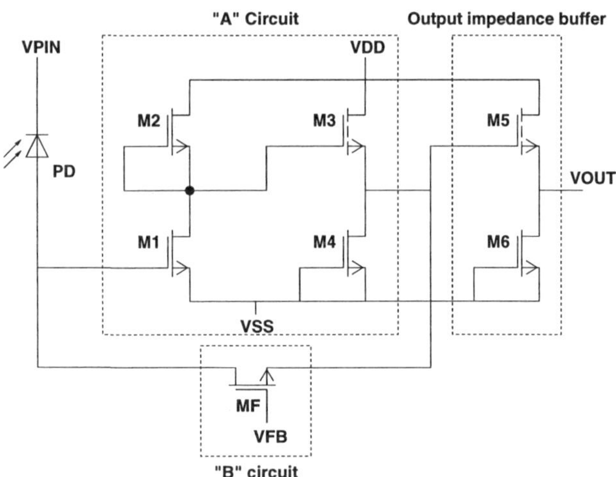

Fig. 12.35. Schematic of an NMOS fiber receiver OEIC [75]

The authors of [75] meanwhile improved their photoreceiver [481]. An N-type Si substrate with a resistivity of 1000–3000 Ωcm was taken and an interdigitated lateral PIN structure with a finger width of 2 μm and a finger spacing of 10 μm instead of the ring structure (see Fig. 3.18) was implemented.

The total area of the photodiode was $50 \times 50\,\mu m^2$. A dark current of 1.3 pA at 5 V and of 63 nA at 30 V was found. The quantum efficiency was increased to 84 and 74 % at 800 and 870 nm, respectively, due to a SiO_2 antireflection coating with a thickness of 150 nm. A bit-rate of 500 Mb/s was achieved from the interdigitated PIN photodiode at 30 V, however, its frequency response showed a diffusion tail below 100 MHz [481].

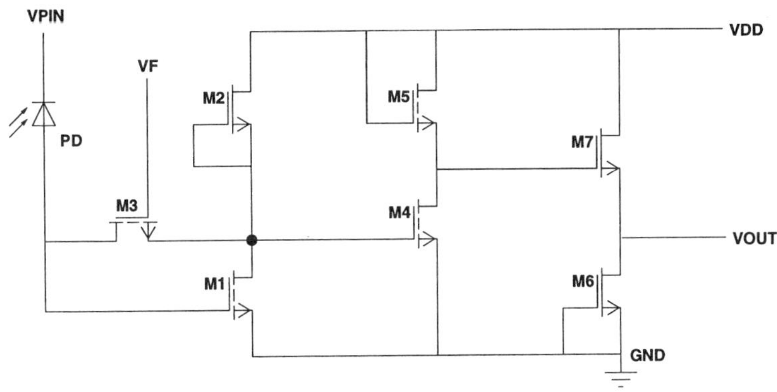

Fig. 12.36. Circuit diagram of an improved NMOS fiber receiver OEIC [481]

The preamplifier has also been modified. Figure 12.36 shows the improved three-stage preamplifier. The feedback via M3 is only across the first stage with the common-source amplifier M1 and the depletion load M2. The second stage with the enhancement mode MOSFETs M4 and M5 further amplifies the signal and is used as a buffer to drive the depletion source follower M7 with an output impedance of 50 Ω. At the optimum feedback, $V_F = 1.25$ V, the transimpedance was 6.5 kΩ and a 45 μA dynamic range of the photocurrent was obtained. The bandwidth of the photoreceiver was 130 MHz for 870 nm light, when biased with VDD = 8 V, $V_F = 1.25$ V, and $V_{PIN} = 30$ V. Open-eye operation under these conditions was demonstrated up to 300 Mb/s. The sensitivity of the photoreceiver was −33 dBm at 155 Mb/s and −25.5 dBm at 300 Mb/s at a bit error rate (BER) of 10^{-9} for a pseudo-random bit sequence (PRBS) of $2^{23} - 1$ under the same conditions. At a bias of VDD = 8 V, the power dissipation was 44 mW, with 2 mW from the first two stages. A redesigned circuit [482] achieved sensitivities of −22.8, −15, and −9.3 dBm at bit rates of 622, 900, and 1000 Mb/s, respectively. This redesigned preamplifier had a bandwidth of 500 MHz and dissipated only 10.8 mW at a power supply voltage of 1.8 V. $V_{PIN} = 30$ V, however, was still necessary for a −3dB frequency of 150 MHz and a −6dB frequency of about 750 MHz for the lateral PIN photodiode.

It can be concluded that the approach of [481,482] results in a rather good performance of the photoreceiver at the cost, however, of a rather large supply voltage of 30 V for the lateral PIN photodiode. This voltage is usually not present in modern electronic systems and advanced microelectronic circuits, which operate at 5 V, 3.3 V, 2.5 V, 1.8 V or even lower voltages.

BiCMOS Receivers. Results of BiCMOS-OEICs, consisting of a PIN photodiode and an amplifier, which exploited only MOSFETs and no bipolar transistors, have been published [236, 237]. The BiCMOS technology, therefore, was chosen merely for the integration of the PIN photodiode. A standard BiCMOS technology with a minimum effective channel length of 0.45 μm was used without any modifications [236, 237]. This effective channel length corresponds to a drawn or nominal channel length of about 0.6 μm. The buried N^+ collector in Fig. 3.87 was used for the cathode of the PIN photodiode, the P^+-source/drain island served for the anode, and the intrinsic zone of the PIN photodiode was formed by the N well (see Sect. 3.7).

The circuit diagram of the CMOS preamplifier used in the BiCMOS OEIC is shown in Fig. 12.37. The input stage of this amplifier with the NMOS transistors N1, N2, and N3 is a single-ended transimpedance amplifier featuring DC input coupling. The feedback resistor R3 had a value of 1.4 kΩ. However, with the additional voltage gain of the following circuit elements, the effective transimpedance of the amplifier was about 3.0 kΩ. The circuit with the transistors N5 and N6 produces a reference voltage for the gate of P5. This voltage is close to the midpoint of the voltage swing at the gate of P3 and can be considered as a kind of decision threshold. P3 and P5 are source followers, which are biased by the current sources P2 and P4, respectively. N4–N6 and P1–P5 perform a single-ended to differential conversion. N7 and N8 form a differential amplifier. N10–N13 are source-follower drivers with an output impedance of 50 Ω.

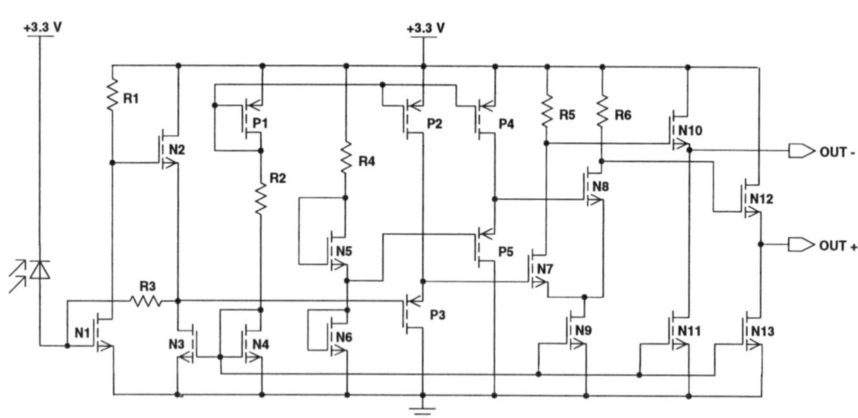

Fig. 12.37. Circuit diagram of an OEIC in BiCMOS technology with a buried N^+ collector as cathode of a PIN photodiode [236]

The power dissipation of the core amplifier circuit was 30 mW from a 3.3 V supply, with an additional 57 mW in the output source followers. The −3dB bandwidth of the receiver was 300 MHz. The OEIC reached a bit rate of 531 Mb/s with a bit error rate of 10^{-9} and a sensitivity of −14.8 dBm for $\lambda = 850$ nm.

In [237] a laser with a wavelength of 670 nm was used for the characterization of the same OEIC as in [236]. The data rate was increased to 622 Mb/s for this wavelength. This bit rate was limited by the capacitance of the photodiode and the feedback resistor of 1.4 kΩ in the amplifier transimpedance input stage.

Another BiCMOS fiber receiver OEIC has been described containing a double photodiode (Fig. 3.90) and a bipolar preamplifier (Fig. 12.38). This OEIC has been fabricated in a 0.8 μm BiCMOS technology.

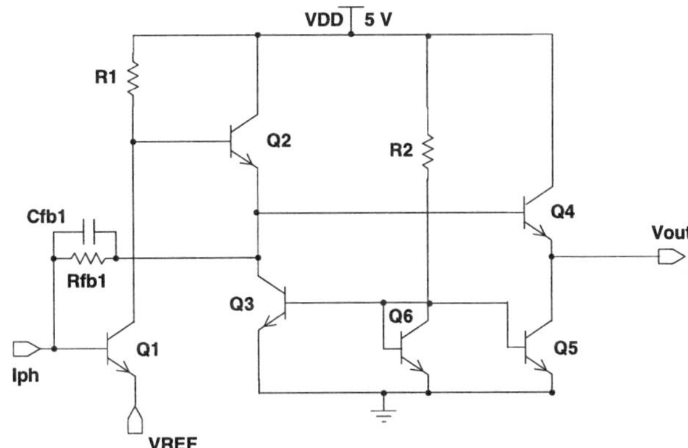

Fig. 12.38. Schematic of fiber receiver OEIC fabricated in BiCMOS technology [483]

The transistor Q1 is used in common-emitter configuration. Together with the emitter follower Q2 and the feedback resistor R_{fb}, Q1 forms a transimpedance input stage. The emitter follower Q4 is used for level shifting and for decoupling the feedback loop from the output. A reference voltage V_{REF} of 2.5 V has been applied to the emitter of Q1 resulting in a reverse bias of about 3.3 V for the double photodiode lying between the input and ground.

The transient response of a double photodiode with an area of 530 μm^2 shown in Fig. 3.94 has been measured with this amplifier. A bandwidth of 367 MHz has been determined for the OEIC. Wide-open eye patterns with a turn-on delay of 0.2 ns at a bit rate of 531 Mb/s for $\lambda = 638$ nm (Fig. 12.39) were obtained with this BiCMOS OEIC containing a double photodiode with an area of 530 μm^2 in a standard technology without any modifications [483].

CMOS Receivers. Another field of application for OEICs in addition to fiber receivers is optical interconnect technology, because electrical interconnects on a board or system level are becoming a problem due to steadily

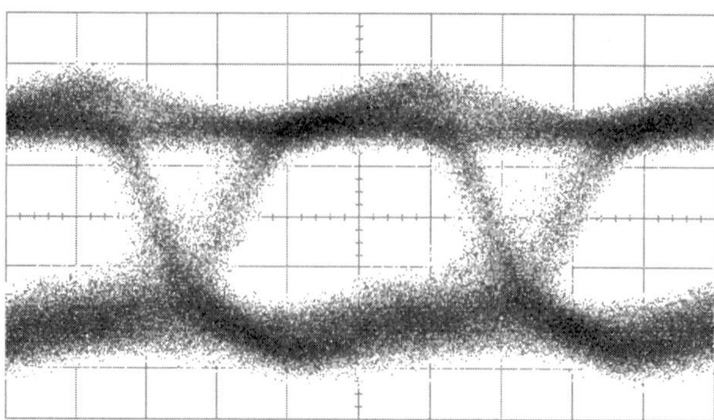

Fig. 12.39. Eye diagram of a BiCMOS fiber receiver OEIC recorded at 531 Mb/s with a PRBS word length of $2^{23} - 1$ (time scale: 400 ps/div.; amplitude: 100 mV/div.) [483]

increasing clock frequencies. In particular, signal reflections on long interconnects on large boards and the cross-talk on high-density electronic boards with conductor widths of 0.1 mm and conductor spacings of 0.1 mm are critical. Optical interconnects, e. g. via waveguide-in-board or fiber-in-board and optical backplanes, avoid the problems of electrical interconnects.

For the application in optical interconnect technology, a CMOS preamplifier OEIC was developed choosing the innovative monolithic integration of vertical PIN photodiodes in a twin-well CMOS-process (Fig. 3.33) [79], which uses epitaxial wafers. The integration of PIN photodiodes in such a CMOS technology requires much less additional process complexity than the published approaches to standard-buried-collector (SBC) based bipolar OEICs [38, 59]. Three additional masks were necessary for the PIN-bipolar integration and for the avoidance of the Kirk effect [38]. Only one photodiode protection mask is added for the PIN-CMOS integration in order to block out an originally unmasked threshold implantation from the photodiode area. A reduction of the standard doping concentration C_e of approximately $1 \times 10^{15}\,\text{cm}^{-3}$ in the epitaxial layer was necessary in order to obtain fast integrated PIN photodiodes. In contrast to a reduction of the current gain and of the transit frequency of bipolar transistors in bipolar OEICs due to the Kirk effect, the electrical performance of the N- and P-channel MOSFETs is not degraded when the doping level in the epitaxial layer is reduced, because these MOSFETs are placed in wells [79]. Reach-through and electrostatic discharge (ESD) aspects in the CMOS OEICs can be dealt with using appropriate design measures.

In contrast to the OEIC of [481], here, only a single power supply of 5 V is needed. The circuit of the preamplifier OEIC is shown in Fig. 12.40.

The amplifier consists of three stages. The input stage with M1–M3 is a transimpedance configuration. The source followers M2, M5, and M8 as well as the current sources M3, M6, and M9 are used for level shifting. Without

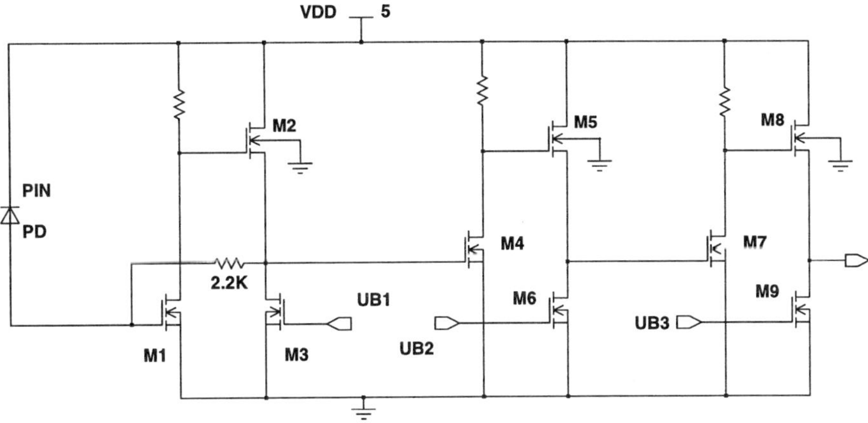

Fig. 12.40. Circuit diagram of a CMOS preamplifier OEIC [82]

these transistors, V_{GS} of transistor M1 would be larger and a lower voltage across the PIN photodiode would result and would increase the rise and fall times of its photocurrent. The threshold implant in Fig. 3.19g was omitted in order to reduce the threshold voltage of transistors M2, M5, and M8 intentionally to about 0.4 V by using the photodiode protection mask in order to obtain lower V_{GS} values and to realize the optimum level shifting in such a way. Due to the feedback across the 2.2 kΩ transimpedance resistor and to the identical dimensions of the transistors in the different stages, a good independence from process deviations within the relatively large specified process tolerances of the used digital CMOS process is obtained.

The CMOS receiver OEIC was fabricated in a 1.0 μm industrial CMOS process [83]. The microphotograph of the CMOS preamplifier OEIC can be seen in Fig. 12.41.

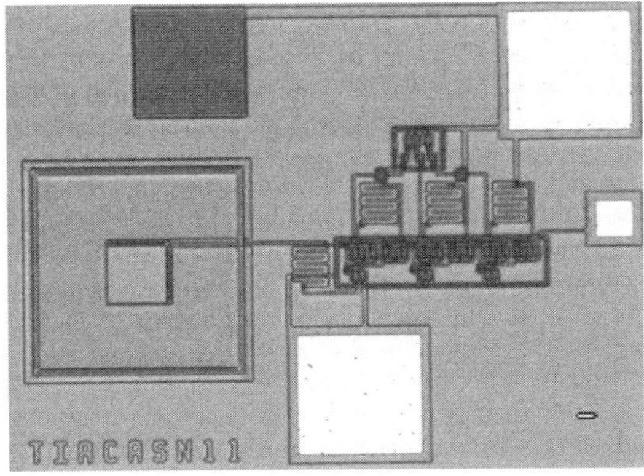

Fig. 12.41. Microphotograph of a CMOS receiver OEIC [82]

The photodiode, having a light sensitive area of 2700 μm^2, together with its metal shield around covers an area of approximately 150 × 150 μm^2. The preamplifier occupies an active area of less than 130 × 200 μm^2. The sensitivity of the PIN CMOS preamplifier OEIC was 4.7 mV/μW without ARC and its power consumption was 19 mW at 5.0 V.

The oscilloscope extracted a rise time $t_\mathrm{r}^{\mathrm{osc,disp}} = 15.5$ ns and a fall time $t_\mathrm{f}^{\mathrm{osc,disp}} = 17.6$ ns for the OEIC with the doping concentration of 1×10^{15} cm^{-3} in the epitaxial layer for a laser wavelength of 638.3 nm. These large values for t_r and t_f are due to the slow carrier diffusion in the standard epitaxial layer of the photodiode. The corresponding values for the concentration of 2×10^{13} cm^{-3} in the epitaxial layer, where the depletion region spreads through the whole epitaxial layer and carrier diffusion in the photodiode is eliminated, were 1.05 ns and 1.26 ns, respectively [82].

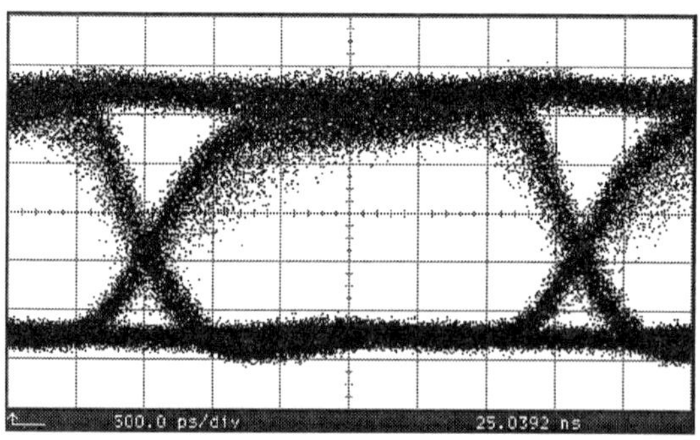

Fig. 12.42. Measured eye diagram of a CMOS preamplifier OEIC for the application in a fiber and interconnect receiver (time: 500 ps/div, amplitude: 0.2 V/div) [82]

Figure 12.42 shows the eye diagram of the CMOS preamplifier OEIC in Figs. 12.40 and 12.41. The eye diagram was measured on the wafer level with a picoprobe (input capacitance: 0.1 pF) at the output of the OEIC, with an HP54750/51 digital sampling oscilloscope, and with an ECL bit pattern generator, which modulated a red semiconductor laser with a wavelength of 638.3 nm. A pseudo-random bit sequence (PRBS) of $2^{23} - 1$ in a non-return-to-zero (NRZ) bit rate of 320 Mb/s was used. The doping concentration in the epitaxial layer of the wafer which was used for the fabrication of the OEIC was 2×10^{13} cm^{-3}.

Simulations showed that shorter rise and fall times can be expected for an improved preamplifier and bit rates in excess of 600 Mb/s seem feasible for a wavelength of 638 nm. For wavelengths of 780 and 850 nm bit rates of 500 Mb/s were estimated by simulations [82]. Meanwhile, the OEIC has been characterized by eye diagram measurements at a bit rate of 622 Mb/s with a wavelength of 638 nm [87].

The fastest fully integrated single-beam optical bulk CMOS receiver (Fig. 12.43) was realized in the Lucent 0.35 μm production process [74]. This receiver contained the photodetector shown in Fig. 3.17. The amplifier of this

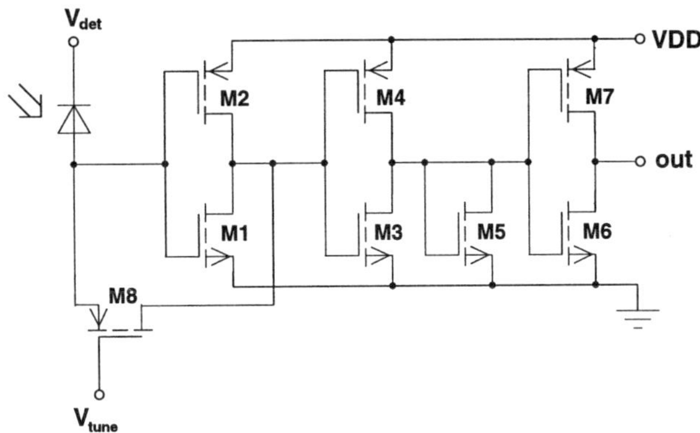

Fig. 12.43. CMOS inverter receiver OEIC [74]

receiver is formed by three inverter stages. The first stage with M1 and M2 is an inverter-based transimpedance stage with the P-channel MOSFET M8 as the feedback element. The gate voltage of M8 could be adjusted for optimum performance (V_{tune}) at a given optical power and bit rate. This first stage converts the photocurrent, flowing from the anode of the photodiode through the active resistor M8, into a voltage. The second stage with the inverter M3, M4 amplifies this voltage. The N-channel MOSFET load M5 reduces the gain and increases the bandwidth. With M5, the technology dependence of the amplifier gain is reduced and the switching threshold of this stage is stabilized [484]. The third stage with M6 and M7 supplies a high gain. This stage acts as an asynchronous decision circuit resulting in a fully digital logic output level. For a correct and optimum performance of this DC-coupled three-stage preamplifier circuit, the dimensions of the transistors M5 and M8 have to be chosen carefully [74].

The supply voltage of the preamplifier was varied between 1.8 and 3.3 V. The best sensitivity was obtained at a supply bias of VDD = 2.2 V. A bit error rate of 10^{-9} for a bit rate of 1 Gb/s was obtained with an average optical input power of −6.3 dBm and with a detector bias $V_{\text{det}} = 10$ V. This low sensitivity results from the low responsivity of the photodiode of less than 0.04 A/W. Another disadvantage of this OEIC is the high detector bias of 10 V.

For an OEIC fabricated in a 0.25 µm fully-depleted CMOS SOI process technology, a single 2 V supply was sufficient [247]. The photodetector of this OEIC was a lateral PIN photodiode exploiting the avalanche effect to achieve a high responsivity of 0.4 A/W, although the OEIC was fabricated in a very thin SOI layer (see Fig. 4.7). The amplifier circuit of this OEIC is shown in Fig. 12.44.

The preamplifier circuit is based on a transimpedance amplifier with a feedback resistor R_{fb} of 5 kΩ. The N-channel MOSFET N1 is used in a common-source circuit. The PMOS transistor P1 forms the active load for

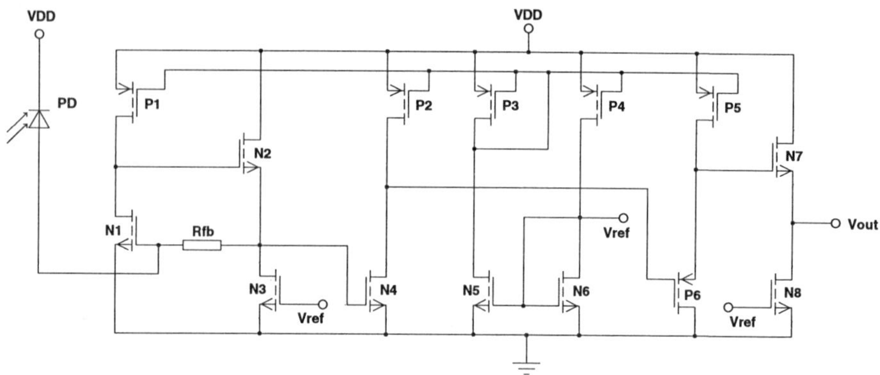

Fig. 12.44. CMOS receiver OEIC on SOI [247]

N1. N2 is used as source follower in order to obtain a low output impedance of the transimpedance input stage. N3 is a constant-current source. N4 amplifies the output voltage of the transimpedance input stage. P6 is used as a source follower for shifting the signal level towards VDD and reducing the output impedance. This shifting towards VDD is necessary to allow the implementation of the second source follower N7 for further reducing the output impedance. P5 and N8 are constant-current sources. P3, P4, N5, and N6 supply the reference voltage V_{ref} for biasing the amplifier. The constant-current sources P1, P2, and P5 are set by P3. These four transistors form current mirrors.

Using a 850 nm wavelength, a bandwidth of 1 GHz was measured for a supply voltage of 2 V for an average optical input power of −13 dBm (50 μW) [247]. The OEIC occupied an area of $650 \times 400\,\mu m^2$.

Based on the circuit shown in Fig. 12.40, an improved CMOS photoreceiver (Fig. 12.45), which contains a vertical PIN photodiode and which combines a data rate of 622 Mb/s with a quantum efficiency of 94 %, has been developed here. The innovative monolithic integration of vertical PIN photodiodes in a twin-well CMOS process (Fig. 3.33) [79] which uses epitaxial wafers, has been demonstrated to combine both high speed and large quantum efficiency of the photodiode [82]. In contrast to the OEICs of [74, 481], there, only a single power supply of 3.3 V was needed. The circuit of the high-bandwidth preamplifier with an integrated PIN photodiode is shown in Fig. 12.45. It is a typical high-frequency amplifier. Only N-channel MOSFETs are used in order to obtain a high bandwidth. The input stage with the transistors M1–M4 is a transimpedance configuration, which converts the photocurrent change in the integrated PIN photodiode to a voltage change. The cascode transistors M1, M5, and M9 reduce the Miller effect and increase the bandwidth correspondingly. The source followers M3, M7, and M11 as well as the current sources M4, M8, and M12 are used for level shifting. The threshold voltage of transistors M3, M7, and M11 has been reduced

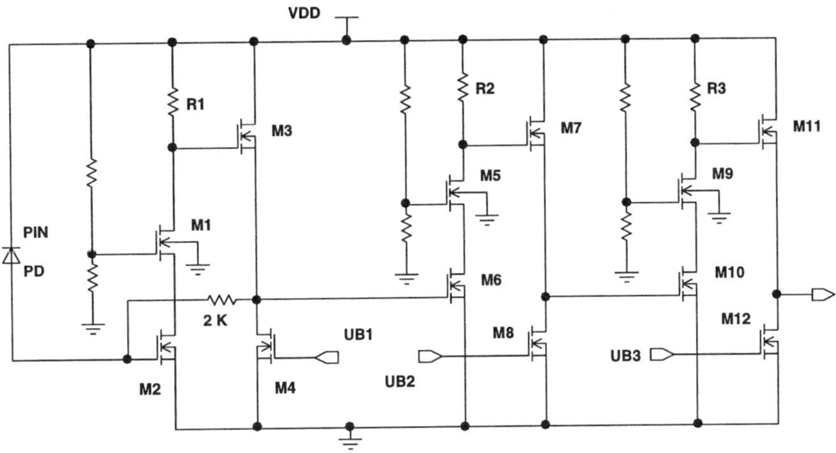

Fig. 12.45. Circuit diagram of a CMOS preamplifier OEIC with a possible data rate of 622 Mb/s for application in a fiber and interconnect receiver [82]

intentionally to about 0.4 V by the photodiode protection mask in order to obtain lower V_{GS} values. Polysilicon resistors were employed as load elements, since depletion transistors were not available in the digital CMOS process. Due to the feedback across the 2 kΩ resistor and to the identical dimensions of the transistors in the different stages, a good independence from process deviations within the relatively large specified process tolerances of the digital CMOS process used has been obtained. Three identical biasing circuits (UB1=UB2=UB3) are used instead of one in order to minimize parasitic coupling between the stages. The sensitivity of the PIN preamplifier OEIC was 4.7 mV/μW for $\lambda = 638$ nm increasing to 9.0 mV/μW with ARC, which corresponds to an overall transimpedance of 18.4 kΩ. Its power consumption was 44 mW at 5.0 V reducing to 17 mW at 3.3 V.

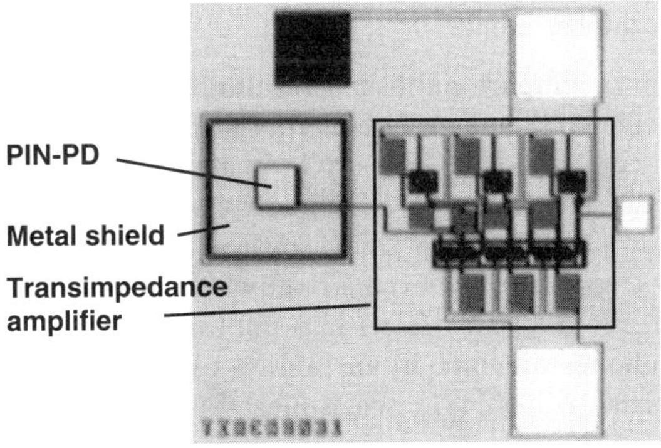

Fig. 12.46. Microphotograph of a high-speed CMOS receiver OEIC [86]

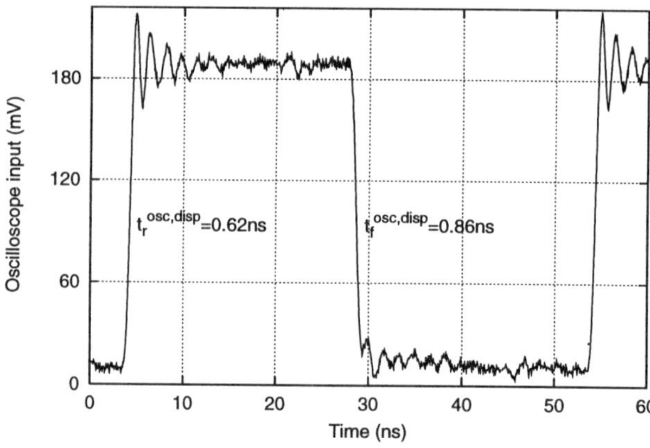

Fig. 12.47. Waveform at the output of the CMOS preamplifier OEIC with a possible data rate of 622 Mb/s for application in a fiber and interconnect receiver

The photodiode, together with its metal shield around, covers an area of approximately $150 \times 150\,\mu\text{m}^2$ (see Fig. 12.46). The preamplifier occupies an active area of less than $190 \times 200\ \mu\text{m}^2$.

Values of 0.62 ns and 0.86 ns for the rise and fall times, respectively, at the output of the preamplifier (Fig. 12.47) were extracted by a digital sampling oscilloscope for a concentration of $2 \times 10^{13}\,\text{cm}^{-3}$ in the epitaxial layer, where the depletion region spreads through the whole epitaxial layer already with a supply voltage of 3.3 V.

The correction of the $t_{\text{r/f}}$ values for the laser and picoprobe rise and fall times results in $t_{\text{r}}^{\text{OEIC}} = 0.53$ ns and $t_{\text{f}}^{\text{OEIC}} = 0.69$ ns. These values indicate that CMOS OEICs with a reduced doping concentration in the epitaxial layer having an appropriate output buffer can be used as receivers for optical data transmission via fibers or for optical interconnects on a board level up to a bit rate BR of 622 Mb/s in the non-return-to-zero (NRZ) mode, verified by a measured eye diagram with a pseudo-random bit sequence (PRBS) of $2^{23}-1$ (Fig. 12.48). The data rate of the OEIC has been limited by the amplifier in a 1.0 µm technology. With sub-micrometer PIN-CMOS-OEICs, however, data rates in excess of 1 Gb/s are possible.

Comparison. It is worthwhile to compare published results on silicon OEICs in Table 12.2. A hybrid receiver-transmitter circuit consisting of a 0.8 µm CMOS amplifier and of a flip-chip bonded GaAs-AlGaAs multi-quantum-well modulator, which also could be used as a PIN photodiode, has been reported to operate at 625 Mb/s [345]. A SiGe OEIC with a PIN photodiode and heterojunction bipolar transistors achieved a bandwidth of 460 MHz corresponding to a bit rate of about 690 Mb/s with a photodiode bias of 9 V [314]. A lateral PIN photodiode was used in an NMOS receiver OEIC, which operated at a bit rate of 300 Mb/s [481], when biased at VDD = 8 V and with a photodiode bias $V_{\text{PD}} = 30$ V. In a redesigned version, the NMOS OEIC achieved a data rate of 1 Gb/s at a sensitivity of −9.3 dBm [482].

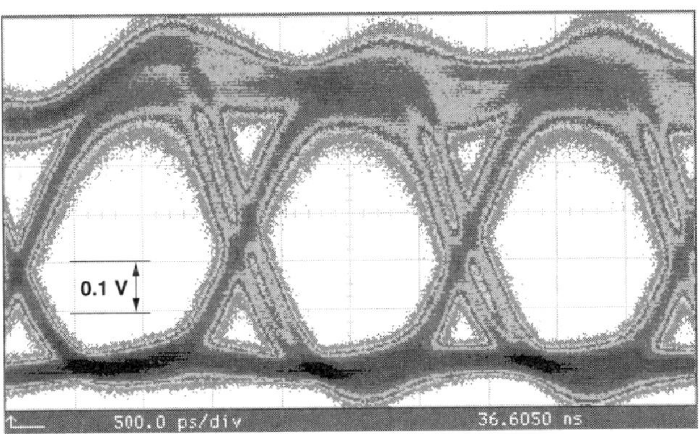

Fig. 12.48. Measured eye diagram of a CMOS preamplifier OEIC with a possible data rate of 622 Mb/s for application in a fiber and interconnect receiver (time: 500 ps/div, amplitude: 0.1 V/div) [82]

With a bipolar OEIC [40] using the BEST-process of AT&T with 1.5 μm design rules [485], a data rate of 150 Mb/s was achieved. An N^+/P-substrate photodiode limited the speed to this data rate. In [38] and [59], vertical PIN photodiodes have been integrated with bipolar transistor technology. The optical receivers described in these two articles demonstrated a data rate of 50 Mb/s [38] and a frequency response of 147 MHz [59] at a supply voltage of 5 V although the PIN photodiodes showed bandwidths of ≈ 300 MHz at a bias of 3 V. The thin P^+/N-collector/N^+-buried collector PIN photodiode in a 0.6 μm BiCMOS process was used together with a MOS amplifier in [236,237] with a rather low responsivity of the photodiode. A double photodiode in a standard 0.8 μm BiCMOS technology enabled a bit rate of the OEIC with a bipolar amplifier in excess of 531 Mb/s [72, 483]. A very low responsivity

Table 12.2. Comparison of silicon receiver OEICs (BR = bit rate; PR = photoreceiver; PD = photodiode; [1)]: with ARC)

Process (μm)	VDD (V)	λ (nm)	BR_{PR} (Mb/s)	V_{PD} (V)	BR_{PD} or f_{3dB}	R_{PD} (A/W)	η_{PD} (%)
0.8 CMOS-MQW [345]	5	850	625	2.5	-	0.5	75
Bipolar SiGe [314]	6	850	690	9	460 MHz	0.3[1)]	43[1)]
1.0 NMOS [481]	8	870	300	30	500 Mb/s	0.52[1)]	74[1)]
1.0 NMOS [482]	1.8	850	1000	30	1.0 Gb/s	0.54[1)]	80[1)]
1.5 Bipolar [40]	5	850	150	4.2	150 Mb/s	0.21	30
Bipolar [38]	5	780	50	3.0	300 MHz	0.35	56
Bipolar [59]	5	830	150	3.0	280 MHz	0.5[1)]	75[1)]
0.6 BiCMOS [236]	3.3	850	531	2.5	700 MHz	0.07	10
0.6 BiCMOS [237]	3.3	670	622	2.5	700 MHz	0.16	29
0.8 BiCMOS [483]	5	638	531	3.3	531 Mb/s	0.49	95[1)]
0.35 CMOS [74]	3.3	850	1000	10	1.0 Gb/s	0.04	5.9
0.25 CMOS SOI [247]	2.0	850	1500	1.5	1.0 GHz	0.4	59
1.0 CMOS [82]	5	638	622	3.0	1.7 GHz	0.25	49
1.0 CMOS [82]	3.3	638	622	1.8	1.4 GHz	0.48[1)]	94[1)]

of 0.04 A/W has been reported for the P^+/N-well photodiode of a 1 Gb/s OEIC in a 0.35 μm CMOS technology [74]. The most sophisticated avalanche photodiode approach in a SOI CMOS technology enabled an OEIC with a bandwidth of 1 GHz corresponding to a bit rate of about 1.5 Gb/s at a supply voltage of only 2 V [247]. Summarizing, a low responsivity has been present in order to achieve a high data rate in the investigations [40, 74, 236, 237], or a high voltage for the photodiode was needed [314, 481, 482], or a high additional process complexity has been needed for the integration of fast and highly efficient photodiodes as in [38, 59]. Even compared to the sophisticated detectors in [74] and [247], the vertically integrated PIN photodiode developed in the CMOS DVD project and described in [82] achieved the highest speed and the highest quantum efficiency with an antireflection coating. The CMOS-integrated vertical PIN photodiode, therefore, shows the largest speed-responsivity product of all integrated silicon photodiodes reported so far whereby standard or near-standard processes have been used.

It can be concluded that with CMOS receiver OEICs comparable or even better data rates can be obtained than with the bipolar OEICs described in [38, 59]. In contrast to the integration of PIN photodiodes in bipolar circuits, the integration of vertical PIN photodiodes in CMOS circuits requires little additional process complexity. The vertical PIN photodiodes combine a high data rate and a high quantum efficiency at a low reverse bias with a single power supply voltage for the OEIC. Low cost PIN-CMOS receiver OEICs for optical data transmission and for optical interconnects on boards and between boards via optical backplanes seem feasible.

12.5 Summary

Digital optical CMOS receivers based on sense-amplifier flip-flops, which require small die areas and are therefore very interesting for the application in massively parallel optical interconnects, were described. OEICs with current mirror amplifiers and current comparators were shown to be appropriate for the wafer-level test of digital CMOS and BiCMOS circuits with considerably increased frequencies.

Many analog photoreceiver circuits were included in this chapter. For instance, a bipolar circuit for optical flame detection with a very high transimpedance and the pixel circuit of an amorphous-silicon-CMOS imager with a dynamical range of 100 dB were described.

The most important results obtained so far within the two projects for the development of OEICs for optical storage systems are: PIN CMOS OEICs for DVD (video) applications with bandwidths in excess of 33 MHz were realized in full custom design. This bandwidth exceeds that of Japanese OEICs for twofold DVD speed [486]. A monolithic integrated CMOS OEIC was shown to exceed the bandwidth of a wire-bonded optoelectronic receiver circuit by a factor of 2.5 due to the avoidance of the bondpad capacitances. A monolithic integrated CMOS OEIC with a reduced doping concentration in the epitaxial layer increased the bandwidth by another factor of two.

The integrated PIN photodiodes can handle much higher data rates (>1 Gb/s) than are necessary for DVD applications. Therefore, a 1.0 μm PIN CMOS OEIC for fiber receiver applications with a non-return-to-zero (NRZ) data rate of 622 Mb/s was demonstrated. This data rate was limited by the amplifier in 1.0 μm CMOS technology. OEICs with data rates in excess of 1 Gb/s, therefore, are possible with submicrometer CMOS processes. Possible applications for PIN CMOS OEICs are Local Area Networks (LANs), such as the Gigabit Ethernet or the Fiber Channel.

A BiCMOS full custom OPTO-ASIC for optical storage systems (CD-ROM, DVD-ROM, DVD-RAM) with bandwidths in excess of 90 MHz has been developed. This bandwidth exceeds that of Japanese OEICs for fourfold DVD speed [479] by almost a factor of two. Meanwhile, an even faster OEIC with a bandwidth of 147 MHz has been demonstrated within the projects for the development of OEICs for optical storage systems [480].

References

1. K. J. Ebeling, *Integrated optoelectronics*, Springer, Berlin, Heidelberg, 1993.
2. E. D. Palik, *Handbook of optical constants of solids*, pp. 547–569, Academic Press, Inc., Orlando, 1985.
3. D. E. Aspnes and A. A. Studna, Phys. Rev. B **27**(2), pp. 985–1009, 1983.
4. S. Selberherr, *Analysis and simulation of semiconductor devices*, Springer, Wien, 1984.
5. Technology Modeling Association, Inc., Palo Alto, CA, *MEDICI manual*, 1994.
6. B. Meinerzhagen and W. L. Engl, "The influence of the thermal equilibrium approximation on the accuracy of classical two-dimensional numerical modeling of silicon submicrometer MOS transistors", IEEE Transactions on Electron Devices **35**(5), pp. 689–697, 1988.
7. A. Forghieri, R. Guerreri, P. Ciampolini, A. Gnudi, M. Rudan, and G. Baccarani, "A new discretization strategy of the semiconductor equations comprising momentum and energy balance", IEEE Trans. Computer-Aided Design **7**(2), pp. 231–242, 1988.
8. V. Agostinelli, T. Bordelon, X. Wang, C. Yeap, C. Maziar, and A. Tasch, "An energy-dependent two-dimensional substrate current model for the simulation of submicron MOSFET's", IEEE Electron Device Lett. **13**(11), pp. 554–556, 1992.
9. M. Shur, *GaAs devices and circuits*, Plenum Press, New York, 1987.
10. D. M. Caughey and R. E. Thomas, "Carrier mobilities in silicon empirically related to doping and field", Proc. IEEE **55**, pp. 2192–2193, 1967.
11. S. Selberherr, "Process and device modeling for VLSI", Microelectron. Reliability **24**(2), pp. 225–257, 1984.
12. S. M. Sze, *Physics of semiconductor devices*, Wiley, New York, 1981.
13. J. M. Senior, *Optical fiber communications*, p. 436, Prentice Hall, New York, 1992.
14. J. Muller, "Photodiode for optical communication", Advances in Electronics and Electron Physics **55**, pp. 192–216, 1981.
15. S. M. Sze, *Physics of semiconductor devices*, p. 758, Wiley, New York, 1981.
16. H. Zimmermann, "Improved CMOS-integrated photodiodes and their application in OEICs", in *IEEE Int. Workshop on High Performance Electron Devices for Microwave & Optoelectronic Applications*, pp. 346–351, 1997.
17. R. N. Hall, "Electron-hole recombination in germanium", Phys. Rev. **87**, p. 387, 1952.
18. W. Shockley and W. T. Read, "Statistics of the recombination of holes and electrons", Phys. Rev. **87**, p. 835, 1952.
19. C. T. Sah, R. N. Noyce, and W. Shockley, "Carrier generation and recombination in p-n junction and p-n junction characteristics", Proc. IRE **45**, p. 1228, 1957.

20. D. M. Chapin, C. S. Fuller, and G. L. Pearson, "A new silicon p-n junction photocell for converting solar radiation into electrical power", J. Appl. Phys. **25**, pp. 676–677, 1954.
21. A. Goetzberger, J. Knobloch, and B. Voss, *Crystalline silicon solar cells*, Wiley, Chichester, New York, 1998.
22. J. A. Mazer, *Solar cells: an introduction to crystalline photovoltaic technology*, Kluwer Academic Publishers, Boston, 1997.
23. L. Partain, *Solar cells and their applications*, Wiley, New York, 1995.
24. H. J. Möller, *Semiconductors for solar cells*, Artech House, Boston, 1993.
25. P. A. H. Hart, "Bipolar transistors and integrated circuits", in C. Hilsum, ed., *Handbook on Semiconductors*, vol. 4, (Device Physics), pp. 99–264, North-Holland, Amsterdam, 1993.
26. J. M. Senior, *Optical fiber communications*, p. 437, Prentice Hall, New York, 1992.
27. G. Winstel and C. Weyrich, *Optoelektronik II*, p. 76, Springer, Berlin, Heidelberg, 1986.
28. Landolt-Börnstein, *Numerical data and functional relationships in science and technology*, vol. 17c, p. 474, Springer, Berlin, 1984.
29. M. Takagi, K. Nakayama, C. Terada, and H. Kamoka, Proc. 4th Conf. Solid-State Devices, Suppl. Jpn. Soc. Appl. Phys. **42**, p. 101, 1973.
30. D. D. Tang, T. H. Ning, R. D. Isaac, G. C. Feth, S. K. Wiedmann, and H.-N. Yu, "Subnanosecond self-aligned I^2L/MTL Circuits", IEEE J. Solid-Sate Circuits **15**(4), pp. 444–449, 1980.
31. E. F. Labuda and J. T. Clemens, "Integrated circuit technology", in R. E. Kirk and D. F. Othmer, eds., *Encyclopedia of Chemical Technology*, Wiley, New York, 1980.
32. J. A. Appels, E. Kooi, M. M. Paffen, J. J. H. Schlorje, and W. H. C. G. Verkuylen, "Local oxidation of silicon and its application in semiconductor technology", Philips Res. Rep. **25**, p. 118, 1970.
33. M. Grossman, "Recessed-oxide isolation hikes IBM's LSI density and speed", Electron Design **12**(6), pp. 26–28, 1979.
34. R. J. Blumberg and S. Brenner, "A 1500 gate, random logic, large-scale integrated (LSI) masterslice", IEEE J. Solid-State Circuits **14**(5), pp. 818–822, 1979.
35. D. D. Tang, P. M. Soloman, T. H. Ning, R. D. Issac, and R. E. Burger, "1.25–μm deep-groove-isolated self-aligned bipolar circuits", IEEE J. Solid-Sate Circuits **17**(5), pp. 925–931, 1982.
36. A. Hayasaka, Y. Takami, M. Kawamura, K. Ogiue, and S. Ohwaki, "U-groove isolation technique for high speed bipolar VLSI's", in *IEDM Digest Technical Papers*, pp. 62–65, 1982.
37. H. Goto, T. Takada, R. Abe, Y. Kawabe, K. Oami, and M. Tanaka, "An isolation technology for high performance bipolar memories: IOP-II", in *IEDM Digest Technical Papers*, pp. 58–61, 1982.
38. M. Yamamoto, M. Kubo, and K. Nakao, "Si-OEIC with a built-in PIN-photodiode", IEEE Transactions on Electron Devices **42**(1), pp. 58–63, 1995.
39. H.-M. Rein and R. Ranfft, *Integrierte Bipolarschaltungen*, p. 50, Springer, Berlin, Heidelberg, 1987.
40. H. H. Kim, R. G. Swartz, Y. Ota, T. K. Woodward, M. D. Feuer, and W. L. Wilson, "Prospects for silicon monolithic opto-electronics with polymer light emitting diodes", IEEE J. Lightwave Technology **12**(12), pp. 2114–2121, 1994.
41. J. Popp and H. v. Philipsborn, "10 Gbit/s on-chip photodetection with self-aligned silicon bipolar transistors", in *ESSDERC*, pp. 571–574, 1990.

42. J. Wieland, H. Duran, and A. Felder, "Two-channel 5Gbit/s silicon bipolar monolithic receiver for parallel optical interconnects", Electronics Letters **30**(4), p. 358, 1994.
43. H. Kabza, K. Ehinger, T. F. Meister, H.-W. Meul, P. Weger, I. Kerner, M. Miura-Mattausch, R. Schreiter, D. Hartwig, M. Reisch, M. Ohnemus, R. Köpl, J. Weng, H. Klose, H. Schaber, and L. Treitinger, "A 1-μm polysilicon self-aligned bipolar process for low-power high-speed integrated circuits", IEEE Electron Device Letters **10**(8), pp. 344–346, 1989.
44. J. Wieland, H. Melchior, M. Q. Kearley, C. R. Morris, A. M. Moseley, M. J. Goodwin, and R. C. Goodfellow, "Optical receiver array in silicon bipolar technology with selfaligned, low parasitic III/V detectors for DC-1 Gbit/s parallel links", Electronics Letters **24**(27), pp. 2211–2213, 1994.
45. G. Winstel and C. Weyrich, *Optoelektronik II*, p. 97, Springer, Berlin, Heidelberg, 1986.
46. J. Lindemayer and C. Y. Wrigley, "Beta cuttoff frequencies of junction transistors", Proc. IRE **50**, pp. 194–198, 1962.
47. D. Bolliger, P. Malcovati, A. Häberli, H. Baltes, P. Sarro, and F. Maloberti, "Integrated ultraviolet sensor system with on-chip 1 GΩ transimpedance amplifier", in *ISSCC*, pp. 328–329, 1996.
48. D. Bolliger, R. S. Popovic, and H. Baltes, "Integration of a smart selective UV detector", in *Transducers'95 and Eurosensors IX, Digest of Technical Papers 2 (8th Int. Conf. Solid-State Sensors and Actuators)*, pp. 144–147, 1995.
49. L. K. Nanver, E. J. G. Goudena, and H. W. van Zeijl, "DIMES-01, a baseline BIFET process for smart sensor experimentation", Sensors and Actuators A **36**, pp. 139–147, 1993.
50. C. T. Kirk, "A theory of transistor cutoff frequency (f_T) falloff at high current densities", IRE Trans. Electron Devices **ED-9**, pp. 164–174, 1962.
51. L. A. Hahn, "The saturation characteristics of high-voltage transistors", Proc. IEEE **55**(8), pp. 1384–1388, 1967.
52. J. R. A. Beale and J. A. G. Slatter, "The equivalent circuit of a transistor with a lightly doped collector operating in saturation", Solid-State Electron. **11**, pp. 241–252, 1968.
53. J. A. Pals and H. C. de Graaff, "On the behaviour of the base-collector junction of a transistor at high collector current densities", Philips Res. Rep. **24**, pp. 53–69, 1969.
54. R. J. Whittier and D. A. Tremere, "Current gain and cutoff frequency falloff at high currents", IEEE Trans. Electron Devices **ED-16**(1), pp. 39–57, 1969.
55. H. C. Poon and H. K. Gummel, "Modeling of emitter capacitance", Proc. IRE **57**, pp. 2181–2182, 1969.
56. H. C. de Graaff, "Collector models for bipolar transistors", Solid-State Electron. **16**, pp. 587–600, 1973.
57. G. Rey and J. P. Bailbe, "Some aspects of current gain variations in bipolar transistors", Solid-State Electron. **17**, pp. 1045–1057, 1974.
58. T. H. Ning, D. D. Tang, and P. M. Solomon, "Scaling properties of bipolar devices", in *IEEE Int. Electron Device Meeting*, pp. 61–64, Wash., D.C., 1980.
59. M. Kyomasu, "Development of an integrated high speed silicon PIN photodiode sensor", IEEE Transactions on Electron Devices **42**(6), pp. 1093–1099, 1995.
60. R. F. Wolffenbuttel, "A simple integrated color indicator", IEEE J. Solid-State Circuits **22**(3), pp. 350–356, 1987.
61. G. Schumicki and P. Seegebrecht, *Prozesstechnologie*, pp. 370–380, Springer, Berlin, Heidelberg, 1991.

62. L. C. Parrillo, R. S. Payne, R. E. Davies, G. W. Rentlinger, and R. L. Field, "Twin-tub CMOS - a technology for VLSI circuits", in *IEDM Digest Technical Papers*, pp. 752–755, 1980.
63. T. Hori, *Gate dielectrics and MOS ULSIs*, Springer, New York, 1997.
64. C.-Y. Lu, J. J. Sung, H. C. Kirsch, N.-S. Tsai, R. Liu, A. S. Manocha, and S. J. Hillenius, "High-perfomance salicide shallow-junction CMOS devices for submicrometer VLSI application in twin-tub VI", IEEE Trans. Electron Devices **36**(11), pp. 2530–2536, 1989.
65. R. A. Chapman, C. C. Wei, D. A. Bell, S. Aur, G. A. Brown, and R. A. Haken, "0.5 micron CMOS for high performance at 3.3 V", in *IEDM Digest Technical Papers*, pp. 52–55, 1988.
66. B. Davari, W. H. Chang, M. Wordeman, C. Ș. Oh, Y. Taur, K. E. Petrillo, D. Moy, J. J. Bucchignano, H. Y. Ng, M. G. Rosenfield, F. J. Hohn, and M. D. Rodriguez, "A high performance 0.25 μm CMOS technology", in *IEDM Digest Technical Papers*, pp. 56–59, 1988.
67. E. Braß, U. Hilleringmann, and K. Schumacher, "System integration of optical devices and analog CMOS amplifiers", IEEE J. Solid-State Circuits **29**(8), pp. 1006–1010, 1994.
68. U. Hilleringmann and K. Goser, "Optoelectronic system integration on silicon: waveguides, photodetectors, and VLSI CMOS circuits on one chip", IEEE Trans. Electron Devices **42**(5), pp. 841–846, 1995.
69. E. Fullin, G. Voirin, M. Chevroulet, A. Lagos, and J.-M. Moret, "CMOS-based technology for integrated optoelectronics: a modular approach", in *IEDM Digest Technical Papers*, pp. 527–530, 1994.
70. R. Kauert, W. Budde, and A. Kalz, "A monolithic field segment photo sensor system", IEEE J. Solid-State Circuits **30**(7), pp. 807–811, 1995.
71. P. Lee, A. Simoni, A. Sartori, and G. Torelli, "A photosensor array for spectrophotometry", Sensors and Actuators A **46–47**, pp. 449–452, 1995.
72. H. Zimmermann, K. Kieschnick, T. Heide, and A. Ghazi, "Integrated high-speed, high-responsivity photodiodes in CMOS and BiCMOS technology", in *Proc. 29th European Solid-State Device Conference (ESSDERC), Sept. 13–15*, pp. 332–335, 1999.
73. M. L. Simpson, M. N. Ericson, G. E. Jellison, W. B. Dress, A. L. Wintenberg, and M. B. Bobrek, "Application specific spectral response with CMOS compatible photodiodes", IEEE Trans. Electron Devices **46**(5), pp. 905–913, 1999.
74. T. K. Woodward and A. V. Krishnamoorthy, "1 Gbit/s CMOS photoreceiver with integrated detector operating at 850 nm", Electronics Letters **34**(12), pp. 1252–1253, 1998.
75. L. D. Garrett, J. Qi, C. L. Schow, and J. C. Campbell, "A silicon-based integrated NMOS-p-i-n photoreceiver", IEEE Trans. Electron Devices **43**(3), pp. 411–416, 1996.
76. S. He, L. D. Garrett, K.-H. Lee, and J. C. Campbell, "Monolithic integrated silicon NMOS PIN photoreceiver", Electronics Letters **30**(22), pp. 1887–1888, 1994.
77. S. M. Sze, *VLSI Technology*, McGraw-Hill, New York, 1988.
78. T. Hori, J. Hirase, Y. Odake, and T. Yasui, "Deep-submicrometer large-angle-tilt implanted drain (LATID) technology", IEEE Trans. Electron Devices **39**(10), pp. 2312–2324, 1992.
79. H. Zimmermann, "Monolithic Bipolar-, CMOS-, and BiCMOS-receiver OEICs", in *Proc. Int. Semicond. Conference (CAS'96)*, pp. 31–40, Sinaia, Romania, 1996.
80. H.-J. Voss, private communication, 1996.

81. M.-J. Zhou, J. Holleman, and H. Wallinga, "Elimination or minimisation of optoelectronic crosstalk between photodiodes and electronic devices in OEIC on Si", Electronics Letters **30**(11), pp. 895–897, 1994.
82. H. Zimmermann, T. Heide, and A. Ghazi, "Monolithic high-speed CMOS-photoreceiver", IEEE Photonics Technology Letters **11**(2), pp. 254–256, 1999.
83. H. Zimmermann, U. Müller, R. Buchner, and P. Seegebrecht, "Optoelectronic receiver circuits in CMOS-technology", in *Mikroelektronik'97, GMM-Fachbericht 17*, pp. 195–202, VDE-Verlag, Berlin, Offenbach, 1997.
84. H. Zimmermann, T. Heide, A. Ghazi, and K. Kieschnick, "PIN-CMOS-receivers for optical interconnects", in *Ext. Abstr. 2nd IEEE Workshop on Signal Propagation on Interconnects, Travemünde, Germany*, pp. 88–89, 1998.
85. A. Ghazi, T. Heide, and H. Zimmermann, "PIN CMOS OEIC for DVD systems", in *Proc. 43rd Int. Scientific Colloquium, TU Ilmena, Germany*, vol. 2, pp. 380–385, 1998.
86. T. Heide, A. Ghazi, and H. Zimmermann, "High speed optical PIN-CMOS-receiverss", in *IEEE Int. Workshop on High Performance Electron Devices for Microwave & Optoelectronic Applications*, pp. 72–76, 1998.
87. H. Zimmermann, A. Ghazi, T. Heide, R. Popp, and R. Buchner, "Advanced photo integrated circuits in CMOS technology", in *Proc. 49th Electronic Components and Technology Conference (ECTC)*, pp. 1030–1035, 1999.
88. R. A. Chapman, R. A. Haken, D. A. Bell, C. C. Wei, R. H. Havemann, T. E. Tang, T. C. Holloway, and R. J. Gale, "An 0.8 μm CMOS technology for high performance logic applications", in *IEDM Digest Technical Papers*, pp. 362–365, 1987.
89. P. Pavan, G. Spiazzi, E. Zanoni, M. Muschitiello, and M. Cecchetti, "Latch-up DC triggering and holding characteristics of n-well, twin-tub and epitaxial CMOS technologies", IEE Proceedings-G **138**(5), pp. 605–612, 1991.
90. H. Zimmermann, T. Heide, A. Ghazi, and P. Seegebrecht, "PIN-CMOS-receivers for optical interconnects", in H. Grabinski, ed., *Signal Propagation on Interconnects*, vol. II, Kluwer, Amsterdam, 1999.
91. F. Yokogawa, S. Ohsawa, T. Iida, Y. Araki, K. Yamamoto, and Y. Moriyama, "The path from a digital versatile disc (DVD) using a red laser to a DVD using a blue laser", Jpn. J. Appl. Phys. **37**(4B), pp. 2176–2178, 1998.
92. S. Nakamura, M. Senoh, and Y. Sugimoto, "InGaN multi-wuantum-well-structure laser diodes with cleaved mirror cavity facets", Jpn. J. Appl. Phys. **35**(2B), p. L217, 1996.
93. S. Nakamura and G. Fasol, *The blue laser diode*, Springer, Heidelberg, 1998.
94. W. Snoeys, J. D. Plummer, S. Parker, and C. Kenney, "PIN detector arrays and integrated readout circuitry on high-resistivity float-zone silicon", IEEE Transactions on Electron Devices **41**(6), pp. 903–912, 1994.
95. W. S. Boyle and G. E. Smith, "Charge coupled semiconductor devices", Bell Syst. Tech. **49**(4), pp. 587–593, 1970.
96. E. G. Stevens, B. C. Burkey, D. N. Nicols, Y. S. Yee, D. L. Losee, T.-H. Lee, T. J. Tredwell, and R. P. Khosla, "A 1-megapixel, progressive-scan image sensor with antiblooming control and lag-free operation", IEEE Transactions on Electron Devices **38**(5), pp. 981–988, 1991.
97. T. Kuriyama, H. Kodama, T. Kozono, Y. Kitahama, Y. Morita, and Y. Hiroshima, "A 1/3-in 270000 pixel CCD image sensor", IEEE Transactions on Electron Devices **38**(5), pp. 949–953, 1991.
98. S. Kawai, N. Mutoh, and N. Teranishi, "Thermionic-emission-based barrier height analysis for precise estimation of charge handling capacity in CCD registers", IEEE Trans. Electron Devices **44**(10), pp. 1588–1592, 1997.

99. J. P. Lavine and E. K. Banghart, "The effect of potential obstacles on charge transfer in image sensors", IEEE Trans. Electron Devices **44**(10), pp. 1593–1598, 1997.
100. R. Miyagawa and T. Kanade, "CCD-based range-finding sensor", IEEE Transactions on Electron Devices **44**(10), pp. 1648–1652, 1997.
101. T. Spirig, M. Marley, and P. Seitz, "The multitap lock-in CCD with offset subtraction", IEEE Transactions on Electron Devices **44**(10), pp. 1643–1647, 1997.
102. B. E. Burke, J. A. Gregory, M. W. Bautz, G. Y. Prigozhin, S. E. Kissel, B. B. Kosiki, A. H. Loomis, and D. J. Young, "Soft-X-ray CCD imagers for AXAF", IEEE Transactions on Electron Devices **44**(10), pp. 1633–1642, 1997.
103. K. Itakura, T. Nobusada, Y. Saitou, N. Kokusenya, R. Nagayoshi, M. Ozaki, Y. Sugawara, K. Mitani, and Y. Fujita, "A 2/3-in 2.0 M-pixel CCD imager with an advanced M-FIT architecture capable of progressive scan", IEEE Transactions on Electron Devices **44**(10), pp. 1625–1632, 1997.
104. K. Tachikawa, T. Umeda, Y. Oda, and T. Kuroda, "Device design with automatic simulation system for basic CCD characteristics", IEEE Transactions on Electron Devices **44**(10), pp. 1611–1616, 1997.
105. T. Yamada, Y. Kawakami, T. Nakano, N. Mutoh, K. Orihara, and N. Teranishi, "Driving voltage reduction in a two-phase CCD by suppression of potential pockets in inter-electrode gaps systems", IEEE Trans. Electron Devices **44**(10), pp. 1580–1587, 1997.
106. B. E. Burke, R. K. Reich, J. A. Gregory, W. H. McGonagle, A. M. Waxman, E. D. Savoye, and B. B. Kosiki, "640×480 back-illuminated CCD imager with improved blooming control for night vision", in *IEDM Digest Technical Papers*, pp. 33–36, 1998.
107. J. T. Bosiers, A. C. Kleinman, A. van der Sijde, L. Korthout, D. W. Verbugt, H. L. Peek, E. Roks, A. Heringa, F. F. Vledder, and P. Opmeer, "A 2/3" 2-M pixel progressive scan FT-CCD for digital still camera applications", in *IEDM Digest Technical Papers*, pp. 37–40, 1998.
108. A. Tanabe, Y. Kudoh, Y. Kawakami, K. Masubuchi, S. Kawai, T. Yamada, M. Morimoto, K. Arai, K. Hatano, M. Furumiya, Y. Nakashiba, N. Mutoh, K. Orihara, and N. Teranishi, "Dynamic range improvement by narrow-channel effect suppression and smear reduction technologies in small pixel IT-CCD image sensors", in *IEDM Digest Technical Papers*, pp. 41–44, 1998.
109. H. Fiedler and K. Knupfer, "Market overview: charge-coupled devices", in H. Baltes, W. Göpel, and J. Hesse, eds., *Sensors update*, vol. 1, pp. 223–271, VCH, Weinheim, 1996.
110. P. Seitz and K. Knop, "Image sensors", in H. Baltes, W. Göpel, and J. Hesse, eds., *Sensors update*, vol. 2, pp. 85–103, VCH, Weinheim, 1996.
111. Y. Hagiwara, "High-density and high-quality frame transfer CCD imager with very low smear, low dark current, and very high blue sensitivity", IEEE Trans. Electron Devices **43**(12), pp. 2122–2130, 1996.
112. J. Hynecek, "Low-noise and high-speed charge detection in high-resolution CCD image sensors", IEEE Transactions on Electron Devices **44**(10), pp. 1679–1688, 1997.
113. N. Tanaka, N. Nakamura, Y. Matsunaga, S. Manabe, H. Tango, and O. Yoshida, "A low driving voltage CCD with single layer electrode structure for area image sensor", IEEE Transactions on Electron Devices **44**(11), pp. 1869–1874, 1997.
114. S. M. Sze, *Physics of semiconductor devices*, p. 412, Wiley, New York, 1981.

115. R. H. Walden, R. H. Krambeck, R. J. Strain, J. McKenna, N. L. Schryer, and G. E. Smith, "The buried channel charge coupled device", Bell Syst. Tech. **51**, pp. 1635–1640, 1972.
116. M. M. Blouke, J. R. Janesick, J. E. Hall, M. W. Cowens, and P. J. May, "800 × 800 charge coupled device image sensor", Opt. Eng. **22**(5), pp. 607–614, 1983.
117. Y. Ishihara, E. Oda, H. Tanigawa, A. Kohno, N. Teranishi, E.-I. Takeuchi, I. Akiyama, and T. Kamata, "Interline CCD image sensor with an antiblooming structure", IEEE Trans. Electron Devices **31**(1), pp. 83–88, 1984.
118. N. Teranishi and Y. Ishihara, "Smear reduction in interline CCD image sensor", IEEE Trans. Electron Devices **34**(5), pp. 1052–1056, 1987.
119. E. Oda, K. Orihara, T. Tanaka, T. Kamata, and Y. Ishihara, "1/2-in 768(H) × 492(V) pixel CCD image sensor", IEEE Trans. Electron Devices **36**(1), pp. 46–53, 1989.
120. K. Knop, "Image sensors", in W. Göpel, J. Hesse, and J. N. Zemel, eds., *Optical sensors*, pp. 233–252, VCH, Weinheim, 1992.
121. P. Centen, "CCD on-chip amplifiers: noise performance versus MOS transistor dimensions", IEEE Transactions on Electron Devices **38**(5), pp. 1206–1216, 1991.
122. J. Hojo, Y. Naito, H. Mori, K. Fujikawa, N. Kato, T. Wakayama, E. Komatsu, and M. Itasaka, "A 1/3-in 510(H) × 492(V) CCD image sensor with mirror image function", IEEE Trans. Electron Devices **38**(5), pp. 954–959, 1991.
123. N. Mutoh, "Simulation for 3-D optical and electrical analysis of CCD", IEEE Trans. Electron Devices **44**(10), pp. 1604–1610, 1997.
124. M. Yamagishi, M. Negishi, H. Yamada, T. Tsunakawa, K. Shinohara, T. Ishimaru, Y. Kamide, Y. Yamazaki, H. Abe, H. Kanbe, Y. Tomiya, K. Yonemoto, T. Iizuka, S. Nakamura, K. Harada, and K. Wada, "A 2 million pixel FIT-CCD image sensor for HDTV camera systems", IEEE Trans. Electron Devices **38**(5), pp. 976–980, 1991.
125. G. Williams and J. Janesick, "Cameras with CCD's capture new markets", Laser Focus World, Detector Handbook pp. S5–S9, 1996.
126. A. L. Lattes, S. C. Munroe, and M. M. Seaver, "Ultrafast shallow-buried-channel CCD's with built-in drift fields", IEEE Electron Device Lett. **12**(2), pp. 104–107, 1991.
127. T. Satoh, N. Mutoh, M. Furumiya, I. Murakami, S. Suwazono, C. Ogawa, K. Hatano, H. Utsumi, S. Kawai, K. Arai, M. Morimoto, K. Orihara, T. Tamura, N. Teranishi, and Y. Hokari, "Optical limitations to cell size reduction in IT-CCD image sensors", IEEE Trans. Electron Devices **44**(10), pp. 1599–1603, 1997.
128. R. Dawson, J. Preisig, J. Carnes, and J. Pridgen, "CMOS/buried-N-channel CCD compatible process for analog signal processing applications", RCA Review **38**, pp. 406–435, 1977.
129. D. Ong, "An all-implanted CCD/CMOS process", IEEE Transactions on Electron Devices **28**(1), pp. 6–12, 1981.
130. E. R. Fossum, "CMOS image sensors: electronic camera-on-a-chip", IEEE Transactions on Electron Devices **44**(10), pp. 1689–1698, 1997.
131. A. Simoni, A. Sartori, M. Gottardi, and A. Zorat, "A digital vision sensor", Sensors and Actuators A **46–47**, pp. 439–443, 1995.
132. P. Noble, "Self-scanned silicon image detector arrays", IEEE Transactions on Electron Devices **15**(4), pp. 202–209, 1968.
133. F. Andoh, K. Taketoshi, J. Yamazaki, M. Sugawara, Y. Fujita, F. Mitani, Y. Matuzawa, K. Miyata, and S. Araki, "A 250000 pixel image sensor with

FET amplification at each pixel for high-speed television cameras", in *ISSCC Dig. Tech. Papers*, pp. 212–213, 1990.
134. H. Kawashima, F. Andoh, N. Murata, K. Tanaka, M. Yamawaki, and K. Taketoshi, "A 1/4 inch format 250000 pixel amplifier MOS image sensor using CMOS process", in *IEDM Digest Technical Papers*, pp. 575–578, 1993.
135. M. Sugawara, H. Kawashima, F. Andoh, N. Murata, Y. Fujita, and M. Yamawaki, "An amplified MOS imager suited for image processing", in *ISSCC Dig. Tech. Papers*, pp. 228–229, 1994.
136. E. Oba, K. Mabuchi, Y. Iida, N. Nakamura, and H. Miura, "A 1/4 inch 330 k square pixel progressive scan CMOS active pixel image sensor", in *ISSCC Dig. Tech. Papers*, pp. 180–181, 1997.
137. C. Aw and B. Wooley, "A 128 × 128 pixel standard CMOS image sensor with electronic shutter", IEEE J. Solid-State Circuits **31**(12), pp. 1922–1930, 1996.
138. R. H. Nixon, S. E. Kemeny, B. Pain, C. O. Staller, and E. R. Fossum, "256 × 256 CMOS active pixel sensor camera-on-a-chip", IEEE J. Solid-State Circuits **31**(12), pp. 2046–2050, 1996.
139. R. H. Nixon, S. E. Kemeny, R. C. Gee, B. Pain, Q. Kim, and E. R. Fossum, "CMOS active pixel image sensors for highly integrated imaging systems", IEEE J. Solid-State Circuits **32**(2 (Feb.)), pp. 187–197, 1997.
140. O. Yadid-Pecht and E. R. Fossum, "Wide intrascene dynamic range CMOS APS using dual sampling", IEEE Trans. Electron Devices **44**(10), pp. 1721–1723, 1997.
141. Z. Zhou, B. Pain, and E. R. Fossum, "CMOS active pixel sensor with on-chip successive approximation analog-to-digital converter", IEEE Trans. Electron Devices **44**(10), pp. 1759–1763, 1997.
142. Z. Zhou, B. Pain, and E. R. Fossum, "Frame-transfer CMOS active pixel sensor with pixel binning", IEEE Trans. Electron Devices **44**(10), pp. 1764–1768, 1997.
143. G. Yang, O. Yadid-Pecht, C. Wrigley, and B. Pain, "A snap-shot CMOS active pixel imager for low-noise, high-speed imaging", in *IEDM Digest Technical Papers*, pp. 45–48, 1998.
144. D. Scheffer, B. Dierickx, and G. Meynants, "Random addressable 2048 × 2048 active pixel image sensor", IEEE Trans. Electron Devices **44**(10), pp. 1716–1720, 1997.
145. F. Pardo, B. Dierickx, and D. Scheffer, "CMOS foveated image sensor: signal scaling and small geometry effects", IEEE Trans. Electron Devices **44**(10), pp. 1731–1737, 1997.
146. A. Dickinson, B. Auckland, E.-S. Eid, D. Inglis, and E. R. Fossum, "A 256x256 CMOS active pixel image sensor with motion detection", in *ISSCC*, pp. 226–227, 1995.
147. R. M. Guidash, T.-H. Lee, P. P. K. Lee, D. H. Sackett, C. I. Drowley, M. S. Swenson, L. Arbaugh, R. Hollstein, F. Shapiro, and S. Domer, "A 0.6 μm CMOS pinned photodiode color imager technology", in *IEDM Digest Technical Papers*, pp. 927–929, 1997.
148. S. Mendis, S. E. Kemeny, and E. R. Fossum, "CMOS active pixel image sensor", IEEE Trans. Electron. Devices **41**, pp. 452–453, 1994.
149. M.-H. Chi, "Technologies for high performance CMOS active pixel imaging system-on-a-chip", in *Proc. 5th Int. Conf. on Solid-State and Integrated-Circuit Technology*, pp. 180–183, 1998.
150. P. Gillingham, B. Hold, I. Mes, C. O'Connell, P. Schofield, K. Skjaveland, R. Torrance, T. Wojcicki, and H. Chow, "A 768 k embedded DRAM for 1.244 Gb/s ATM switch in a 0.8 μm logic process", in *ISSCC*, pp. 267–268, 1996.

151. B. El-Kareh, G. B. Bronner, and S. E. Schuster, "The evolution of DRAM cell technology", Solid-State Technology (May), pp. 89–101, 1997.
152. K. Itoh, Y. Nakagome, S. Kimura, and T. Watanabe, "Limitations and challenges of multigigabit DRAM chip design", IEEE J. Solid-State Circuits **32**(5), pp. 624–634, 1997.
153. H. Ishiuchi, T. Yoshida, H. Takato, K. Tomioka, H. Momose, S. Sawada, K. Yamzaki, and K. Maeguchi, "Embedded DRAM technologies", in *IEDM Digest Technical Papers*, pp. 33–36, 1997.
154. J. Nakamura, Y. Gomi, M. Uno, and H. Hayashi, "Nondestructive readout mode static induction transistor (SIT) photo sensors", IEEE Trans. Electron. Devices **40**(2), pp. 334–341, 1993.
155. A. Yusa, J. Nishizawa, M. Imai, H. Yamada, J. Nakamura, T. Mizoguchi, Y. Ohta, and M. Takayama, "SIT image sensor: design consideration and characteristics", IEEE Transactions on Electron Devices **33**(6), pp. 735–742, 1986.
156. E. H. Rhoderick, "Transport processes in Schottky diodes", in K. M. Pepper, ed., *Inst. Phys. Conf. Ser.*, vol. 22, pp. 3–19, Institute of Physics, Manchester, UK, 1974.
157. E. H. Rhoderick, *Metal-semiconductor contacts*, Clarendon, Oxford, 1978.
158. V. L. Rideout, "A review of the theory, technology and applications of metal-semiconductor rectifiers", Thin Solid Films **48**, p. 261, 1978.
159. C. K. Chen, B. Nechay, and B.-Y. Tsaur, "Ultraviolet, visible, and infrared response of PtSi Schottky-barrier detectors operated in the front-illuminated mode", IEEE Trans. Electron Devices **38**(5), pp. 1094–1103, 1991.
160. M.-A. Nicolet and S. S. Lau, "VLSI electronics microstructure science", in *Materials and process characterization*, vol. 6, pp. 329–464, Academic Press, New York, 1983.
161. S. P. Murarka, *Silicides for VLSI applications*, Academic Press, New York, 1983.
162. S. A. Audet, A. E. White, K. T. Short, Y.-F. Hsieh, F. M. Ross, and C. S. Rafferty, "30 nm $CoSi_2$ surface layers for contact metallization in complementary metal-oxide-semiconductor processes", Appl. Phys. Lett. **61**(19), pp. 2311–2313, 1992.
163. N. Brun, A. Kalnitsky, and A. Brun, "$TiSi_2$ integration in a submicron CMOS process, I. salicide formation", J. Electrochem. Soc. **142**(6), pp. 1992–1996, 1995.
164. A. Kalnitsky, N. Brun, A. Brun, and J.-P. Gonchond, "$TiSi_2$ integration in a submicron CMOS process, II. integration issues", J. Electrochem. Soc. **142**(6), pp. 1992–1996, 1995.
165. W. F. Kosonocky, H. Elabd, H. G. Erhardt, F. V. Shallcross, G. M. Meray, T. S. Villani, J. V. Groppe, R. Miller, V. L. Frantz, M. J. Cantella, J. Klein, and N. Roberts, "Design and performance of 64×128 element PtSi Schottky-barrier infrared charge-coupled device (IRCCD) focal plane array", Proc. SPIE **344**, pp. 66–77, 1982.
166. H. Elabd, T. Villani, and W. F. Kosonocky, "Palladium-silicide Schottky-barrier IR-CCD for SWIR applications at intermediate temperatures", IEEE Electron Device Letters **EDL-3**(4), pp. 89–90, 1982.
167. M. Denda, M. Kimata, and N. Yutani, "A PtSi Schottky-barrier infrared MOS area imager with large fill factor", in *IEDM Digest Technical Papers*, pp. 722–725, 1983.
168. M. Kimata, M. Denda, N. Yutani, N. Tsubouchi, H. Shibata, H. Kurebayashi, S. Uematsu, R. Tsunoda, and T. Kanno, "A 256×256-elements Si monolithic IR-CCD sensor", in *ISSCC*, pp. 254–255, 1983.

169. M. Denda, M. Kimata, S. Iwade, N. Yutani, T. Kondo, and N. Toubouchi, "4-band × 4096-element Schottky-barrier infrared linear image sensor", IEEE Trans. Electron Devices **38**(5), pp. 1131–1135, 1991.
170. A. Czernik, H. Palm, W. Cabanski, M. J. Schulz, and U. Suckow, "Infrared photoemission of holes from ultrathin (3–20 nm) Pt/Ir-compound silicide films into silicon", Appl. Phys. A **55**, pp. 180–191, 1992.
171. M. Chi and J. Mooney, "Infrared optical absorption of iridium silicide films on silicon", J. Appl. Phys. **76**(5), pp. 3028–3031, 1994.
172. W. F. Kosonocky, "Review of Schottky-barrier imager technology", Proc. SPIE **1308**, pp. 2–26, 1990.
173. F. D. Shepherd and A. C. Yang, "Silicon Schottky retinas for infrared imaging", in *IEDM Digest Technical Papers*, pp. 310–313, 1973.
174. V. E. Vickers, "Model of Schottky barrier hot-electron-mode photodetection", Applied Optics **10**(9), pp. 2190–2192, 1971.
175. V. L. Dalal, "Analysis of photoemissive Schottky barrier photodetectors", J. Appl. Phys. **42**(5), pp. 2280–2284, 1971.
176. H. Elabd and W. F. Kosonocky, "Theory and measurements of photoresponse for thin film Pd_2Si and PtSi infrared Schottky-barrier detectors with optical cavity", RCA Review **43**, pp. 569–589, 1982.
177. W. A. Cabanski and M. J. Schulz, "Electronic and IR-optical properties of silicide/silicon interfaces", Infrared Phys. **32**, pp. 29–44, 1991.
178. J. A. Pals, "Properties of Au, Pt, Pd and Rh levels in silicon measured with a constant capacitance technique", Solid-State Electronics **17**, pp. 1139–1145, 1974.
179. B. J. Baliga and E. Sun, "Comparison of gold, platinum, and electron irradiation for controlling lifetime in power rectifiers", IEEE Trans. Electron Devices **24**(6), pp. 685–688, 1977.
180. K. Maex, A. Lauwers, P. Besser, E. Kondoh, M. de Potter, and A. Steegen, "Self-aligned $CoSi_2$ for 0.18 μm and below", IEEE Trans. Electron Devices **46**(7), pp. 1545–1550, 1999.
181. W. Kuiper, private communication, 1999.
182. E. Roca, K. Kyllesbech-Larsen, S. Kolodinski, and R. Mertens, "Increase in the infrared response of silicide Schottky barrier diodes by grain boundary scattering", Appl. Phys. Lett. **67**(10), pp. 1372–1374, 1995.
183. D. L. Rogers, "Integrated optical receivers using MSM detectors", J. Lightwave Technology **9**(12), pp. 1635–1638, 1991.
184. B. W. Mullins, S. F. Soares, K. A. McArdle, C. M. Wilson, and S. R. J. Brueck, "A simple high-speed Si photodiode", IEEE Photonics Technology Letters **3**(4), pp. 360–362, 1991.
185. C. Buchal and M. Löken, "Silicon-based metal-semiconductor-metal detectors", MRS Bulletin (April), pp. 55–59, 1998.
186. J. Burm, K. I. Litvin, W. J. Schaff, and L. F. Eastman, "Optimization of high-speed metal-semiconductor-metal photodetectors", IEEE Photonics Technology Letters **6**(6), pp. 722–724, 1994.
187. M. Y. Liu and S. Y. Chou, "Picosecond silicon metal-semiconductor-metal photodiode", in *Proc. SPIE*, vol. 2022, p. 76, 1993.
188. S. Alexandrou, C.-C. Wang, T. Y. Hsiang, M. Y. Liu, and S. Y. Chou, Appl. Phys. Lett. **62**(20), pp. 2507–2509, 1993.
189. C.-C. Wang, S. Alexandrou, D. Jacobs-Perkins, and T. Y. Hsiang, "Comparison of the picosecond characterisitics of silicon and silicon-on-sapphire metal-semiconductor-metal photodiodes", Appl. Phys. Lett. **64**(26), pp. 3578–3580, 1994.

190. P. K. Vasudew and P. M. Zeitzoff, "Si-ULSI with a scaled-down future", IEEE Circuits and Devices (March), pp. 19–29, 1998.
191. M. Y. Liu, E. Chen, and S. Y. Chou, "140 GHz metal-semiconductor-metal photodetectors on silicon-on-insulator substrate with a scaled active layer", Appl. Phys. Lett. **65**(7), pp. 887–888, 1994.
192. E. Chen and S. Y. Chou, "High-efficiency and high-speed metal-semiconductor-metal photodetectors operating in the infrared", Appl. Phys. Lett. **70**(6), pp. 753–755, 1997.
193. J. Y. L. Ho and K. S. Wong, "Bandwidth enhancement in silicon metal-semiconductor-metal photodetector by trench formation", IEEE Photonic Technology Letters **8**(8), pp. 1064–1066, 1996.
194. F. Rüders, J. Kim, M. Hacke, S. Mesters, C. Buchal, and S. Mantl, "Vertical MSM photodiodes in silicon based on epitaxial Si/$CoSi_2$/Si", Thin Solid Films **294**, pp. 351–353, 1997.
195. S. Mantl, "Ion beam synthesis of epitaxial silicides: fabrication, characterization and applications", Mater. Sci. Rep. **8**, pp. 1–95, 1992.
196. M. Löken, L. Kappius, S. Mantl, and C. Buchal, "Fabrication of ultrafast Si based MSM photodetector", Electronics Letters **34**(10), pp. 1027–1028, 1998.
197. M. Kunii, K. Hasegawa, H. Oka, Y. Nakazawa, T. Takeshita, and H. Kurihara, "Performance of a high-resolution contact-type linear image sensor with a-Si:H/a-SiC:H heterojunction photodiodes", IEEE Transactions on Electron Devices **36**(12), pp. 2877–2881, 1989.
198. H. Kakinuma, M. Sakamoto, Y. Kasuya, and H. Sawai, "Characterisitics of Cr Schottky amorphous silicon photodiodes and their application in linear image sensors", IEEE Transactions on Electron Devices **37**(1), pp. 128–133, 1990.
199. L. E. Antonuk, J. Boudry, Y. El-Mohri, W. Huang, J. Siewerdsen, J. Yorkston, and R. A. Street, "A high-resolution, high frame rate flatpanel TFT array for digital X-Ray imaging", SPIE, Phys. Med. Imag. **2163**, pp. 118–127, 1994.
200. N. C. Bird, C. J. Curling, and C. van Berkel, "Large-area image sensing using amorphous silicon NIP diodes", Sensors and Actuators **46–47**, pp. 444–448, 1995.
201. X. D. Wu, R. A. Street, R. Weisfield, S. Ready, and S. Nelson, "Page sized a-Si:H two-dimensional array as imaging devices", in *Proc. 4th Int. Conf. Solid-State and Integrated-Circuit Technology*, pp. 724–726, 1995.
202. E. D. Palik, *Handbook of optical constants of solids*, pp. 571–586, Academic Press, Inc., Orlando, 1985.
203. R. H. Bube, "Solar cells", in C. Hilsum, ed., *Handbook on Semiconductors*, vol. 4, (Device Physics), pp. 825–826, North-Holland, Amsterdam, 1993.
204. M. Böhm, F. Blecher, A. Eckhardt, B. Schneider, S. Benthien, H. Keller, T. Lule, P. Rieve, M. Sommer, R. C. Lind, L. Humm, M. Daniels, and N. Wu, "High dynamic range image sensors in thin film on ASIC - technology for automotive applications", in D. E. Ricken and W. Gessner, eds., *Advanced Microsystems for Automotive Applications*, pp. 157–172, Springer, Berlin, Heidelberg, 1998.
205. B. Schneider, H. Fischer, S. Benthien, H. Keller, T. Lule, P. Rieve, M. Sommer, J. Schulte, and M. Böhm, "TFA image sensors: from the one transistor cell to a locally adaptive high dynamic range sensor", in *IEDM Digest Technical Papers*, pp. 209–212, 1997.
206. H. Fischer, J. Schulte, P. Rieve, and M. Böhm, "Technology and performance of TFA (Thin Film on ASIC)-sensors", in *Mat. Res. Soc. Symp. Proc.*, vol. 336, pp. 867–872, 1994.

207. J. Schulte, H. Fischer, T. Lule, Q. Zhu, and M. Böhm, "Properties of TFA (Thin Film on ASIC) sensors", in H. Reichl and A. Heuberger, eds., *Micro System Technologies*, pp. 783–790, 1994.
208. H. Fischer, J. Schulte, J. Giehl, Böhm, and J. P. M. Schmitt, "Thin film on ASIC - a novel concept of intelligent image sensors", in *Mat. Res. Soc. Symp. Proc.*, vol. 258, pp. 1139–1144, 1992.
209. J. Schulte, H. Fischer, Q. Zhu, J. Giehl, H. Stiebig, Z. P. Xu, and M. Böhm, "a-Si:H on ASIC - a new approach to intelligent image sensing", in H. Reichl and A. Heuberger, eds., *Micro System Technologies*, pp. 265–275, 1992.
210. Q. Zhu, H. Stiebig, P. Rieve, and M. Böhm, "A novel a-Si(C):H color sensor array", in *Mat. Res. Soc. Symp. Proc.*, vol. 336, pp. 843–845, 1994.
211. Q. Zhu, S. Coors, B. Schneider, P. Rieve, and M. Böhm, "Bias sensitive a-Si(C):H multispectral detectors", IEEE Trans. Electron Devices **45**(7), pp. 1393–1398, 1998.
212. R. Crandall, "Modeling of thin film solar cells: uniform field approximation", J. Appl. Phys. **54**(12), pp. 7176–7186, 1983.
213. Q. Zhu, J. Sterzel, B. Schneider, S. Coors, and M. Böhm, "Transient behavior of a-Si(C):H bulk barrier color detectors", J. Appl. Phys. **83**(7), pp. 3906–3910, 1998.
214. K. Aflatooni, A. Nathan, R. Hornsey, I. A. Cunningham, and S. G. Chamberlain, "a-Si:H Schottky diode direct detection pixel for large area X-ray imaging", in *IEEE Int. Electron Device Meeting*, pp. 197–200, 1997.
215. K. Aflatooni, A. Nathan, R. I. Hornsey, and I. A. Cunningham, "A novel detection scheme for large area imaging of low energy X-rays using amorphous silicon technology", in *9th Int. Conf. on Solid State Sensors and Actuators*, pp. 1299–1302, 1997.
216. M. P. Vidal, M. Bafleur, J. Buxo, and G. Sarrabayrouse, "A bipolar photodetector compatible with standard CMOS technology", Solid-State Electron. **34**(8), pp. 809–814, 1991.
217. E. A. Vittoz, "MOS transistors operated in the lateral bipolar mode and their application in CMOS technology", IEEE J. Solid-State Circuits **18**(6), pp. 273–279, 1983.
218. R. W. Sandage and J. A. Connelly, "A fingerprint opto-detector using lateral bipolar phototransistors in a standard CMOS process", in *IEDM Digest Technical Papers*, pp. 171–174, 1995.
219. W. T. Holman and J. A. Connelly, "A compact low-noise operational amplifier for a 1.2 μm digital CMOS technology", IEEE J. Solid-State Circuits **30**(6), pp. 710–714, 1995.
220. H. Beneking, "Gain and bandwidth of fast near-infrared photodetectors: a comparison of diodes, phototransistors, and photoconductive devices", IEEE Trans. Electron Devices **29**(9), pp. 1420–1430, 1982.
221. M. Schanz, W. Brockherde, R. Hauschild, B. J. Hosticka, and M. Schwarz, "Smart CMOS image sensor arrays", IEEE Trans. Electron Devices **44**(10), pp. 1699–1705, 1997.
222. M. J. M. Pelgrom, A. C. J. Duinmaijer, and A. P. G. Welbers, "Matching properties of MOS transistors", IEEE J. Solid-State Circuits **24**(10), pp. 1433–1440, 1989.
223. W. Zhang, M. Chan, and P. K. Ko, "A novel high-gain CMOS image sensor using floating N-well/gate tied PMOSFET", in *IEDM Digest Technical Papers*, pp. 1023–1025, 1998.
224. S. Wolf, *Silicon processing for the VLSI era, Vol. 2 – Process integration*, Lattice Press, Sunset Beach, 1990.

225. R. H. Havemann and R. H. Eklund, "Process integration issues for submicron BiCMOS technology", Solid State Technology (6), pp. 71–76, 1992.
226. T. Ikeda, A. Watanabe, Y. Nishio, I. Masuda, N. Tamba, M. Odaka, and K. Ogiue, "High-speed BiCMOS technology with a buried twin well structure", IEEE Trans. Electron Devices **34**(6), pp. 1304–1310, 1987.
227. S. S. Ahmed, W. W. Asakawa, M. T. Bohr, S. S. Chambers, T. Deeter, M. Denham, J. K. Greason, W. W. Holt, R. R. Taylor, and I. Young, "A triple diffused approach for high performance 0.8 μm BiCMOS technology", Solid State Technology (10), pp. 33–40, 1992.
228. R. A. Chapman, D. A. Bell, R. H. Eklund, R. H. Havemann, M. G. Harward, and R. A. Haken, "Submicrometer BiCMOS well design for optimum circuit performance", in *IEDM Digest Technical Papers*, pp. 756–759, 1988.
229. H. Iwai, G. Sasaki, Y. Unno, Y. Niitsu, M. Norishima, Y. Sugimoto, and K. Kanzaki, "0.8-μm Bi-CMOS technology with high f_T ion-implanted emitter bipolar transistor", in *IEDM Digest Technical Papers*, pp. 28–31, 1987.
230. T.-Y. Chiu, G. M. Chin, M. Y. Lau, R. C. Hanson, M. D. Morris, K. F. Lee, A. M. Voshchenkov, R. G. Swartz, V. D. Archer, M. T. Y. Liu, S. N. Finegan, and M. D. Feuer, "Non-overlapping super self-aligned BiCMOS with 87 ps low power ECL", in *IEDM Digest Technical Papers*, pp. 752–755, 1988.
231. W. R. Burger, C. Lage, T. Davies, M. DeLong, D. Haueisen, J. Small, G. Huglin, A. Landau, F. Whitwer, and B. Bastani, "An advanced self-aligned BICMOS technology for high performance 1-megabit ECL i/O SRAMs", in *IEDM Digest Technical Papers*, pp. 421–424, 1989.
232. Y. Kobayashi, C. Yamaguchi, Kobayashi, Y. Amemiya, and T. Sakai, "High perfomance LSI process technology: SST CBi-CMOS", in *IEDM Digest Technical Papers*, pp. 760–763, 1988.
233. K. Sakaue, Y. Shobatake, M. Motoyama, Y. Kumaki, S. Takatsuka, S. Tanaka, H. Hara, K. Matsuda, S. Kitaoka, M. Noda, Y. Niitsu, M. Norishima, H. Momose, K. Maeguchi, M. Ishibe, S. Shimizu, and T. Kodama, "A 0.8-μm BiCMOS ATM switch on an 800-Mb/s asynchronous buffered banyan network", IEEE J. Solid-State Circuits **26**(8), pp. 1133–1144, 1991.
234. M. El-Diwany, J. Borland, J. Chen, S. Hu, P. v. Wijnen, C. Vorst, V. Akylas, M. Brassington, and R. Razouk, "An advanced BiCMOS process utilizing ultra-thin silicon epitaxy over arsenic buried layers", in *IEDM Digest Technical Papers*, pp. 245–248, 1989.
235. M. Norishima, Y. Niitsu, G. Sasaki, H. Iwai, and K. Maeguchi, "Bipolar transistor design for low process-temperature 0.5 μm BI-CMOS", in *IEDM Digest Technical Papers*, pp. 237–240, 1989.
236. P. J.-W. Lim, A. Y. C. Tzeng, H. L. Chuang, and S. A. S. Onge, "A 3.3 V monolithic photodetector/CMOS preamplifier for 531 Mb/s optical data link applications", in *ISSCC*, pp. 96–97, 1993.
237. D. M. Kuchta, H. A. Ainspan, F. J. Canora, and R. P. Schneider, "Performance of fiber-optic data links using 670 nm CW VCSELs and a monolithic Si photodetector and CMOS preamplifier", IBM J. Res. Develop. **39**(1/2), pp. 63–72, 1995.
238. H. Zimmermann, K. Kieschnick, M. Heise, and H. Pless, "BiCMOS OEIC for optical storage systems", Electronics Letters **34**(19), pp. 1875–1876, 1998.
239. H. Zimmermann, "Full custom CMOS and BiCMOS OPTO-ASICs", in *Proc. 5th Int. Conf. on Solid-State and Integrated-Circuit Technology*, pp. 344–347, 1998.
240. K. Kieschnick, H. Zimmermann, and P. Seegebrecht, "Silicon-based optical receivers in BiCMOS technology for advanced optoelectronic integrated cir-

cuits", in *Proc. European Materials Research Society Meeting (E-MRS), Strasbourg, June 1–4*, 1999.

241. J.-P. Colinge, *Silicon-on-insulator technology: materials to VLSI*, Kluwer Academic Publishers, Boston, 1991.
242. J.-P. Colinge, "p-i-n photodiodes made in laser-recrystallized silicon-on-insulator", IEEE Transactions on Electron Devices **33**(2), pp. 203–205, 1986.
243. M. Ghioni, F. Zappa, V. P. Kesan, and J. Warnock, "A VLSI-compatible high-speed silicon photodetector for optical data link applications", IEEE Transactions on Electron Devices **43**(7), pp. 1054–1060, 1996.
244. J. Y. L. Ho and K. S. Wong, "High-speed and high-sensitivity silicon-on-insulator metal-semiconductor-metal photodetector with trench structure", Appl. Phys. Lett. **69**(1), pp. 16–18, 1996.
245. B. F. Levine, J. D. Wynn, F. P. Klemens, and G. Sarusi, "1 Gb/s Si high quantum efficiency monolithically integrable $\lambda = 0.88\,\mu\text{m}$ detector", Appl. Phys. Lett. **66**(22), pp. 2984–2986, 1995.
246. E. Yablonovitch, "Statistical ray optics", J. Opt. Soc. Am. **72**(7), pp. 899–907, 1982.
247. T. Yoshida, Y. Ohtomo, and M. Shimaya, "A novel p-i-n photodetector fabricated on SIMOX for 1 GHz 2 V CMOS OEICs", in *IEDM Digest Technical Papers*, pp. 29–32, 1998.
248. W. N. Grant, "Electron and hole ionization rates in epitaxial silicon at high electric fields", Solid-State Electron. **16**, pp. 1189–1203, 1973.
249. W. Zhang, M. Chan, S. K. H. Fung, and P. K. Ko, "Performance of a CMOS compatible lateral bipolar photodetector on SOI substrate", IEEE Electron Device Lett. **19**(11), pp. 435–437, 1998.
250. S. Verdonckt-Vandebroek, S. S. Wong, J. C. S. Woo, and P. K. Ko, "High-gain lateral bipolar action in a MOSFET structure", IEEE Trans. Electron Devices **38**(11), pp. 2487–2496, 1991.
251. S. A. Parke, C. Hu, and P. K. Ko, "Bipolar-FET hybrid-mode operation of quarter-micrometer SOI MOSFETs", IEEE Electron Device Lett. **14**(5), pp. 234–236, 1993.
252. S. A. Parke, F. Assaderaghi, J. Chen, J. King, C. Hu, and P. K. Ko, "A versatile SOI BiCMOS technology with complementary lateral BJT's", in *IEDM Digest Technical Papers*, pp. 453–456, 1992.
253. H. Yamamoto, K. Taniguchi, and C. Hamaguchi, "High-sensitivity SOI MOS photodetector with self-amplification", Jpn. J. Appl. Phys. **35**(2B), pp. 1382–1386, 1996.
254. X. Xiao, J. C. Sturm, P. V. Schwartz, and K. K. Goel, "Vertical 1.3 µm optical modulator in silicon-on-insulator", in *Proc. IEEE SOS/SOI Technology Conference*, pp. 171–172, 1990.
255. O. Mikami and H. Nakagome, "Waveguided optical switch in InGaAsP/InP using free carrier plasma dispersion", Electron. Lett. **20**, pp. 228–229, 1984.
256. J. L. Moll, M. Tanenbaum, J. M. Goldey, and N. Holonyak, "p-n-p-n-transistor switches", Proc. IRE **44**, p. 1174, 1956.
257. S. K. Ghandhi, *Semiconductor power devices*, Wiley, New York, 1977.
258. B. J. Baliga, *Modern power devices*, Wiley, New York, 1986.
259. W. Gerlach, "Light-activated power thyristors", Inst. Phys. Conf. Ser. **32**, pp. 111–133, 1977.
260. M. Watanabe, Y. Takahashi, K. Endo, H. Kakigi, and O. Hashimoto, "A 115-mm ϕ 6-kV 2500-A light-triggered thyristor", IEEE Transactions on Electron Devices **37**(1), pp. 285–289, 1990.

261. Y. Shimizu, R. Iyotanu, N. Konishi, and T. Yatsuo, "A high-voltage light-activated thyristor with a novel over-voltage self-protection structure", IEEE Transactions on Electron Devices **36**(5), p. 355, 1989.
262. F. E. Gentry, R. I. Scace, and J. K. Flowers, "Bidirectional triode p-n-p-n switches", Proc. IEEE **53**(4), pp. 355–369, 1965.
263. J. F. Essom, "Bidirectional triode thyristor applied voltage rate effect following conduction", Proc. IEEE **55**, pp. 1312–1317, 1967.
264. S.-Q. Zhao, Z.-Y. Wang, and D.-S. Gao, "A high power light triggered triac with a novel light sensitive structure", IEEE Transactions on Electron Devices **42**(4), pp. 773–779, 1995.
265. E. Kasper, H. J. Herzog, and H. Kibbel, "A one-dimensional SiGe superlattice grown by UHV epitaxy", Appl. Phys. **8**, pp. 199–205, 1975.
266. H. M. Manasevit, I. S. Gergis, and A. B. Jones, "The properties of $Si/Si_{1-x}Ge_x$ films grown on Si substrates by chemical vapor deposition", J. Electron. Mater. **12**(4), pp. 637–651, 1983.
267. J. C. Bean, T. T. Sheng, L. C. Feldman, A. T. Fiory, and R. T. Lynch, "Pseudomorphic growth of Ge_xSi_{1-x} on silicon by molecular beam epitaxy", Appl. Phys. Lett. **44**(1), pp. 102–104, 1984.
268. R. People, J. C. Bean, D. V. Lang, A. M. Sergent, H. L. Stormer, K. W. Wecht, R. T. Lynch, and K. Baldwin, "Modulation doping in Ge_xSi_{1-x}/Si strained layer heterostructures", Appl. Phys. Lett. **45**(11), pp. 1231–1233, 1984.
269. R. People, "Indirect band gap of coherently strained Ge_xSi_{1-x} bulk alloys on <001> silicon substrates", Phys. Rev. B **32**(2), pp. 1405–1408, 1985.
270. D. C. Ahlgren, M. Gilbert, D. Greenberg, S. J. Jeng, J. Malinowski, D. Nguyen-Ngoc, K. Schonenberg, K. Stein, R. Groves, K. Walter, G. Hueckel, D. Colavito, G. Freeman, D. Sunderland, D. L. Harame, and B. Meyerson, "Manufacturability demonstration of an integrated SiGe HBT technology for the analog and wireless marketplace", in *IEEE Int. Electron Device Meeting*, pp. 859–862, Wash., D.C., 1996.
271. S. Subbanna, D. Ahlgren, D. Harame, and B. Meyerson, "How SiGe evolved into a manufacturable semiconductor production process", in *IEEE Int. Solid-State Circuits Conference*, pp. 66–67, 1999.
272. D. J. Paul, "Silicon-germanium strained layer materials in microelectronics", Advanced Materials **11**(3), pp. 191–204, 1999.
273. S. Luryi, T. P. Pearsall, H. Tempkin, and J. C. Bean, "Waveguide infrared photodetectors on a silicon chip", IEEE Electron Device Lett. **7**(2), pp. 104–107, 1986.
274. A. Splett, T. Zinke, K. Petermann, E. Kasper, H. Kibbel, H.-J. Herzog, and H. Presting, IEEE Photonics Technology Letters **6**, pp. 59–, 1994.
275. S. B. Samavedam, M. T. Currie, T. A. Langdo, and E. A. Fitzgerald, "High-quality germanium photodiodes integrated on silicon substrates using optimized relaxed graded buffers", Appl. Phys. Lett. **73**(15), pp. 2125–2127, 1998.
276. T. L. Lin and J. Maserjian, "Characterization of wafer bonded photodetectors fabricated using various annealing temperatures and ambients", Appl. Phys. Lett. **57**(14), pp. 1422–1424, 1990.
277. T. L. Lin, E. W. Jones, A. Ksendzov, S. M. Dejewski, R. W. Fathauer, T. N. Krabach, and J. Maserjian, "A novel Si-based LWIR detector: the SiGe/Si heterojunction internal photoemission detector", in *IEDM Digest Technical Papers*, pp. 641–644, 1990.
278. R. P. G. Karunasiri, J. S. Park, and K. L. Wang, "$Si_{1-x}Ge_x/Si$ multiple quantum well infrared detector", Appl. Phys. Lett. **59**(20), pp. 2588–2590, 1991.
279. R. A. Soref, "Silicon-based photonic devices", in *ISSCC*, pp. 66–67, 1995.

280. Q. Mi, X. Xiao, J. C. Sturm, L. Lenchyshyn, and M. L. W. Thewalt, "Room-temperature 1.3 μm electroluminescence from strained $Si_{1-x}Ge_x$/Si quantum wells", Appl. Phys. Lett. **60**(25), pp. 3177–3179, 1992.
281. H. Presting, T. Zinke, A. Splett, H. Kibbel, and M. Jaros, Appl. Phys. Lett. **69**, pp. 2376–, 1996.
282. R. A. Soref, "Silicon-based optoelectronics", Proc. IEEE **81**(12), pp. 1687–1706, 1993.
283. G. Abstreiter, "Engineering the future of electronics", Phys. World **5**, pp. 36–39, 1992.
284. J. C. Sturm, "Advanced column-IV epitaxial materials for silicon-based optoelectronics", MRS Bulletin (April), pp. 60–64, 1998.
285. D. C. Houghton, "Strain relaxation kinetics in $Si_{1-x}Ge_x$/Si heterostructures", J. Appl. Phys. **70**(4), pp. 2136–2151, 1991.
286. D. C. Houghton, C. Gibbings, C. Tuppen, M. Lyons, and M. Halliwell, "Equilibrium critical thickness for $Si_{1-x}Ge_x$ strained layers on (100) Si", Appl. Phys. Lett. **56**(5), pp. 460–462, 1990.
287. F. K. LeGoues, B. S. Meyerson, and J. Morar, "Anomalous strain relaxation in SiGe thin films and superlattices", Phys. Rev. Lett. **66**(22), pp. 2903–2906, 1991.
288. Y. J. Mii, Y. H. Xie, E. A. Fitzgerald, D. Monroe, F. A. Thiel, B. E. Weir, and L. C. Feldman, "Extremely high electron mobility in Si/Ge_xSi_{1-x} structures grown by molecular beam epitaxy", Appl. Phys. Lett. **59**(13), pp. 1611–1613, 1991.
289. K. Ismail, S. Rishton, J. O. Chu, K. Chan, and B. S. Meyerson, "High-performance Si/SiGe n-type modulation-doped transistors", IEEE Electron Device Lett. **14**(7), pp. 348–350, 1993.
290. G. Kissinger, T. Morgenstern, G. Morgenstern, and H. Richter, "Stepwise equilibrated graded Ge_xSi_{1-x} buffer with very low threading dislocation density on Si(001)", Appl. Phys. Lett. **66**(16), pp. 2083–2085, 1995.
291. E. J. Prinz, P. M. Garone, P. V. Schwartz, X. Xiao, and J. C. Sturm, "The effects of base dopant outdiffusion and undoped $Si_{1-x}Ge_x$ junction spacer layers in Si/$Si_{1-x}Ge_x$/Si heterojunction bipolar transistors", IEEE Electron Device Lett. **12**(2), pp. 42–44, 1991.
292. J. W. Slotboom, B. Streutker, A. Pruijimboom, , and D. J. Gravesteijn, "Parasitic energy barriers in SiGe HBTs", IEEE Electron Device Lett. **12**(9), pp. 486–488, 1991.
293. U. Gnutzmann and K. Clausecker, "Theory of direct optical transitions in an optical indirect semiconductor with a superlattice structure", Appl. Phys. **3**, pp. 9–14, 1974.
294. E. D. Palik, *Handbook of optical constants of solids II*, pp. 607–636, Academic Press, Inc., Boston, 1991.
295. J. C. Bean, "Recent developments in the strained layer epitaxy of germanium-silicon alloys", J. Vac. Sci. Technol. **B** 4(6), pp. 1427–1429, 1986.
296. B. Jalali, A. F. J. Levi, F. Ross, and E. A. Fitzgerald, Electronics Letters **28**, pp. 269–270, 1992.
297. B. Schuppert, J. Schmidtchen, A. Splett, U. Fischer, T. Zinke, R. Moosburger, and K. Petermann, "Integrated optics in silicon and SiGe-heterostructures", IEEE J. Lightwave Technol. **14**(10), pp. 2311–2323, 1996.
298. A. Vonsovici, "Room temperature photocurrent spectroscopy of SiGe/Si p-i-n photodiodes grown by selective epitaxy", IEEE Trans. Electron Devices **45**(2), pp. 538–542, 1998.

299. E. A. Fitzgerald, Y. H. Xie, M. L. Green, D. Brasen, A. R. Kortan, J. Michel, Y. J. Mii, and B. E. Weir, "Totally relaxed Ge_xSi_{1-x} layers with low threading dislocation densities grown on Si substrates", Appl. Phys. Lett. **59**(7), pp. 811–813, 1991.
300. M. T. Currie, S. B. Samavedam, T. A. Langdo, C. W. Leitz, and E. A. Fitzgerald, "Controlling threading dislocation densities in Ge on Si using graded SiGe layers and chemical-mechanical polishing", Appl. Phys. Lett. **72**(14), pp. 1718–1720, 1998.
301. S. Luryi, A. Katalsky, and J. C. Bean, "New infrared detector on a silicon chip", IEEE Trans. Electron Devices **31**(9), pp. 1135–1139, 1984.
302. J. B. Posthill, R. A. Rudder, S. V. Hattangaddy, C. G. Fountain, and R. J. Markunas, "On the feasibility of growing dilute C_xSi_{1-x} epitaxial alloys", Appl. Phys. Lett. **56**(8), pp. 734–736, 1990.
303. K. Eberl, S. S. Iyer, S. Zollner, J. C. Tsang, and F. K. LeGoues, "Growth and strain compensation effects in the ternary $Si_{1-x-y}Ge_xC_y$ alloy system", Appl. Phys. Lett. **60**(24), pp. 3033–3035, 1992.
304. C. L. Chang, A. S. Amour, and J. C. Sturm, "Effect of carbon on the valence band offset of $Si_{1-x-y}Ge_xC_y$/Si heterojunctions", in *IEDM Digest Technical Papers*, pp. 257–260, 1996.
305. Landolt-Börnstein, *Numerical data and functional relationships in science and technology*, vol. 22a, Springer, Berlin, Heidelberg, 1989.
306. R. A. Soref, "Optical band gap of the ternary semiconductor $Si_{1-x-y}Ge_xC_y$", J. Appl. Phys. **70**(4), pp. 2470–2472, 1991.
307. L. C. Chang97, A. S. Amour, and J. C. Sturm, "The effect of carbon on the valence band offset of compressively strained $Si_{1-x-y}Ge_xC_y$/(100)Si heterojunctions", Appl. Phys. Lett. **70**(12), pp. 1557–1559, 1997.
308. P. A. Stolk, D. J. Eaglesham, H.-J. Gossmann, and J. M. Poate, "Carbon incorporation in silicon for suppressing interstitial-enhanced boron diffusion", Appl. Phys. Lett. **66**(11), pp. 1370–1372, 1995.
309. L. D. Lanzerotti, J. C. Sturm, E. Stach, R. Hull, T. Buyuklimanli, and C. Magee, "Suppression of boron transient enhanced diffusion in SiGe heterojunction bipolar transistors by carbon incorporation", Appl. Phys. Lett. **70**(23), pp. 3125–3127, 1997.
310. T. Tashiro, T. Tatsumi, M. Sugiyama, T. Hashimoto, and T. Morikawa, "A selective epitaxial SiGe/Si planar photodetector for Si-based OEIC's", IEEE Trans. Electron Devices **44**(4), pp. 545–550, 1997.
311. K. Aketagawa, T. Tatsumi, M. Hiroi, T. Niino, and J. Sakai, "Si/$Si_{1-x}Ge_x$ selective epitaxial growth by ultra-high-vacuum chemical vapor deposition using Si_2H_6, GeH_4", in *Proc. Solid State Devices and Materials*, pp. 719–722, 1991.
312. R. T. Carline, D. A. O. Hope, V. Nayar, D. J. Robbins, and M. B. Stanaway, "A vertical cavity longwave infrared SiGe/Si photodetector using a buried silicide mirror", in *IEEE Int. Electron Device Meeting*, pp. 891–894, Wash., D.C., 1997.
313. A. Schüppen, U. Erben, A. Gruhle, H. Kibbel, H. Schumacher, and U. König, "Enhanced SiGe heterojunction bipolar transistors with 160 GHz-f_{max}", in *IEDM Digest Technical Papers*, pp. 743–746, 1995.
314. J.-S. Rieh, D. Klotzkin, O. Qasaimeh, L.-H. Lu, K. Yang, L. P. B. Katehi, P. Bhattacharya, and E. T. Croke, "Monolithically integrated SiGe-Si PIN-HBT front-end photoreceiver", IEEE Photonics Technology Letters **10**(3), pp. 415–417, 1998.
315. M. S. Ünlü and S. Strite, "Resonant cavity enhanced photonic devices", J. Appl. Phys. **78**(2), pp. 607–639, 1995.

316. P. Bhattacharya, *Semiconductor optoelectronic devices*, p. 426, Prentice Hall, Upper Saddle River, NJ, 1997.
317. G. W. Neudeck, J. Denton, J. Qi, J. D. Schaub, R. Li, and J. C. Campbell, "Selective epitaxial growth Si resonant-cavity photodetector", IEEE Photonics Technology Letters **10**(1), pp. 129–131, 1998.
318. H. C. Chao and G. W. Neudeck, "Crystal silicon Fabry–Perot cavities deposited with dichlorosilane in a reduced pressure chemical vapor deposition reactor for thermal sensing", Electron. Lett. **31**(13), pp. 1101–1102, 1995.
319. G. W. Neudeck, J. Spitz, J. C. Chang, J. P. Denton, and N. Gallagher, "Precision crystal corner cube arrays for optical gratings formed by (100) silicon planes with selective epitaxial growth", J. Appl. Opt. **35**(19), pp. 3466–3470, 1996.
320. J. C. Bean, J. Qi, L. Schow, R. Li, H. Nie, J. Schaub, and J. C. Campbell, "High-speed polysilicon resonant-cavity photodiodes with SiO_2/Si Bragg reflectors", IEEE Photon. Technol. Lett. **9**(6), pp. 806–808, 1997.
321. E. A. Fitzgerald, Y.-H. Xie, D. Monroe, P. J. Silverman, J.-M. Kuo, A. R. Kortan, F. A. Thiel, B. E. Weir, and L. C. Feldman, "Relaxed Ge_xSi_{1-x} structures for III-V integration with Si and high mobility two-dimensional electron gases in Si", J. Vac. Sci. Technol. B **10**(4), pp. 1807–1819, 1992.
322. T. Egawa, T. Soga, T. Jimbo, and M. Umeno, "Room-temperature continuous-wave operation of AlGaAs-GaAs single-quantum-well lasers on Si by metalorganic chemical-vapor deposition using AlGaAs-AlGaP intermediate layers", IEEE J. Quantum Electron. **27**(6), pp. 1798–1803, 1991.
323. E. A. Fitzgerald and L. C. Kimerling, "Silicon-based microphotonics and integrated optoelectronics", MRS Bulletin (April), pp. 39–47, 1998.
324. E. A. Fitzgerald, "The effect of substrate growth area on misfit and threading dislocation densities in mismatched heterostructures", J. Vac. Sci. Technol. B **7**(4), pp. 782–788, 1989.
325. M. Koyanagi, "Optically coupled three-dimensional memory system", in *Proc. SPIE, Int. Conf. on Advances in Interconnection and Packaging*, vol. 1390, pp. 467–476, 1990.
326. M. Koyanagi, H. Takata, H. Mori, and J. Iba, "Design of 4-kbit×4-layer optically coupled three-dimensional common memory for parallel processor system", IEEE J. Solid-State Circuits **25**(1), pp. 109–116, 1990.
327. H. Shichijo and R. J. Matyi, "GaAs-on-silicon integrated circuits; savior for GaAs or for Si?", in *IEDM Digest Technical Papers*, pp. 88–91, 1987.
328. I. Suemune, Y. Kunitsugu, Y. Tanaka, Y. Kan, and M. Yamanishi, "New low-temperature process for growth of GaAs on Si with metalorganic molecular beam epitaxy assisted by hydrogen plasma", Appl. Phys. Lett. **53**(22), pp. 2173–2175, 1988.
329. T. Hamaguchi, N. Endo, M. Kimura, and M. Nakamae, "Novel LSI/SOI wafer fabrication using device layer transfer technique", in *IEDM Digest Technical Papers*, pp. 688–691, 1985.
330. H. K. Choi, G. W. Turner, T. H. Windhorn, and B.-Y. Tsaur, "Monolithic integration of GaAs/AlGaAs double-heterostructure LEDs and Si MOSFETs", IEEE Electron Device Lett. **7**(9), pp. 500–502, 1986.
331. T. Egawa, T. Jimbo, and M. Umeno, "Monolithically integrated AlGaAs/InGaAs laser diode, p-n photodetector and GaAs MESFET grown on Si substrate", Jpn. J. Appl. Phys., Part 1 **32**(1B), pp. 650–653, 1993.
332. T. Egawa, T. Jimbo, and M. Umeno, "Monolithic integration of AlGaAs/GaAs MQW laser diode and GaAs MESFET grown on Si using selective regrowth", IEEE Photon. Technol. Lett. **4**(6), pp. 612–614, 1992.

333. H. K. Choi, C. A. Wang, and N. H. Karam, "GaAs-based diode laser on Si with increased lifetime obtained by using strained InGaAs active layer", Appl. Phys. Lett. **59**(21), pp. 2634–2635, 1991.
334. K. Iga, F. Koyama, and S. Kinoshita, "Surface emitting semiconductor lasers", IEEE J. Quantum Electron. **24**(9), pp. 1845–1855, 1988.
335. D. G. Deppe, N. Chand, J. P. van der Ziel, and G. J. Zydzik, "$Al_xGa_{1-x}As$-GaAs vertical-cavity surface-emitting laser grown on Si substrate", Appl. Phys. Lett. **56**(8), pp. 740–742, 1990.
336. J. L. Jewell, J. P. Harbison, A. Schere, Y. H. Lee, and L. T. Florez, "Vertical-cavity surface-emitting lasers: design, growth, fabrication, characterization", IEEE J. Quantum Electron. **27**(6), pp. 1332–1346, 1991.
337. T. Egawa, T. Jimbo, and M. Umeno, "Low-threshold AlGaAs/GaAs vertical-cavity surface-emitting laser on Si substrate at 300 K", in *IEDM Digest Technical Papers*, pp. 597–600, 1993.
338. R. S. Sussmann, "Ultra-low-capacitance flip-chip-bonded GaInAs PIN photodetector for long-wavelength high-data-rate fibre-optic systems", Electron. Lett. **21**, pp. 593–595, 1985.
339. K. Katsura, T. Hayashi, F. Ohira, S. Hata, and K. Iwashita, "A novel flip-chip interconnection technique using solder bumps for high-speed photoreceivers", IEEE J. Lightwave Technology **8**(9), pp. 1323–1326, 1990.
340. L. S. Goldmann, "Self-alignment capability of controlled-collapse chip joining", in *Proc. 22nd Electronic Components Conf.*, pp. 332–339, 1972.
341. W. Hunziker, W. Vogt, H. Melchior, P. Buchmann, and P. Vettiger, "Passive self-aligned low-cost packaging of semiconductor laser arrays on Si motherboard", IEEE Photonics Technology Letters **7**(11), pp. 1324–1326, 1995.
342. A. V. Krishnamoorthy and D. A. B. Miller, "Scaling optoelectronic-VLSI circuits into the 21st century: a technology roadmap", IEEE J. Selected Topics in Quantum Electronics **2**(1), pp. 55–76, 1996.
343. K. W. Goossen, J. A. Walker, L. A. D'Asaro, S. P. Hui, B. Tseng, R. Leibenguth, D. Kossives, D. D. Bacon, D. Dahringer, L. M. F. Chirovsky, A. L. Lentine, and D. A. Miller, "GaAs MQW modulators integrated with silicon CMOS", IEEE Photonics Technology Letters **7**(4), pp. 360–362, 1995.
344. T. K. Woodward, A. L. Lentine, A. V. Krishnamoorthy, K. W. Goossen, J. A. Walker, J. E. Cunningham, W. J. Yan, B. T. Tseng, S. P. Hui, and R. E. Leibenguth, "Parallel operation of 50 element two-dimensional CMOS smart-pixel receiver array", Electronics Letters **34**(10), pp. 936–937, 1998.
345. A. L. Lentine, K. W. Goossen, J. A. Walker, J. E. Cunningham, W. Y. Jan, T. K. Woodward, A. V. Krishnamoorthy, B. J. Tseng, S. P. Hui, R. E. Leibenguth, L. M. F. Chirovsky, R. A. Novotny, D. B. Buchholz, and R. L. Morrison, "Optoelectronic VLSI switching chip with greater than 1 Tbit/s potential optical I/O bandwidth", Electronics Letters **33**(10), pp. 894–895, 1997.
346. A. V. Krishnamoorthy, L. M. F. Chirovsky, W. S. Hobson, R. E. Leibenguth, S. P. Hui, G. J. Zydzik, K. W. Goossen, J. D. Wynn, B. J. Tseng, J. Lopata, J. A. Walker, J. E. Cunningham, and L. A. D'Asaro, "Vertical-cavity surface-emitting lasers flip-chip bonded to gigabit-per-second CMOS circuits", IEEE Photonics Technology Letters **11**(1), pp. 128–130, 1999.
347. A. V. Krishnamoorthy and K. W. Goossen, "Progress in optoelectronic-VLSI smart pixel technology based on GaAs/AlGaAs MQW modulators", Int. J. Optoelectron. **11**, pp. 181–198, 1997.
348. L. M. F. Chirovsky, R. E. Leibenguth, W. S. Hobson, S. P. Hui, G. J. Zydzik, B. J. Tseng, J. D. Wynn, J. Lopata, and L. A. D'Asaro, "VCSELs with ion-implanted current apertures and index guiding", in *CLEO'98*, 1998.

349. J. Haisma, B. Spierings, U. Biermann, and A. van Gorkum, "Diversity and feasibility of direct bonding: a survey of a dedicated optical technology", Appl. Opt. **33**(7), pp. 1154–1169, 1994.
350. Y. H. Lo, R. Bhat, D. M. Hwang, C. Chua, and C.-H. Lin, "Semiconductor lasers on Si substrates using the technology of bonding by atomic rearrangement", Appl. Phys. Lett. **62**(10), pp. 1038–1040, 1993.
351. K. Mori, K. Tokutome, K. Nishi, and S. Sugou, "High-quality InGaAs/InP multiple-quantum-well structures on Si fabricated by direct bonding", Electron. Lett. **30**, pp. 1008–1009, 1994.
352. F. E. Ejeckam, C. L. Chua, Z. H. Zhu, Y. H. Lo, M. Hong, and R. Bhat, "High-performance InGaAs photodetectors on Si and GaAs substrates", Appl. Phys. Lett. **67**(26), pp. 3936–3938, 1995.
353. H. Wada and T. Kamijoh, "Room-temperature CW operation of InGaAsP lasers on Si fabricated by wafer bonding", IEEE Photonics Technology Letters **8**(2), pp. 173–175, 1996.
354. A. R. Hawkins, T. E. Reynolds, D. R. England, D. I. Babic, M. J. Mondry, K. Streubel, and J. E. Bowers, "Silicon heterointerface photodetector", Appl. Phys. Lett. **68**(26), pp. 3692–3694, 1996.
355. A. R. Hawkins, W. Wu, P. Abraham, K. Streubel, and J. E. Bowers, "High gain-bandwidth-product silicon heterointerface photodetector", Appl. Phys. Lett. **70**(3), pp. 303–305, 1997.
356. Q.-Y. Tong and U. Gösele, *Semiconductor wafer bonding*, Wiley, New York, 1999.
357. Q.-Y. Tong, private communication, 1999.
358. V. Lehmann, K. Mitani, R. Stengl, T. Mii, and U. Gösele, "Bubble-free wafer bonding of GaAs and InP on silicon in a microcleanroom", Jpn. J. Appl. Phys. **28**, p. L2141, 1989.
359. A. Yamada, M. Oasa, H. Nagabuchi, and M. Kawashima, "Direct adhesion of single-crystal GaAs wafers", Mater. Lett. **6**, p. 167, 1988.
360. E. Jalaguier, B. Aspar, S. Pocas, J. F. Michaud, M. Zussy, A. M. Papon, and M. Bruel, "Transfer of 3-in GaAs film on silicon substrate by proton implantation process", Electronics Letters **34**(4), pp. 408–409, 1998.
361. M. Bruel, "Silicon on insulator material technology", Electronics Letters **31**(14), pp. 1201–1202, 1995.
362. B. F. Levine, A. R. Hawkins, S. Hiu, B. J. Tseng, C. A. King, L. A. Gruezke, R. W. Johnson, D. R. Zolnowski, and J. E. Bowers, "20 GHz high performance planar Si/InGaAs P-I-N photodetector", Appl. Phys. Lett. **70**(18), pp. 2449–2451, 1997.
363. B. F. Levine, A. R. Hawkins, S. Hiu, B. J. Tseng, J. P. Reilley, C. A. King, L. A. Gruezke, R. W. Johnson, D. R. Zolnowski, and J. E. Bowers, "Characterization of wafer bonded photodetectors fabricated using various annealing temperatures and ambients", Appl. Phys. Lett. **71**(11), pp. 1507–1509, 1997.
364. M. Sugo, H. Mori, Y. Sakai, and Y. Itoh, "Stable CW operation at room temperature of a 1.5–μm wavelength multiple quantum well laser on a Si substrate", Appl. Phys. Lett. **60**, pp. 472–473, 1992.
365. K. Mori, K. Tokutome, and S. Sugou, "Low-threshold pulsed operation of long-wavelength lasers on Si fabricated by direct bonding", Electron. Lett. **31**(4), pp. 284–285, 1995.
366. G. Deboy and J. Kölzer, "Fundamentals of light emission from silicon devices", Semicond. Sci. Technol. **9**, pp. 1017–1032, 1993.
367. J. Bude, N. Sano, and A. Yoshii, "Hot-carrier luminescence in Si", Phys. Rev. B **45**(11), pp. 5848–5856, 1992.

368. I. V. Grekhov, A. F. Shulekin, M. I. Vexler, and H. Zimmermann, "Concepts of light emission from a silicon MOS tunnel emitter Auger transistor", Solid-State Electron. **43**, pp. 417–426, 1999.
369. K. Imai and H. Unno, "FIPOS (Full Isolation by Porous Oxidized Silicon) technology and its applications to LSIs", IEEE Trans. Electron Devices **31**(3), pp. 297–302, 1984.
370. V. Lehmann and U. Gösele, "Porous silicon formation: a quantum wire effect", Appl. Phys. Lett. **58**(8), pp. 856–858, 1991.
371. L. T. Canham, "Silicon quantum wire array fabrication by electrochemical and chemical dissolution of wafers", Appl. Phys. Lett. **57**(10), pp. 1046–1048, 1990.
372. A. Richter, P. Steiner, F. Kozlowski, and W. Lang, "Current-induced light emission from a porous silicon device", IEEE Electron Device Lett. **12**(12), pp. 691–692, 1991.
373. N. Koshida and H. Koyama, "Visible electroluminescence from porous silicon", Appl. Phys. Lett. **60**(3), pp. 347–349, 1992.
374. N. Koshida, H. Koyama, Y. Yamamoto, and G. J. Collins, "Visible electroluminescence from porous silicon diodes with an electropolymerized contact", Appl. Phys. Lett. **63**(19), pp. 2655–2657, 1993.
375. A. Loni, A. J. Simons, T. I. Cox, and P. D. J. C. L. T. Canham, "Electroluminescent porous silicon device with an external quantum efficiency greater than 0.1 % under CW operation", Electronics Letters **31**(15), pp. 1288–1289, 1995.
376. P. M. Fauchet, "Progress toward nanoscale silicon light emitters", IEEE J. Selected Topics in Quantum Electronics **4**(6), pp. 1020–1028, 1998.
377. D. I. Kovalev, I. D. Yaroshetzkii, T. Muschik, V. Petrova-Koch, and F. Koch, "Fast and slow visible luminescence bands of oxidized porous Si", Appl. Phys. Lett. **64**(2), pp. 214–216, 1994.
378. S. Lazarouk, P. Jaguiro, S. Katsouba, G. Masini, S. L. Monica, G. Maiello, and A. Ferrari, "Stable electroluminesccence from reverse biased n-type porous silicon-aluminum Schottky junction device", Appl. Phys. Lett. **68**(12), pp. 1646–1648, 1996.
379. A. G. Nassiopoulou, "Low dimensional silicon for integrated optoelectronics", in *Proc. Int. Semicond. Conference (CAS'98)*, pp. 417–426, Sinaia, Romania, 1998.
380. C. Delerue, G. Allen, and M. Lannoo, "Theoretical aspects of the luminescence of porous silicon", Phys. Rev. B **48**(15), pp. 11024–11036, 1993.
381. S. B. Zhang and A. Zunger, "Prediction of unusual electronic properties of Si quantum films", Appl. Phys. Lett. **63**(10), pp. 1399–1401, 1993.
382. M. S. Hybertsen, "Absorption and emission of light in nanoscale silicon structures", Phys. Rev. Lett. **72**(10), pp. 1514–1517, 1994.
383. A. G. Nssiopoulou, V. Ioannou-Sougleridis, P. Photopoulos, A. Travlos, V. Tsakiri, and D. Papadimitriou, "Stable visible photo- and electroluminescence from nanocrystalline silicon thin films fabricated on thin SiO_2 layers by low pressure chemical vapour deposition", Physica Status Solidi A **165**, pp. 79–85, 1998.
384. L. Tsybeskov, K. D. Hirschman, S. P. Duttagupta, P. M. Fauchet, M. Zacharias, J. P. McCaffrey, and D. J. Lockwood, "Fabrication of nanocrystalline silicon superlattices by controlled thermal recrystallization", Physica Status Solidi A **165**, pp. 69–77, 1998.
385. K. D. Hirschman, L. Tsybeskov, S. P. Duttagupta, and P. M. Fauchet, "Silicon-based visible light-emitting devices integrated into microelectronic circuits", Nature **34**, pp. 338–341, 1996.

386. L. Tsybeskov, S. P. Duttagupta, K. D. Hirschman, and P. M. Fauchet, "Stable and efficient electroluminescence from a porous silicon-based bipolar device", Appl. Phys. Lett. **68**(15), pp. 2058–2060, 1996.
387. L. Tsybeskov, S. P. Duttagupta, K. D. Hirschman, and P. M. Fauchet, "Efficient electroluminescence from oxidized porous silicon", in D. J. Lockwood, P. M. Fauchet, and N. Koshida, eds., *Advanced Luminescent Materials*, pp. 34–47, Electrochemical Society, Pennington, NJ, 1996.
388. P. Boucaud, C. Francis, F. H. Julien, J.-M. Lourtioz, D. Bouchier, S. Bodnar, B. Lambert, and J. L. Regolini, "Band-edge and deep level photoluminescence of pseudomorphic $Si_{1-x-y}Ge_xC_y$ alloys", Appl. Phys. Lett. **64**(7), pp. 875–877, 1994.
389. A. S. Amour, C. W. Lu, J. C. Sturm, Y. Lacroix, and M. L. W. Thewalt, "Defect-free band-edge photoluminescence and band gap measurement of pseudomorphic $Si_{1-x-y}Ge_xC_y$ alloy layers on Si (100)", Appl. Phys. Lett. **67**(26), pp. 3915–3917, 1995.
390. K. Brunner, K. Eberl, and W. Winter, "Near-band-edge photoluminescence from pseudomorphic $Si_{1-y}C_y$/Si quantum well structures", Phys. Rev. Lett. **76**(2), pp. 303–306, 1996.
391. J. C. Sturm, H. Manoharan, L. C. Lenchyshyn, M. L. W. Thewalt, N. L. Rowell, J.-P. Noel, and D. C. Houghton, "Well-resolved band-edge photoluminescence of excitons confined in strained $Si_{1-x}Ge_x$ quantum wells", Phys. Rev. Lett. **66**(10), pp. 1362–1365, 1991.
392. J. Weber and M. I. Alonso, "Near-band-gap photoluminescence of Si-Ge alloys", Phys. Rev. B **40**(8), pp. 5683–5693, 1989.
393. R. Apetz, L. Vescan, A. Hartmann, C. Dieker, and H. Lüth, "Photoluminescence and electroluminescence of SiGe dots fabricated by island growth", Appl. Phys. Lett. **66**(4), pp. 445–447, 1995.
394. Y. S. Tang, W.-X. Ni, C. M. S. Torres, and G. V. Hansson, "Fabrication and characterisation of Si-$Si_{0.7}Ge_{0.3}$ quantum dot light emitting diodes", Electronics Letters **31**(16), pp. 1385–1386, 1995.
395. R. Soref, "Applications of silicon-based optoelectronics", MRS Bulletin (April), pp. 20–24, 1998.
396. G. He and H. A. Atwater, "Interband transistions in Sn_xGe_{1-x} alloys", Phys. Rev. **79**(10), pp. 1937–1940, 1997.
397. J. Michel, L. C. Kimerling, J. L. Benton, D. J. Eaglesham, F. A. Fitzgerald, D. C. Jacobson, J. M. Poate, Y.-H. Xie, and R. F. Ferrante, "Dopant enhancement of the 1.54 μm emission of erbium implanted silicon", in *Mater. Sci. Forum, Vols. 83–87*, pp. 653–658, Trans Tech Publications, Switzerland, 1992.
398. S. Coffa, F. Priolo, G. Franzo, A. Polman, S. Libertino, M. Saggio, and A. Carnera, "Materials issues and device performances for light emitting Er-implanted Si", Nuclear Instruments and Methods in Physics Research B **106**, pp. 386–392, 1995.
399. H. Ennen, G. Pomrenke, A. Axmann, K. Eisele, W. Hayde, and J. Schneider, "1.54-μm electroluminescence of erbium-doped silicon grown by molecular beam epitaxy", Appl. Phys. Lett. **46**(4), pp. 381–383, 1985.
400. B. Zheng, J. Michel, F. Y. G. Ren, L. C. Kimerling, D. C. Jacobson, and J. M. Poate, "Room-temperature sharp line electroluminescence at λ=1.54 μm from erbium-doped, silicon light-emitting diode", Appl. Phys. Lett. **64**(21), pp. 2842–2844, 1994.
401. J. Michel, B. Zheng, J. Palm, E. Ouellette, F. Gan, and L. C. Kimerling, "Erbium doped silicon for light emitting devices", in *Mat. Res. Soc. Symp. Proc., Vol. 422*, pp. 317–324, 1996.

402. J. Michel, F. Y. G. Ren, B. Zheng, D. C. Jacobson, J. M. Poate, and L. C. Kimerling, "The physics and application of Si:Er for light emitting diodes", in *Mater. Sci. Forum, Vols. 143–147*, pp. 707–713, Trans Tech Publications, Switzerland, 1994.
403. S. Lombardo, S. U. Campisano, G. N. van den Hoven, A. Cacciato, and A. Polman, "Room-temperature luminescence from Er-implanted semi-insulating polycrystalline silicon", Appl. Phys. Lett. **63**(14), pp. 1942–1944, 1993.
404. S. Lombardo, S. U. Campisano, G. N. van den Hoven, and A. Polman, "Erbium in oxygen-doped silicon: electroluminescence", J. Appl. Phys. **77**(12), pp. 6504–6510, 1995.
405. H. Lange, "Electronic properties of semiconducting silicides", Phys. Stat. Sol. (b) **201**(1), pp. 3–65, 1997.
406. A. B. Filonov, D. B. Migas, V. L. Shaposhnikov, V. E. Borisenko, W. Henrion, M. Rebien, P. Strauss, H. Lange, and G. Behr, "Theoretical and experimental study of interband optical transitions in semiconducting iron disilicide", J. Appl. Phys. **83**(8), pp. 4410–4414, 1998.
407. H. Lange, "Physical properties of semiconducting transition metal silicides and their prospects in Si-based device applications", in *Proc. 5th Int. Conf. on Solid-State and Integrated-Circuit Technology*, pp. 247–250, 1998.
408. D. Leong, M. Harry, K. J. Reeson, and K. P. Homewood, "A silicon/iron-disilicide light-emitting diode operating at a wavelength of 1.5 μm", Nature **387**(6), pp. 686–688, 1997.
409. C. W. Tang and S. A. VanSlyke, "Organic electroluminescent diodes", Appl. Phys. Lett. **51**(12), pp. 913–915, 1987.
410. J. H. Burroughes, D. D. C. Bradley, A. R. Brown, R. N. Marks, K. Mackay, R. H. Friend, P. L. Burns, and A. B. Holmes, "Light-emitting diodes based on conjugated polymers", Nature **347**, pp. 539–541, 1990.
411. Y. Hamada, "The development of chelate metal complexes as an organic electroluminescent material", IEEE Trans. Electron Devices **44**(8), pp. 1208–1217, 1997.
412. T. Wakimoto, Y. Fukuda, K. Nagayama, A. Yokoi, H. Nakada, and M. Tsuchida, "Organic EL cells using alkaline metal compounds as electron injection materials", IEEE Trans. Electron Devices **44**(8), pp. 1245–1248, 1997.
413. L. J. Rothberg, M. Yan, E. W. Kwock, T. M. Miller, M. E. Galvin, S. Son, and F. Papadimitrakopoulos, "Fundamental and practical issues for phenylenevinylene-based polymer light-emitting diodes", IEEE Trans. Electron Devices **44**(8), pp. 1258–1261, 1997.
414. P. E. Burrows, G. Gu, Z. Shen, S. R. Forrest, and M. E. Thompson, "Achieving full-color organic light-emitting devices for lightweight, flat-panel displays", IEEE Trans. Electron Devices **44**(8), pp. 1188–1203, 1997.
415. K. Itano, T. Tsuzuki, H. Ogawa, S. Appleyard, M. R. Willis, and Y. Shirota, "Fabrication and performance of a double-layer organic electroluminescent device using a novel starburst molecule, 1,3,5-Tris[N-(4-diphenylaminophenyl)phenylamino]benzene, as a hole-transport material and Tris(8-quinolinolato)aluminum as an emitting material", IEEE Trans. Electron Devices **44**(8), pp. 1218–1221, 1997.
416. H. Antoniadis, J. N. Miller, D. B. Roitman, and I. H. Cambell, "Effects of hole carrier injection and transport in organic light-emitting diodes", IEEE Trans. Electron Devices **44**(8), pp. 1289–1294, 1997.
417. C. Zhang, D. Braun, and A. Heeger, "Light-emitting diodes from partially conjugated poly(p-phenylene vinylene", J. Appl. Phys. **73**(10), pp. 5177–5180, 1993.

418. L. M. H. Heinrich, J. Müller, U. Hilleringmann, K. F. Goser, A. Holmes, D.-H. Hwang, and R. Stern, "CMOS-compatible organic light-emitting diodes", IEEE Trans. Electron Devices **44**(8), pp. 1249–1252, 1997.
419. J. Ammer, M. Bolotski, and T. F. Knight, "A 160 $\times$ 120 pixel liquid-crystal-on-silicon microdisplay with an adiabatic DACM", in *IEEE Int. Solid-State Circuits Conference*, pp. 212–213, 1999.
420. J. Wilson and J. F. B. Hawkes, *Optoelectronics*, p. 310, Prentice Hall, New York, 1989.
421. D. Peters, K. Fischer, and J. Müller, "Integrated optics based on silicon oxynitride thin films deposited on silicon substrates for sensor applications", Sensors and Actuators A **25–27**, pp. 425–431, 1991.
422. T. Tamir, *Integrated optics*, pp. 62–66, Springer, Berlin, Heidelberg, 1991.
423. T. Nagata, T. Tanaka, K. Miyake, H. Kurotaki, S. Yokohama, and M. Koyanagi, "Micron-size optical waveguide for optoelectronic integrated cicuits", in *Ext. Abstr. 1993 Int. Conf. on Solid State Devices and Materials*, pp. 1047–1049, 1993.
424. U. Fischer, T. Zinke, J.-R. Kropp, F. Arndt, and K. Petermann, "0.1 dB/cm waveguide losses in singlemode SOI rib waveguides", IEEE Photonics Technology Letters **8**(5), pp. 647–648, 1996.
425. S. F. Pesarcik, G. V. Treyz, S. S. Iyer, and J. M. Halbout, "Silicon germanium optical waveguides with 0.5 dB/cm losses for singlemode fiber optic systems", Electronics Letters **28**, pp. 59–60, 1992.
426. R. A. Soref, F. Namawar, N. M. Kalkhoran, and D. M. Koker, "Silicon optical waveguides with buried-$CoSi_2$ cladding layers", Optics Letters **19**(17), pp. 1319–1321, 1994.
427. X.-P. Feng and K. Ujihara, "Quantum theory of spontaneous emission in a one-dimensional optical cavity with two-side output coupling", Phys. Rev. A **41**(5), pp. 2668–2676, 1990.
428. J. D. Joannopoulos, P. R. Villeneuve, and S. Fan, "Photonic crystals: putting a new twist on light", Nature **386**, pp. 143–149, 1997.
429. C. M. Soukoulis, *Photonic band gap materials*, Kluwer, Dordrecht, 1996.
430. J. D. Joannopoulos, R. D. Meade, and J. N. Winn, *Photonics crystals*, Princetown, New York, 1995.
431. E. Yablonovitch, "Photonic band-gap structures", J. Opt. Soc. Am. B **10**(2), pp. 283–295, 1993.
432. J. S. Foresi, P. R. Villeneuve, J. Ferrera, E. R. Thoen, G. Steinmeyer, S. Fan, J. D. Joannopoulos, L. C. Kimerling, H. I. Smith, and E. P. Ippen, "Photonic-bandgap microcavities in optical waveguides", Nature **390**(11), pp. 143–145, 1997.
433. S. Fan, J. N. Winn, A. Devenyi, J. C. Chen, R. D. Meade, and J. D. Joannopoulos, "Guided and defect modes in periodic dielectric waveguides", J. Opt. Soc. Am. B **12**(7), pp. 1267–1272, 1995.
434. I. Y. Yang, S. Silverman, J. Ferrera, K. Jackson, J. M. Carter, D. A. Antoniadis, and H. I. Smith, "Combining and matching optical, electron-beam, and X-ray lithographies in the fabrication of Si complementary metal-oxide-semiconductor circuits with 0.1 and sub-0.1 μm features", J. Vac. Sci. Technol. B **13**(6), pp. 2741–2744, 1995.
435. H. Yokoyama and S. D. Brorson, "Rate equation analysis of microcavity lasers", J. Appl. Phys. **66**(10), pp. 4801–4805, 1989.
436. E. F. Schubert, A. M. Vredenberg, N. E. J. Hunt, Y. H. Wong, P. C. B. J. M. Poate, D. C. Jacobson, L. C. Feldman, and G. J. Zydzik, "Giant enhancement of luminescence intensity in Er-doped Si/SiO_2 resonant cavities", Appl. Phys. Lett. **61**(12), pp. 1381–1383, 1992.

437. Y. S. Liu, "Lighting the way in computer design", Circuits and Devices (Jan.), pp. 23–31, 1998.
438. Y. S. Liu, R. J. Wojnarowski, W. A. Hennessy, J. P. Bristow, Y. Liu, A. Peczalski, J. Rowlette, A. Plotts, J. Stack, J. Yardley, L. Eldada, R. M. Osgood, R. Scarmozzino, S. H. Lee, and V. Uzguz, "Polymer optical interconnect technology (POINT) - optoelectronic packaging and interconnect for board and backplane applications", in *Proc. SPIE, Critical reviews of optical science and technology*, vol. 62, pp. 405–414, 1996.
439. Y. S. Liu, R. J. Wojnarowski, W. A. Hennessy, J. P. Bristow, Y. Liu, A. Peczalski, J. Rowlette, A. Plotts, J. Stack, J. Yardley, L. Eldada, R. M. Osgood, R. Scarmozzino, S. H. Lee, and V. Uzguz, "High density optical interconnect for board and backplane applications using VCSEL and polymer waveguides", in *Proc. 47th Electronic Components and Technology Conference (ECTC)*, pp. 391–398, 1997.
440. T. R. Haller, B. S. Whitmore, P. J. Zabinski, and B. K. Gilbert, "High frequency performance of GE high density interconnect modules", IEEE Trans. Components, Hybrids, and Manufacturing Tech. **16**(1), pp. 21–27, 1993.
441. R. F. Carson, M. L. Lovejoy, K. L. Lear, M. E. Warren, P. K. Seigal, D. C. Craft, S. P. Kilcoyne, G. A. Patrizi, and O. Blum, "Low-power approaches for parallel, free-space photonic interconnects", in *Proc. Optoelectronic Interconnects and Packaging*, pp. 35–63, 1996.
442. T. Uno, M. Mitsuda, M. Kito, H. Asano, M. Ishino, G. Tohmon, and Y. Matsui, "Optical transceiver formed with fiber-embedded lightwave circuit on silicon substrate", in *ISSCC Dig. Tech. Papers*, pp. 388–389, 1999.
443. G. de Graaf, J. H. Correia, and R. F. Wolffenbuttel, "On-chip integrated CMOS optical microspectrometer with light-to-frequency converter and bus interface", in *IEEE Int. Solid-State Circuits Conference*, pp. 208–209, 1999.
444. A. T. T. D. Tran, Y. Z. Lo, Z. H. Zhu, D. Haroinan, and E. Mozdy, "Surface micromachined Fabry–Perot tuneable filter", IEEE Photonics Technology Letters **8**(3), pp. 393–395, 1996.
445. J. H. Correi, E. Cretu, M. Bartek, and R. F. Wolfenbuttel, "Bulk-micromachined tuneable Fabry–Perot microinterferometer for the visible spectral range", in *Proc. Eurosensors XII*, Southampton, UK, 1998.
446. C. J. Aswell, J. Berlien, E. Dierschke, and M. Hassan, "A monolithic light-to-frequency converter with a scalable sensor array", in *IEEE Int. Solid-State Circuits Conference*, pp. 158–159, 1994.
447. J. H. Correi, E. Cretu, M. Bartek, and R. F. Wolfenbuttel, "A local bus for MCM-based microinstrumentation systems", Sensors and Actuators A **68**(1–3), pp. 460–465, 1998.
448. S. Ura, "An integrated-optic disk pickup device", J. Lightwave Technology **4**(7), pp. 913–918, 1986.
449. G. Bouwhuis and J. J. M. Braat, "Video disk player optics", Appl. Opt. **17**(13), pp. 1993–2000, 1978.
450. S. Ura, M. Shinohara, T. Suhara, and H. Nishihara, "An integrated-optic parallel data pickup device", in *Technical Digest of Int. Symposium on Optical Memory*, pp. 27–28, 1991.
451. H. Nishihara, T. Suhara, and S. Ura, "Integrated photonic functional devices using grating couplers", in T. Sueta and T. Okoshi, eds., *Ultrafast and ultra-parallel optoelectronics*, p. 241, Wiley, Chichester, 1995.
452. J. M. Rabaey, *Digital integrated circuits: a design perspective*, Prentice Hall, New York, 1996.
453. S.-M. Kang and Y. Leblebici, *CMOS digital integrated circuits: analysis and design*, McGraw-Hill, New York, 1996.

454. T. A. DeMassa and Z. Ciccone, *Digital integrated circuits*, Wiley, New York, 1996.
455. K. R. Laker and W. M. C. Sansen, *Design of analog integrated circuits and systems*, McGraw-Hill, New York, 1994.
456. Dept. Electrical Engineering and Computer Sciences, University of California, Berkely, CA 94720, *BSIM3v3 manual*, 1995.
457. H. Shichman and D. Hodges, "Modelling and simulation of insulated-gate field-effect transistor switching circuits", IEEE J. Solid-State Circuits **3**(3), pp. 285–289, 1968.
458. S. Liu and L. W. Nagel, "Small-signal MOSFET models for analog circuit design", IEEE J. Solid-State Circuits **17**(6), pp. 983–998, 1982.
459. P. E. Allen and D. R. Holberg, *CMOS analog circuit design*, Holt, Rinehart, and Winston, New York, 1987.
460. P. R. Gray and R. G. Meyer, *Analysis and design of analog integrated circuits*, p. 39, Wiley, New York, 1993.
461. Technology Modeling Association, Inc., Palo Alto, CA, *SUPREM-3 manual*, 1994.
462. Technology Modeling Association, Inc., Palo Alto, CA, *TSUPREM-4 manual*, 1994.
463. P. Hoppe, *Übertragungsverhalten analoger Schaltungen*, p. 161, B. G. Teubner, Stuttgart, 1994.
464. H. Zimmermann, K. Kieschnick, M. Heise, and H. Pless, "High-bandwidth BiCMOS OEIC for optical storage systems", in *IEEE Int. Solid-State Circuits Conference*, pp. 384–385, 1999.
465. K. Ayadi, M. Kuijk, P. Heremans, G. Bickel, G. Borghs, and R. Vounckx, "A monolithic optoelectronic receiver in standard 0.7 μm CMOS operating at 180 MHz and 176 fJ light input energy", IEEE Photonics Technology Letters **9**(1), pp. 88–90, 1997.
466. J. F. Heanue, M. C. Bashaw, and L. Hesselink, "Volume holographic storage and retrieval of digital data", Science **265**(8), pp. 749–752, 1994.
467. M. E. Schaffer and P. A. Mitkas, "Smart photodetector array for page-oriented optical memory in 0.35–μm CMOS", IEEE Photonics Technology Letters **10**(6), pp. 866–868, 1998.
468. F. Esfahani, K.-O. Hofacker, A. Benedix, and H. H. Berger, "Small area optical inputs for high speed CMOS circuits", in *9th Int. IEEE ASIC Conference*, pp. 7–10, Rochester, N.Y., 1996.
469. H. H. Berger, J. Sturm, F. Esfahani, A. Benedix, S. von Aichberger, B. Müller, and K.-O. Hofacker, "Optical signal injection for high-speed wafer level function test of integrated circuits", in *IEEE Int. Conf. on Microelectronic Test Structures*, pp. 39–42, IEEE, Monterey, CA, 1997.
470. T. C. Banwell, A. C. V. Lehmen, and R. R. Cordell, "VCSE laser transmitters for parallel data links", IEEE J. Quantum Electron. **29**(2), pp. 635–644, 1993.
471. D. L. Mathine, R. Droopad, and G. N. Maracas, "A vertical-cavity surface-emitting laser applied to a 0.8 μm NMOS driver", IEEE Photonics Technology Letters **9**(7), pp. 869–871, 1997.
472. C. Stanescu, S. Porumbescu, A. Hanganu, S. Costea, I. Mirea, G. Aungurecei, M. Furis, and B. Mihalea, "Bipolar preamplifier for monolithic OEIC receiver", in *Proc. Int. Semicond. Conference (CAS'97)*, pp. 567–570, Sinaia, Romania, 1997.
473. M. Seifart, *Analoge Schaltungen*, pp. 315–316, Verlag Technik, Berlin, 1996.
474. T. Lule, H. Fischer, S. Benthien, H. Keller, M. Sommer, J. Schulte, P. Rieve, and M. Böhm, "Image sensor with per-pixel programmable sensitivity in TFA

technology", in H. Reichl and A. Heuberger, eds., *Micro System Technologies*, p. 675, 1996.

475. M. Mansuripur and G. Sincerbox, Proc. IEEE **85**(11), pp. 1780–1796, 1997.
476. R. Gregorian and G. C. Temes, *Analog integrated circuits for signal processing*, Wiley, New York, 1986.
477. B.-E. Kim, M.-S. Jeong, D.-M. Cho, J.-K. KIM, J.-S. Lee, S.-K. KIM, and S.-W. KIM, "0.8 μm CMOS analog front-end processor for CD-ROM", IEEE Trans. Consumer Electronics **42**(3), pp. 826–831, 1996.
478. P. R. Gray and R. G. Meyer, *Analysis and design of analog integrated circuits*, p. 456, Wiley, New York, 1993.
479. T. Takimoto, N. Fukunaga, M. Kubo, , and N. Okabayashi, "High speed Si-OEIC(OPIC) for optical pickup", IEEE Trans. Consumer Electronics **44**(1), pp. 137–142, 1998.
480. K. Kieschnick, T. Heide, A. Ghazi, H. Zimmermann, and P. Seegebrecht, "High-speed photonic CMOS and BiCMOS receiver ICs", in *Proc. 25th European Solid-State Circuits Conference (ESSCIRC), Sept. 21–23*, pp. 398–401, 1999.
481. J. Qi, C. L. Schow, L. D. Garrett, and J. C. Campbell, "A silicon NMOS monolithically integrated optical receiver", IEEE Photonics Technology Letters **9**(5), pp. 663–665, 1997.
482. C. L. Schow, J. D. Schaub, R. Li, J. Qi, and J. C. Campbell, "A monolithically integrated 1-Gb/s silicon photoreceiver", IEEE Photonics Technology Letters **11**(1), pp. 120–121, 1999.
483. K. Kieschnick, H. Zimmermann, H. Pless, and P. Seegebrecht, "BiCMOS receiver OEIC for optical interconnects", in *Ext. Abstr. 3nd IEEE Workshop on Signal Propagation on Interconnects, Neustadt, Germany*, pp. 72–73, 1999.
484. G. Williams, "Lightwave receivers", in L. Tingye, ed., *Topics in lightwave systems*, pp. 79–148, Academic Press, New York, 1991.
485. K. G. Moerschel, T. Y. Chiu, W. A. Possanza, K. S. Lau, R. G. Swartz, R. A. Mantz, T. Y. M. Liu, K. F. Lee, V. D. Archer, G. R. Hower, G. T. Mazsa, R. E. Carsia, J. A. Pavlo, M. P. Ling, J. L. Dolcin, F. M. Erceg, J. J. Egan, C. J. Fassl, J. T. Glick, and M. A. Prozonic, "BEST: a BiCMOS-compatible super-self-aligned ECL technology", in *IEEE Custom Integrated Circuits Conference*, pp. 18.3.1–18.3.4, 1990.
486. N. Fukunaga, M. Yamamoto, M. Kubo, and N. Okabayashi, "Si-OEIC(OPIC) for optical pickup", IEEE Trans. Consumer Electronics **43**(2), pp. 157–164, 1997.

Index

Druck: Strauss Offsetdruck, Mörlenbach
Verarbeitung: Schäffer, Grünstadt